T0281480

Introduction to Statistical Methods, Design of Experiments and Statistical Quality Control

Dharmaraja Selvamuthu · Dipayan Das

Introduction to Statistical Methods, Design of Experiments and Statistical Quality Control

 Springer

Dharmaraja Selvamuthu
Department of Mathematics
Indian Institute of Technology Delhi
New Delhi, India

Dipayan Das
Department of Textile Technology
Indian Institute of Technology Delhi
New Delhi, India

ISBN 978-981-13-4673-6   ISBN 978-981-13-1736-1   (eBook)
https://doi.org/10.1007/978-981-13-1736-1

This Springer imprint is published by the registered company Springer Nature Singapore Pte Ltd.
The registered company address is: 152 Beach Road, #21-01/04 Gateway East, Singapore 189721, Singapore

Foreword

Present-day research either academic or applied is facing entirely a different kind of problem with respect to data; in particular, the volume, velocity, variety, and veracity of data available have changed dramatically. Alongside the validity, variability, veracity, of data have brought in new dimensions to associated risk in using such data. While the value of data-driven decision making is becoming the norm these days, the volatility and vulnerability of data pose a newer cautious approach to using data gathered from social networks and sources. However, the age-old problem of visualizing data and value adding from data analysis remains unchanged.

In spite of the availability of newer approaches to learning from data, like machine learning, deep learning, and other such modern data analytical tools, a student of engineering or management or social and natural sciences still needs a good basic grasp of the statistical methods and concepts (1) to describe a system in a quantitative way, (2) to improve system through experiments, (3) to maintain the system unaffected by external sources of variation, and (4) to analyze and predict the dynamics of the system in the future. Toward this goal, students need a resource (1) that does not require too many prerequisites, (2) that is easy to access, (3) that explains concepts with examples, (4) that explains the validity of the methods without getting lost in rigor, and last but not least, (5) that enhances the learning experience. Professors Dharmaraja Selvamuthu and Dipayan Das have translated their years of teaching and writing experience in the fields of descriptive and inferential statistical methods, the design of experiments, and statistical quality control to come out with a valuable resource that has all the desired features outlined above.

The efforts of the authors can be seen in the depth and breadth of the topics covered with the intention to be useful in different courses that are taught in engineering colleges and technology institutions.

On the other hand, the instructors will enjoy using this resource as it makes their teaching experience enhanced by the learning outcomes that are bound to accrue from the content, structure, and exposition of this book. The exercises in this book add value as assessment tools for instructors and also offer additional practice for students. The levels of difficulty in exercises are designed with such end in mind. The authors will be appreciated by both students and instructors for this valuable addition.

Good textbooks are like caring companions for students. This book has achieved that merit.

Auckland, New Zealand   Prof. Tiru Arthanari
                                                                                                            University of Auckland

Preface

Statistics has great relevance to several disciplines like economics, commerce, engineering, medicine, health care, agriculture, biochemistry, and textiles. A large number of students with varied disciplinary backgrounds need a course in basics of statistics, the design of experiments, and statistical quality control at an introductory level to pursue their discipline of interest. The idea of writing this book emerged several years ago since there is no textbook available which covers all the three areas in one book. In view of the diverse audience, this book addresses these three areas. No previous knowledge of probability or statistics is assumed, but an understanding of calculus is a prerequisite. The main objective of this book is to give an accessible presentation of concepts from probability theory, statistical methods, the design of experiments, and statistical quality control. Practical examples and end-of-chapter exercises are the highlights of the text as they are purposely selected from different fields.

Organized into ten chapters, the book comprises major topics on statistical methods, the design of experiments, and statistical quality control. Chapter 1 is the introductory chapter which describes the importance of statistical methods, design of experiments, and statistical quality control. Chapters 2–6 alone could be used as a text for a one-semester, beginner's level course in statistical methods. Similarly, Chaps. 7–10 alone could be used as a text for a one-semester course in design of experiments. Chapters 2–6 and 10 could be used as a text for a one-semester introductory course in statistical and quality control. The whole book serves as a master-level introductory course in all the three topics, as required in textile engineering or industrial engineering. At the Indian Institute of Technology (IIT) Delhi, the course Introduction to Statistics and Design of Experiments for which this text was developed has been taught for over a decade, chiefly to students majoring in engineering disciplines or mathematics. Chapter 2 introduces the basic concepts of probability theory, conditional probability, the notion of independence, and common techniques for calculating probabilities. To introduce probability concepts and to demonstrate probability calculations, simple probabilistic experiments such as selecting a card from a deck or rolling a die are considered. In addition, the standard distributions, moments, and central limit theorem with examples are also discussed in Chap. 2. Chapter 3 presents the descriptive statistics, which starts with concepts such as data, information, and description. Various descriptive measures, such as central tendency measures, variability measures, and coefficient of variation, are presented in this chapter. Inference in mathematics is based on logic and presumably infallible at least when correctly applied, while statistical inference considers how inference should proceed when the data are subject to random fluctuation. Sampling theory can be employed to obtain information about samples drawn at random from a known population. However, often it is more important to be able to infer information about a population from samples drawn from it. Such problems are dealt with in statistical inference. The statistical inference may be divided into four major areas: theory, estimation, tests of hypothesis, and correlation and regression analysis. This book treats these four areas separately, dealing with the theory of sampling distributions and estimation in Chap. 4, hypothesis testing in Chap. 5, and correlation and regression analysis in Chap. 6. The statistical inference is dealt with in detail with sampling distribution in Chap. 4. The standard sampling distributions such as chi-square, Student's t, and F distributions are presented. The sample mean and sample variance are studied, and their expectations and variances are given. The central limit theorem is applied to determine the probability distribution they follow. Then, this chapter deals with point estimation, a method of moments, maximum likelihood estimator, and interval estimation. The classic methods are used to estimate unknown population parameters such as mean, proportion, and variance by computing statistics from random samples and applying the theory of sampling distributions.

Chapter 5 covers a statistical test of the hypothesis in detail with many examples. The topics such as simple and composite hypotheses, types of error, power, operating characteristic curves, p value, Neyman–Pearson method, generalized likelihood ratio test, use of asymptotic results to construct tests, and generalized ratio test statistic are covered. In this chapter, analysis of variance, in particular, one-way ANOVA, is also introduced, whereas its applications are presented in the later chapters. Chapter 6 discusses the analysis of correlation and regression. This chapter starts by introducing Spearman's correlation coefficient and rank correlation and later on presents simple linear regression and multiple linear regression. Further, in this chapter, nonparametric tests such as Wilcoxon, Smirnov, and median tests are presented. The descriptive statistics, sampling distributions, estimations, statistical inference, testing of hypothesis, and correlation and regression analysis are presented in Chaps. 2–6 and are applied to the design and analysis of experiments in Chaps. 7–9. Chapter 7 gives an introduction to the design of experiments. Starting with the definition of the design of experiments, this chapter gives a brief history of experimental design along with the need for it. It then discusses the principles and provides us with the guidelines of the design of experiments and ends with the illustration of typical applications of statistically designed experiments in process-, product-, and management-related activities. This chapter also deals with a very popular design of experiments, known as a completely randomized design, which describes how to conduct an experiment and discusses the analysis of the data obtained from the experiment. The analysis of the experimental data includes the development of descriptive and regression models, a statistical test of hypothesis based on the one-way classification of analysis of variance, and multiple comparisons among treatment means. This chapter presents many numerical examples to illustrate different methods of data analysis. At the end, the reader is asked to solve many numerical problems to have a full understanding of a completely randomized design.

Chapter 8 discusses two important block designs, namely randomized block design and Latin square design. It describes these designs by using practical examples and discusses the analysis of the data obtained from experiments conducted in accordance with these designs. The data analysis includes the development of descriptive models, statistical tests of a hypothesis based on the two-way and three-way classifications of analysis of variance, and multiple comparisons among treatment mean. Also, in this chapter, many numerical examples are solved, and several numerical problems are given at the end of the chapter as exercises. Chapter 8 deals with an important class of experimental designs, known as factorial designs. This chapter discusses the design and analysis of factorial experiments with two or three factors, where each factor might have the same level or different levels. It also discusses the design and analysis of 2^2 and 2^3 full factorial experiments. This chapter explains two important design techniques, namely blocking and confounding, which are often followed by a factorial experiment. The design and analysis of two-level fractional factorial design and the concept of design resolution are explained. In this chapter, many numerical examples are given to illustrate the concepts of different factorial designs and their methods of analysis. Additional end-of-chapter exercises are provided to assess students' understanding of factorial experiments. Chapter 9 deals with response surface methodology, a collection of mathematical and statistical tools and techniques used in developing, understanding, and optimizing processes and products along with a description of response surface models. It discusses the analysis of first-order and second-order response surface models. It describes popular response surface designs that are suitable for fitting the first-order and second-order models. Also, it describes the multi-factor optimization technique based on the desirability function approach. This chapter reports many numerical examples to illustrate different concepts of response surface methodology. At the end, readers are asked to solve several numerical problems based on the response surface methodology. Chapter 10 deals with statistical quality control. This chapter discusses acceptance sampling techniques used for inspection of incoming and outgoing materials in an industrial environment. It describes single and double sampling plans for attributes and acceptance sampling of variables. Further, this chapter also describes a very important tool in process control, known as a control chart, which is used to monitor a manufacturing process with quality assurance in mind. It provides an introduction to control chart. It describes Shewhart's three-sigma control charts for variables and attributes. It discusses the process capability analysis. Also, it describes an advanced control chart which is very efficient to detect a small shift in the mean of a process. Finally, this chapter discusses many numerical examples to illustrate different concepts of acceptance sampling techniques and quality control charts.

The exposition of the entire book is processed with easy access to the subject matter without sacrificing rigor, at the same time keeping prerequisites to a minimum. A distinctive feature of this text is the "Remarks" following most of the theorems and definitions. In Remarks, a particular result or concept being presented is discussed from an intuitive point of view. A list of references is given at the end of each chapter. Also, at the end of each chapter, there is a list of exercises to facilitate the understanding of the main body of each chapter. Most of the examples and exercises are classroom-tested in the course that we taught over many years. Since the book is the outcome of years of teaching experience continuously improved with students' feedback, it is expected to yield a fruitful learning experience for the students, and the instructors will also enjoy facilitating such creative learning. We hope that this book will serve as a valuable text for students.

We would like to express our gratitude to our organization—Indian Institute of Technology Delhi—and numerous individuals who have contributed to this book. Many former students of IIT Delhi, who took courses, namely MAL140 and TTL773, provided excellent suggestions that we have tried to incorporate in this book. We are immensely thankful to Prof. A. Rangan of IIT Madras for his encouragement and criticism during the writing of this book. We are also indebted to our doctoral research scholars, Dr. Arti Singh, Mr. Puneet Pasricha, Ms. Nitu Sharma, Ms. Anubha Goel, and Mr. Ajay K. Maddineni, for their tremendous help during the preparation of the manuscript in LaTeX and also for reading the manuscript from a student point of view.

We gratefully acknowledge the book grant provided by the office of Quality Improvement Programme of the IIT Delhi. Our thanks are also due to Mr. Shamim Ahmad from Springer for his outstanding editorial work for this book. We are also grateful to those anonymous referees who reviewed our book and provided us with excellent suggestions. On a personal note, we wish to express our deep appreciation to our families for their patience and support during this work.

In the end, we wish to tell our dear readers that we have tried hard to make this book free of mathematical and typographical errors and misleading or ambiguous statements. However, it might be possible that some are still being left in this book. We will be grateful to receive such corrections and also suggestions for further improvement of this book.

New Delhi, India   Dharmaraja Selvamuthu
April 2018   Dipayan Das

Contents

About the Authors

Dharmaraja Selvamuthu is Professor in the Department of Mathematics, Indian Institute of Technology Delhi, India. He also served as Head of the Department of Mathematics, Indian Institute of Technology Delhi. He earned his M.Sc. degree in applied mathematics from Anna University, Chennai, India, in 1994, and Ph.D. degree in mathematics from the Indian Institute of Technology Madras, India, in 1999. He has held visiting positions at Duke University, USA; Emory University, USA; University of Calgary, Canada; University of Los Andes, Bogota, Colombia; National University of Colombia, Bogota, Colombia; University of Verona, Italy; Sungkyunkwan University, Suwon, Korea; and Universita Degli Studi di Salerno, Fisciano, Italy. His research interests include applied probability, queueing theory, stochastic modeling, performance analysis of computer and communications systems, and financial mathematics. He has published over 45 research papers in several international journals of repute and over 20 research papers at various international conferences.

Dipayan Das is Professor in the Department of Textile Technology, Indian Institute of Technology Delhi, India. He obtained his Ph.D. degree from the Technical University of Liberec, the Czech Republic, in 2005. His research interests include modeling of fibrous structures and their properties, product and process engineering using statistical and mathematical techniques, and nonwoven products and processes. He has published four books including two monographs and over 100 research papers in scientific journals and conference proceedings. He is a recipient of the BIRAC-SRISTI Gandhian Young Technological Innovation (GYTI) Appreciation Award (in 2018), IIT Delhi Teaching Excellence Award (in 2017), and Kusuma Trust Outstanding Young Faculty Fellowship (from 2008 to 2013).

Acronyms

CDF   Cumulative distribution function
MGF   Moment-generating function
MLE   Maximum likelihood estimator
MSE   Mean square error
PDF   Probability density function
PGF   Probability generating function
PMF   Probability mass function
RV   Random variable

Chapter 1
Introduction

1.1 Statistics

Statistics is the science of data. The term statistics is derived from the New Latin statisticum collegium ("council of state") and the Italian word statista ("statesman"). In a statistical investigation, it is known that for reasons of time or cost, one may not be able to study each individual element (of population). Consider a manufacturing unit that receives raw material from the vendors. It is then necessary to inspect the raw materials before accepting it. It is practically impossible to check each and every item of raw material. Thus, a few items (sample) are randomly selected from the lot or batch and inspected individually before taking a decision to reject or accept the lot. Consider another situation where one wants to find the retail book value (dependent variable) of a used automobile using the age of the automobile (independent variable). After conducting a study over the past sale of the used automobile, we are left with the set of numbers. The challenge is to extract meaningful information from the behavior observed (i.e., how age of the automobile is related to the retail book value). Hence, statistics deals with the collection, classification, analysis, and interpretation of data. Statistics provide us with an objective approach to do this. There are several statistical techniques available for learning from data. One needs to note that the scope of statistical methods is much wider than only statistical inference problems. Such techniques are frequently applied in different branches of science, engineering, medicine, and management. One of them is known as design of experiments. When the goal of a study is to demonstrate cause and effect, experiment is the only source of convincing data. For example, consider an investigation in which researchers observed individuals and measure variable of interest but do not attempt to influence response variable. But to study cause and effect, the researcher deliberately imposes some treatment on individuals and then observes the response variables. Thus, design of experiment refers to the process of planning and conducting experiments and analyzing the experimental data by statistical methods so that valid and objective conclusions can be obtained with minimum use of resources. Another

D. Selvamuthu and D. Das, *Introduction to Statistical Methods, Design of Experiments and Statistical Quality Control,* https://doi.org/10.1007/978-981-13-1736-1_1

important application of statistical techniques lies in statistical quality control, often abbreviated as SQC. It includes statistical process control and statistical product control. Statistical process control involves certain statistical techniques for measurement and analysis of process variation, while statistical product control involves certain statistical techniques for taking a decision whether a lot or batch of incoming and outgoing materials is acceptable or not.

This book comprises of statistical methods, design of experiment, and statistical quality control. A brief introduction to these three parts is as follows.

1.2   Statistical Methods

The architect of modern statistical methods in the Indian subcontinent was undoubtedly Mahalanobis,[1] but he was helped by a very distinguished scientist C R Rao.[2] Statistical methods are mathematical formulas, models, and techniques that are used in statistical inference of raw data. Statistical inference mainly takes the form of problem of point or interval estimation of certain parameters of the population and of testing various claims about the population parameters known as hypothesis testing problem. The main approaches to statistical inference can be classified into parametric, nonparametric, and Bayesian. Probability is an indispensable tool for statistical inference. Further, there is a close connection between probability and statistics. This is because characteristics of the population under study are assumed to be known in probability problem, whereas in statistics, the main concern is to learn these characteristics based on the characteristics of sample drawn from the population.

1.2.1   Problem of Data Representation

Statistics and data analysis procedures generally yield their output in numeric or tabular forms. In other words, after an experiment is performed, we are left with the set of numbers (data). The challenge is to understand the features of the data and extract useful information. Empirical or descriptive statistics helps us in this. It encompasses both graphical visualization methods and numerical summaries of the data.

---

[1]Prasanta Chandra Mahalanobis (June 29, 1893–June 28, 1972) was an Indian scientist and applied statistician. He is best remembered for the Mahalanobis distance, a statistical measure, and for being one of the members of the first Planning Commission of free India.

[2]Calyampudi Radhakrishna Rao, known as C R Rao (born September 10, 1920) is an Indian-born, naturalized American, mathematician, and statistician. He is currently Professor Emeritus at Penn State University and Research Professor at the University at Buffalo. He has been honored by numerous colloquia, honorary degrees, and festschrifts and was awarded the US National Medal of Science in 2002.

Graphical Representation

Over the years, it has been found that tables and graphs are particularly useful ways for presenting data. Such graphical techniques include plots such as scatter plots, histograms, probability plots, spaghetti plots, residual plots, box plots, block plots, and bi-plots. In descriptive statistics, a box plot is a convenient way of graphically depicting groups of numerical data through their quartiles. A box plot presents a simple but effective visual description of the main features, including symmetry or skewness, of a data set. On the other hand, pie charts and bar graphs are useful in the scenario when one is interested to depict the categories into which a population is categorized. Thus, they apply to categorical or qualitative data. In a pie chart, a circle (pie) is used to represent a population and it is sliced up into different sectors with each sector representing the proportion of a category. One of the most basic and frequently used statistical methods is to plot a scatter diagram showing the pattern of relationships between a set of samples, on which there are two measured variables x and y (say). One may be interested in fitting a curve to this scatter, or in the possible clustering of samples, or in outliers, or in colinearities, or other regularities. Histograms give a different way to organize and display the data. A histogram does not retain as much information on the original data as a stem-and-leaf diagram, in the sense that the actual values of the data are not displayed. Further, histograms are more flexible in selecting the classes and can also be applied to the bivariate data. Therefore, this flexibility makes them suitable as estimators of the underlying distribution of the population.

Descriptive Statistics

Descriptive statistics are broken down into measures of central tendency and measures of variability (spread), and these measures provide valuable insight into the corresponding population features. Further, in descriptive statistics, the feature identification and parameter estimation are obtained with no or minimal assumptions on the underlying population. Measures of central tendency include the mean, median, and mode, while measures of variability include the standard deviation or variance, the minimum and maximum variables, and the kurtosis and skewness. Measures of central tendency describe the center position of a data set. On the other hand, measures of variability help in analyzing how spread-out the distribution is for a set of data. For example, in a class of 100 students, the measure of central tendency may give average marks of students to be 62, but it does not give information about how marks are distributed because there can still be students with 1 and 100 marks. Measures of variability help us communicate this by describing the shape and spread of the data set.

1.2.2 Problem of Fitting the Distribution to the Data

There is a need to learn how to fit a particular family of distribution models to the data; i.e., identify the member of the parametric family that best fits the data.

For instance, suppose one is interested to examine n people and record a value 1 for people who have been exposed to the tuberculosis (TB) virus and a value 0 for people who have not been so exposed. The data will consist of a random vector $X = (X_1, X_2, \ldots, X_n)$ where $X_i = 1$ if the ith person has been exposed to the TB virus and $X_i = 0$ otherwise. A possible model would be to assume that $X_1, X_2, \ldots, X_n$ behave like n independent Bernoulli random variables each of which has the same (unknown) probability p of taking the value 1. If the assumed parametric model is a good approximation to the data generation mechanism, then the parametric inference is not only valid but can be highly efficient. However, if the approximation is not good, the results can be distorted. For instance, we wish to test a new device for measuring blood pressure. We will try it out on n people and record the difference between the value returned by the device and the true value as recorded by standard techniques. The data will consist of a random vector $X = (X_1, X_2, \ldots, X_n)$ where X_i is the difference for the ith person. A possible model would be to assume that $X_1, X_2, \ldots, X_n$ behave like n independent random variables each having a normal distribution with mean 0 and variance σ^2 density where σ^2 is some unknown positive real number. It has been shown that even small deviations of the data generation mechanism from the specified model can lead to large biases. Three methods of fitting models to data are: (a) the method of moments, which derives its name because it identifies the model parameters that correspond (in some sense) to the nonparametric estimation of selected moments, (b) the method of maximum likelihood, and (c) the method of least squares which is most commonly used for fitting regression models.

1.2.3 Problem of Estimation of Parameters

In addition, there is a need to focus on one of the main approaches for extrapolating sample information to the population, called the parametric approach. This approach starts with the assumption that the distribution of the population of interest belongs to a specific parametric family of distribution models. Many such models depend on a small number of parameters. For example, Poisson models are identified by the single parameter λ, and normal models are identified by two parameters, μ and σ^2. Under this assumption (i.e., that there is a member of the assumed parametric family of distributions that equals the population distribution of interest), the objective becomes that of estimating the model parameters, to identify which member of the parametric family of distributions best fits the data.

Point Estimation

Point estimation, in statistics, is the process of finding an approximate value of some parameter of a population from random samples of the population. The method mainly comprises of finding out an estimating formula of a parameter, which is called the estimator of the parameter. The numerical value, which is obtained from the formula on the basis of a sample, is called estimate.

Example 1.1 Let $X_1, X_2, \ldots, X_n$ be a random sample from any distribution F with mean μ. One may need to estimate the mean of the distribution. One of the natural choices for the estimator of the mean is

$$\frac{1}{n} \sum_{i=1}^{n} X_i.$$

Other examples may need to estimate a population proportion, variance, percentiles, and interquartile range (IQR).

Confidence Interval Estimation

In many cases, in contrast to point estimation, one may be interested in constructing an interval that contains the true value (unknown) of the parameter value with a specified high probability. The interval is known as the confidence interval, and the technique of obtaining such intervals is known as interval estimation.

Example 1.2 A retailer buys garments of the same style from two manufacturers and suspects that the variation in the masses of the garments produced by the two makers is different. A sample of size n_1 and n_2 was therefore chosen from a batch of garments produced by the first manufacturer and the second manufacturer, respectively, and weighed. We wish to find the confidence intervals for the ratio of variances of mass of the garments from the one manufacturer with the other manufacturer.

Example 1.3 Consider another example of a manufacturer regularly tests received consignments of yarn to check the average count or linear density (in the text). Experience has shown that standard count tests on specimens chosen at random from a delivery of a certain type of yarn usually have an average linear density of μ_0 (say). A normal 35-tex yarn is to be tested. One is interested to know that how many tests are required to be 95% sure that the value lies in an interval (a, b) where a and b are known constants.

1.2.4 Problem of Testing of Hypothesis

Other than point estimation and interval estimation, one may be interested in deciding which value among a set of values is true for a given distribution. In practice, the functional form of the distribution is unknown. One may be interested in some properties of the population without making any assumption on the distribution. This procedure of taking a decision on the value of the parameter (parametric) or nature of distribution (nonparametric) is known as the testing of hypothesis. The nonparametric tests are also known as distribution-free tests. Some of the standard hypothesis tests are z test, t test (parametric) and KS test, median test (nonparametric).

Example 1.4 Often one wishes to investigate the effect of a factor (independent variable x) on a response (dependent variable y). We then carry out an experiment to compare a treatment when the levels of the factor are varied. This is a hypothesis testing problem where we are interested in testing the equality of treatment means of a single factor x on a response variable y (such problems are discussed in Chap. 7).

Example 1.5 Consider the following problem. A survey showed that a random sample of 100 private passenger cars was driven on an average 9,500 km a year with a standard deviation of 1,650 km. Use this information to test the hypothesis that private passenger cars are driven on the average 9,000 km a year against the alternatives that the correct average is not 9,000 km a year.

Example 1.6 Consider another example of some competitive examination performance of students from one particular institute in this country who took this examination last year. From the sample of n student's score and known average score, we wish to test the claim of an administrator that these students scored significantly higher than the national average.

Example 1.7 Consider one another example of the weekly number of accidents over a 30-week period in Delhi roads. From the sample of n observations, we wish to test the hypothesis that the number of accidents in a week has a Poisson distribution.

Example 1.8 Let $x_1, x_2, \ldots, x_n$ and $y_1, y_2, \ldots, y_n$ be two independent random samples from two unknown distribution functions F and G. One is interested to know whether both samples come from same distribution or not. This is a problem of nonparametric hypothesis testing.

Nonparametric tests have some distinct advantages. Nonparametric tests may be the only possible alternative in the scenarios when the outcomes are ranked, ordinal, measured imprecisely, or are subject to outliers, and parametric methods could not be implemented without making strict assumptions about the distribution of population. Another important hypothesis test is analysis of variance (ANOVA). It is based on the comparison of the variability between factor levels to average variability within a factor level, and it is used to assess differences in factor levels. The applications of ANOVA are discussed in design of experiments.

1.2.5 Problem of Correlation and Regression

Correlation refers to a broad class of relationships in statistics that involve dependence. In statistics, dependence refers to a relationship between two or more random variables or data sets, for instance, the correlation between the age of a used automobile and the retail book value of an automobile, correlation between the price and demand of a product. However, in practice, correlation often refers to linear relationship between two variables or data sets. There are various coefficients of correlation that are used to measure the degree of correlation. Correlations are useful because they can indicate a predictive relationship that can be exploited in practice. On the other hand, regression analysis is a tool to identify the relationship that exists between a dependent variable and one or more independent variables. In this technique, we make a hypothesis about the relationship and then estimate the parameters of the model and hence the regression equation. Correlation analysis can be used in two basic ways: in the determination of the predictive ability of the variable and also in determining the correlation between the two variables given.

The first part of the book discusses statistical methods whose applications in the field of design of experiments and SQC are discussed in second part of the book. For instance, in design of experiments, a well-designed experiment makes it easier to understand different sources of variation. Analysis techniques such as ANOVA and regression help to partition the variation for predicting the response or determining if the differences seen between factor levels are more than expected when compared to the variability seen within a factor level.

1.3 Design of Experiments

1.3.1 History

The concept of design of experiments was originated by Sir R. A. Fisher[3] (Montgomery 2007; Box et al. 2005). This happened when he was working at the Rothamsted Agricultural Experiment Station near London, England. The station had a huge record of happenstance data of crop yield obtained from a large number of plots of land treated every year with same particular fertilizer. It also had the records of rainfall, temperature, and so on for the same period of time. Sir Fisher was asked if he could extract additional information from these records using statistical methods. The pioneering work of Sir Fisher during 1920–1930 led to introduce, for the first time, the concept of design of experiment. This concept was further developed by many statisticians. The catalytic effect of this concept was seen after the introduction of response surface methodology by Box and Wilson in 1951. The design of experiments in conjunction with response surface methodology of analysis was used to develop, improve, and optimize processes and products. Of late, the design of experiments has started finding applications in cost reduction also. Today, a large number of manufacturing and service industries do use it regularly.

---

[3] Sir Ronald Aylmer Fisher FRS (February 17, 1890–July 29, 1962), who published as R. A. Fisher, was a British statistician and geneticist. For his work in statistics, he has been described as "a genius who almost single-handedly created the foundations for modern statistical science" and "the single most important figure in twentieth-century statistics."

1.3.2 Necessity

There are several ways an experiment can be performed. They include best-guess approach (trial-and-error method), one-factor-at-a-time approach, and design of experiment approach. Let us discuss them one by one with the help of a practical example. Suppose a product development engineer wanted to minimize the electrical resistivity of electro-conductive yarns prepared by in situ electrochemical polymerization of an electrically conducting monomer. Based on the experience, he knew that the polymerization process factors, namely polymerization time and polymerization temperature, played an important role in determining the electrical resistivity of the electro-conductive yarns. He conducted an experiment with 20 min polymerization time and 10 °C polymerization temperature and prepared an electro-conductive yarn. This yarn showed an electrical resistivity of 15.8 kΩ/m. Further, he prepared another electro-conductive yarn keeping the polymerization time at 60 min and polymerization temperature at 30 °C. Thus, prepared electro-conductive yarn exhibited electrical resistivity of 5.2 kΩ/m. He thought that this was the lowest resistivity possible to obtain, and hence, he decided not to carry out any experiment further. This strategy of experimentation, often known as best-guess approach or trial-and-error method, is frequently followed in practice. It sometimes works reasonably well if the experimenter has an in-depth theoretical knowledge and practical experience of the process. However, there are serious disadvantages associated with this approach. Consider that the experimenter does not obtain the desired results. He will then continue with another combination of process factors. This can be continued for a long time, without any guarantee of success. Further, consider that the experimenter obtains an acceptable result. He then stops the experiment, though there is no guarantee that he obtains the best solution. Another strategy of experiment that is often used in practice relates to one-factor-at-a-time approach. In this approach, the level of a factor is varied, keeping the level of the other factors constant. Then, the level of another factor is altered, keeping the level of remaining factors constant. This is continued till the levels of all factors are varied. The resulting data are then analyzed to show how the response variable is affected by varying each factor while keeping other factors constant. Suppose the product development engineer followed this strategy of experimentation and obtained the results as displayed in Fig. 1.1. It can be seen that the electrical resistivity increased from 15.8 to 20.3 kΩ/m when the polymerization time increased from 20 to 60 min, keeping the polymerization temperature constant at 10 °C. Further, it can be seen that the electrical resistivity decreased from 15.8 to 10.8 kΩ/m when the polymerization temperature raised from 10 to 30 °C, keeping the polymerization time at 20 min. The optimal combination of process factors to obtain the lowest electrical resistivity (10.8 kΩ/m) would be thus chosen as 20 min polymerization time and 30 °C polymerization temperature.

The major disadvantage of the one-factor-at-a-time approach lies in the fact that it fails to consider any possible interaction present between the factors. Interaction is said to happen when the difference in responses between the levels of one factor is not same at all levels of the other factors. Figure 1.2 displays an interaction between

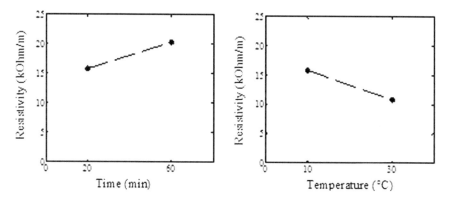

Fig. 1.1   Effects of polymerization process factors on electrical resistivity of yarns

Fig. 1.2   Effect of interaction between polymerization time and temperature

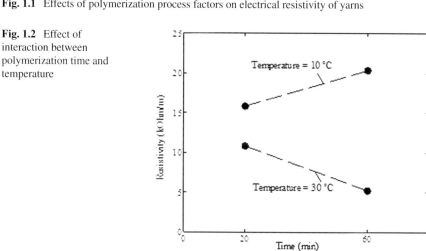

polymerization temperature and polymerization time in determining the electrical resistivity of electro-conductive yarns. It can be observed that the electrical resistivity increased with the increase of polymerization time when the polymerization temperature was kept at a lower level (10 °C). But the resistivity decreased with the increase of polymerization time when the polymerization temperature was kept at a higher level (30 °C). The lowest resistivity (5.2 kΩ/m) was registered at a polymerization time of 60 min and polymerization temperature of 30 °C. Note that the lowest resistivity obtained with the factorial experiment with four runs was much smaller than that obtained with the one-factor-at-a-time experiment with three runs. In practice, interactions between factors happen frequently; hence, the one-factor-at-a-time approach fails to produce the desirable results. The correct approach in dealing with many factors is factorial design of experiment. In this approach, the factors are varied together, instead of one at a time. Let us illustrate this concept with the help of earlier example of electro-conductive yarn. Suppose a factorial design of experiment was carried out with four runs as follows. In run 1, the polymerization time was kept at 20 min and the polymerization temperature was maintained at 10 °C. In run 2, the polymerization time was kept at 60 min and the polymerization temperature was maintained at 10 °C. In run 3, the polymerization time was kept at 20 min and the polymerization temperature was maintained at 30 °C. In run 4, the polymerization time was kept at 60 min and the polymerization temperature was kept at 30 °C. In this way, four specimens of electro-conductive yarns were prepared. This is a two-factor factorial design of experiment with both factors kept at two levels each. Let us denote the factors by symbols A and B, where A represents polymerization time and B refers to polymerization temperature. The levels are called as "low" and "high" and denoted by "−" and "+", respectively. The low level of factor A indicates 20 min, and the high level of factor A refers to 60 min. Similarly, the low level of factor B refers to 10 °C and the high level of factor B indicates 30 °C. The results of electrical resistivity for the four runs are displayed in Table 1.1. It is possible to calculate the main effects of polymerization time and polymerization temperature on the electrical resistivity of yarns. Also, it is possible to calculate the effect of interaction between polymerization time and polymerization temperature on the electrical resistivity of yarns. This is discussed below. The main effect of A is calculated from the difference between the average of the observations when A is at high level and the average of the observations when A is at low level. This is shown below.

$$A = \frac{5.2 + 20.3}{2} - \frac{10.8 + 15.8}{2} = -0.55.$$

Similarly, the main effect of B is calculated from the difference between the average of the observations when B is at high level and the average of the observations when B is at low level. This is shown below.

$$B = \frac{5.2 + 10.8}{2} - \frac{20.3 + 15.8}{2} = -10.05.$$

Similarly, the interaction effect of AB is calculated from the difference between the average of the observations when the product AB is at high level and the average of the observations when the product AB is at low level. This is shown below.

$$AB = \frac{5.2 + 15.8}{2} - \frac{10.8 + 20.3}{2} = -5.05.$$

Table 1.1 Electrical resistivity of yarn

Run	Factor A	Factor B	Product AB	Resistivity (kΩ/m)
1	−	−	+	15.8
2	+	−	−	20.3
3	−	+	−	10.8
4	+	+	+	5.2

The aforesaid analysis reveals several interesting items of information. The minus sign for the main effect of A indicates that the change of levels of A from higher to lower resulted in increase of resistivity. Similarly, the minus sign for the main effect of B indicates that the change of levels of B from higher to lower resulted in increase of resistivity. Similarly, the minus sign for the interaction effect of AB indicates that the change of levels of AB from higher to lower resulted in increase of resistivity. Further, the main effect of B is found to be the largest, followed by the interaction effect of AB and the main effect of A, respectively. One of the very interesting features of the factorial design of experiment lies in the fact that it makes the most efficient use of the experimental data. One can see that in the example of electro-conductive yarn, all four observations were used to calculate the effects of polymerization time, polymerization temperature, and their interaction. No other strategy of experimentation makes so efficient use of the experimental data. This is an essential and useful feature of factorial design of experiment. However, the number of runs required for a factorial experiment increases rapidly as the number of factor increases. For example, a complete replicate of a two-factor factorial design of experiment where each factor is varied at six levels, the total number of runs is 64. In this design, 6 of the 63 degrees of freedom correspond to the main effect, 14 of the 63 degrees of freedom correspond to two-factor interactions, and the remaining 43 degrees of freedom are associated with three-factor and higher-order interactions. If the experimenter can reasonably assume that certain higher-order interactions are negligible, then the information on the main effects and lower-order interaction effects may be obtained by running a fraction of the factorial design of experiment. That is, a one-half fraction of the two-factor factorial design of experiment where each factor is varied at six levels requires 32 runs instead of 64 runs required for the original two-factor factorial design experiment. The fractional factorial design of experiments is advantageous as a factor screening experiment. Using this experiment, the significant main effects of the factors are identified, and the insignificant factors are dropped out. It follows the principle of main effects as proposed by Lucas (1991). According to him, it is the empirical observation that the main effects are more important than the higher-order effects (whether they are two-factor interaction effect or quadratic effect). Taking the significant main effects into account, the observations of screening experiment are analyzed, and an attempt is made to fit a first-order response surface model to the data. The first-order response surface model is a mathematical representation of the linear relationship between the independent variables (factors) and the dependent variable (response). Suppose there are n number of factors $x_1, x_2, \ldots, x_n$, then the first-order response surface model takes the following form

$$\hat{y} = \hat{\beta}_0 + \sum_{i=1}^{n} \hat{\beta}_i x_i$$

where $\hat{y}$ denotes predicted response, and $\hat{\beta}$'s represent estimated coefficients. As this model contains only the main effects, it is sometimes called main effects model. Statistical tests are performed to examine if the first-order model is adequate.

If it is then the first-order model is analyzed to find a direction along which the desired response (higher or lower or target) is lying. If it is not found to be adequate, then the experiment proceeds to a second stage which involves fitting of data to a second-order response surface model as shown below

$$\hat{y} = \hat{\beta}_0 + \sum_{i=1}^{n} \hat{\beta}_i x_i + \sum_{i=1}^{n} \hat{\beta}_{ij} x_i + \sum \sum_{i<j} \hat{\beta}_{ij} x_i x_j.$$

As shown, the second-order response surface model takes into account of linear effect, quadratic effect, and interaction effect of the factors. Generally, the second-order models are determined in conjunction with response surface designs, namely central composite design, Box Behnken design. Experiments are performed following the response surface design, and the data are used to fit a higher-order model. If the model is not found to be adequate, then the experimenter returns to screening experiment with new factor-level combinations. But if the model is found to be adequate, then the second-order model is analyzed to find out the optimum levels of the process factors. This entire approach discussed above is known as sequential experimentation strategy. This works very well with design of experiments and response surface methodology of analysis.

1.3.3 Applications

The statistically designed experiments find applications in almost all kinds of industries. It is often said that wherever there are products and processes, the designed experiments can be applied. Industries like agriculture, chemical, biochemical, pharmaceutical, semiconductor, mechanical, textile, and automobile do use it regularly. Needless to say, there are numerous research articles available that demonstrate widespread applications of statistically designed experiments in many processes, product, and management-related activities, including process characterization, process optimization, product design, product development, and cost reduction. Some examples illustrating the typical applications of statistically designed experiments are given below.

Example 1.9 (Process characterization using statistically designed experiment)
The rotary ultrasonic machining process is used to remove materials from ceramics for the development of advanced ceramic products with precise size and shape. In this process, a higher material removal rate is always desired. It was of interest to Hu et al. (2002) to investigate the effects of machining factors namely static force, vibration amplitude, rotating speed, abrasive grit size, and abrasive grit number on the rate of removal of materials. The authors wished to examine the main effect and the interaction effect of the aforesaid machining factors on the material removal rate. They conducted a factorial design of experiment involving the aforesaid five factors, each varied at two levels. The static force was varied as 100 and 350 N, the vibration

amplitude was varied as 0.02 and 0.04 mm, the rotating speed was varied at 1000 and 3000 rpm, the abrasive grit size was varied as 0.1 and 0.3 mm, and the abrasive grit number was varied as 300 and 900. The experimental data were analyzed to estimate the main effects, two-factor interaction effects, and three-factor interaction effects. Of the main effects, static force, vibration amplitude, and grit size were found to be statistically significant. For two-factor interactions, the interactions between static force and vibration amplitude, between static force and grit size, between vibration amplitude and grit size were significant. For three-factor interactions, the interactions among static force, vibration amplitude, and grit size were significant. The best combination for material removal rate was found with higher static force, larger vibration amplitude, and larger grit size. In addition to this, there are many studies reported on process characterization using designed experiments (Kumar and Das 2017; Das et al. 2012).

Example 1.10 (*Process optimization using statistically designed experiment*)
Corona discharge process is used to apply electrical charge onto fibrous filter media for enhancement of particle capture. Thakur et al. (2014) attempted to optimize this process to achieve higher initial surface potential and higher half-decay time simultaneously. A set of fibrous filter media was prepared by varying the corona charging process factors namely applied voltage, charging time, and distance between electrodes in accordance with a three-factor, three-level factorial design of experiment. The experimental data of initial surface potential and half-decay time were analyzed statistically. The initial surface potential was found to be higher at higher applied voltage, longer duration of charging, and lower distance between electrodes. But the half-decay time was found to be higher at lower applied voltage. Further, the half-decay time increased initially with the increase in charging time and distance between electrodes, but an increase in both the process factors beyond the optimum regions resulted in a decrease in half-decay time. The simultaneous optimization of initial surface potential and half-decay time was carried out using desirability function approach. It was found that the corona charging process set with 15 kV applied voltage, 29.4 min charging time, and 26.35 mm distance between electrodes was found to be optimum, yielding initial surface potential of 10.56 kV and half-decay time of 4.22 min. Also, there are many recent studies reported where the authors attempted to optimize processes using designed experiments (Kumar and Das 2017; Kumar et al. 2017; Thakur and Das 2016; Thakur et al. 2016, 2014; Das et al. 2012a; Pal et al. 2012).

Example 1.11 (*Product design using statistically designed experiment*)
It is well known that the fuel efficiency of automobiles can be achieved better by reduction of vehicle weight. With a view to this, an attempt was made by Park et al. (2015) to design a lightweight aluminum-alloyed automotive suspension link using statistically designed experiment and finite element analysis. Seven design factors of the link were identified, and each factor was varied at two levels. The design factors chosen were number of truss, height of truss, thickness of truss, thickness of upper rib, thickness of lower rib, thickness of vertical beam, and width of link. A 2^7 full factorial design of experiment was carried out, and the weight, stress, and

stiffness of the links were determined. By optimization, the weight of the aluminum suspension link was obtained as 58% of that of the initial steel suspension link, while the maximum von Mises stress was reduced to 37% and stiffness was increased to 174% of those of the steel suspension link. In the literature, many reports are available on product design using designed experiments (Pradhan et al. 2016; Das et al. 2014).

Example 1.12 (*Product development using statistically designed experiment*)
The statistically designed experiments are very popular for research and development in pharmaceutical science. There are many case studies reported on the formulation of tablets using designed experiments by the US Food and Drug Administration (FDA). Besides, a large number of articles are available on this topic in the literature. In one of those articles, Birajdar et al. (2014) made an attempt to formulate fast disintegrating tablets for oral antihypertensive drug therapy. A 2^3 factorial design was applied to investigate the effects of concentration of Isabgol mucilage, concentration of sodium starch glycolate (SSG), and concentration of microcrystalline cellulose (MCC) on the disintegration time of losartan potassium tablets. The analysis of experimental data revealed that the minimum disintegration time was found to be 46 s with 16 mg Isabgol mucilage, 12 mg SSG, and 40 mg MCC. Besides this study, there are many other studies reported on the development of products using designed experiments (Kaur et al. 2013; Das et al. 2012b).

Example 1.13 (*Cost reduction using statistically designed experiment*)
Of late, the statistically designed experiments are started to be used for cost reduction. Phadke and Phadke (2014) made an investigation whether the design of experiments could reduce the IT system testing cost. They organized 20 real end-to-end case studies using orthogonal arrays (OA) for generating test plans at 10 large financial services institutions and compared the results with business-as-usual (BAU) process. It was found that the OA-based testing resulted in an average reduction of total test effort (labor hours) by 41%. Also, in 40% of the cases, the OA-based testing process found more defects than the BAU process.

1.4 Statistical Quality Control

Statistical quality control (SQC) is one of the important applications of statistical techniques in manufacturing industries. Typically, the manufacturing industries receive raw material from the vendors. It is then necessary to inspect the raw material before taking a decision whether to accept them or not. In general, the raw material is available in lots or batches (population). It is practically impossible to check each and every item of the raw material. So a few items (sample) are randomly selected from the lot or batch and inspected individually before taking a decision whether the lot or batch is acceptable or not. Here, two critical questions arise: (1) How many items should be selected? and (2) how many defective items in a sample, if found, would call for rejection of the lot or batch? These questions are answered through acceptance sampling technique. Using this technique, if the raw material is not found to be acceptable, then it may be returned to the vendor. But if it is found to be acceptable, then it may be processed through a manufacturing process and finally converted into products. In order to achieve the targeted quality of the products, the manufacturing process needs to be kept under control. This means that there should not be any assignable variation present in the process. The assignable variation is also known as non-random variation or preventable variation. Examples of assignable variation include defective raw material, faulty equipment, improper handling of machines, negligence of operators, unskilled technical staff. If the process variation is arising due to random variation, the process is said to be under control. But if the process variation is arising due to assignable variation, then the process is said to be out of control. Whether the manufacturing process is under control or out of control can be found through a technique, called control chart. It is therefore clear that control chart helps to monitor a manufacturing process. Once the manufactured products are prepared, they will be again inspected for taking a decision whether to accept or reject the products. The statistical technique used for taking such decisions is known as acceptance sampling technique.

References

Birajdar SM, Bhusnure OG, Mulaje SS (2014) Formulation and evaluation of fast disintegrating losartan potassium tablets by formal experimental design. Int J Res Dev Pharm Life Sci 3:1136–1150

Box GE, Hunter JS, Hunter WG (2005) Statistics for experimenters. Wiley, New York

Das D, Butola BS, Renuka S (2012) Optimization of fiber-water dispersion process using Box-Behnken design of experiments coupled with response surface methodology of analysis. J Dispers Sci Tech 33:1116–1120

Das D, Mukhopadhyay S, Kaur H (2012a) Optimization of fiber composition in natural fiber reinforced composites using a simplex lattice design. J Compos Mater 46:3311–3319

Das D, Thakur R, Pradhan AK (2012b) Optimization of corona discharge process using Box-Behnken design of experiments. J Electrost 70:469–473

Das D, Das S, Ishtiaque SM (2014) Optimal design of nonwoven air filter media: effect of fiber shape. Fibers Polym 15:1456–1461

Hu P, Zhang JM, Pei ZJ, Treadwell C (2002) Modeling of material removal rate in rotary ultrasonic machining: designed experiments. J Mater Process Tech 129:339–344

Kaur H, Das D, Ramaswamy S, Aslam B, Gries T (2013) Preparation and optimization of Flax/PLA yarn performs for development of biocomposites. Melliand Int. 19

Kumar N, Das D (2017) Alkali treatment on nettle fibers. part II: enhancement of tensile properties using design of experiment and desirability function approach, J Text Inst 108:1468–1475

Kumar V, Mukhopadhyay S, Das D (2017) Recipe optimization for sugarcane bagasse fibre reinforced soy protein biocomposite. Indian J Fibre Text Res 42:132–137

Lucas JM (1991) Using response surface methodology to achieve a robust process. In: 45th annual quality congress transactions, vol 45. Milwaukee, WI, pp 383–392

Montgomery DC (2007) Design and analysis of experiments. Wiley, New York

Pal R, Mukhopadhyay S, Das D (2012) Optimization of micro injection molding process with respect to tensile properties of polypropylene. Indian J Fiber Text Res 37:11–15

Park JH, Kim KJ, Lee JW, Yoon JK (2015) Leight-weight design of automotive suspension link based on design of experiment. Int J Automot Tech 16:67–71

Phadke MS, Phadke KM (2014) Utilizing design of experiments to reduce IT system testing cost. In: 2014 Annual reliability and maintenance symposium. IEEE, New York, pp 1–6

Pradhan AK, Das D, Chattopadhyay R, Singh SN (2016) An approach of optimal designing of nonwoven air filter media: effect of fiber fineness. J Ind Text 45(6):1308–1321

Thakur R, Das D (2016) A combined Taguchi and response surface approach to improve electret charge in corona-charged fibrous electrets. Fibers Polym 17:1790–1800

Thakur R, Das D, Das A (2014) Optimization of charge storage in corona-charged fibrous electrets. J Text Inst 105:676–684

Thakur R, Das D, Das A (2016) Optimization study to improve filtration behavior of electret filter media. J Text Inst 107:1456–1462

Chapter 2
Review of Probability

If mathematics is the queen of sciences, then probability is the queen of applied mathematics. The concept of probability originated in the seventeenth century and can be traced to games of chance and gambling. Games of chance include actions like drawing a card, tossing a coin, selecting people at random and noting number of females, number of calls on a telephone, frequency of accidents, and position of a particle under diffusion. Today, probability theory is a well-established branch of mathematics that finds applications from weather predictions to share market investments. Mathematical models for random phenomena are studied using probability theory.

2.1 Basics of Probability

Probability theory makes predictions about experiments whose outcomes depend upon chance.

Definition 2.1 (*Random Experiment*) An experiment is said to be a random experiment if

1. All the possible outcomes of the experiment are known in advance. This implies that before the experiment is executed, we are aware of all the possible outcomes.
2. At any execution of the experiment, the final outcome is not known in advance.
3. The experiment can be repeated under identical conditions any number of times. Here, identical conditions mean that the situation or scenario will not change when the experiment is repeated.

© Springer Nature Singapore Pte Ltd. 2018
D. Selvamuthu and D. Das, *Introduction to Statistical Methods,*
Design of Experiments and Statistical Quality Control,
https://doi.org/10.1007/978-981-13-1736-1_2

Let Ω denotes the set of all possible outcomes of a random experiment. For example,

1. In the random experiment of tossing a coin, $\Omega = \{H, T\}$.
2. In the random experiment of observing the number of calls in a telephone exchange, we have $\Omega = \{0, 1, 2, \ldots\}$.
3. In the random experiment of measuring the lifetime of a light bulb, $\Omega = [0, \infty)$.

From the above examples, one can observe that the elements of Ω can be non-numerical, integers, or real numbers. Also, the set Ω may be finite or countably infinite or uncountable.

Definition 2.2 (*Sample Points, Sample Space, and Events*) An individual element $w \in \Omega$ is called a sample point. Let S denote the collection of all possible subsets of Ω including the null set. The pair (Ω, S) or Ω itself is called the sample space, and any element of S is called an event. That means, if $A \subseteq \Omega$, then A is an event. Note that, the null set, denoted by $\emptyset$, is also a subset of Ω and hence is an event.

For example, in the random experiment of tossing two coins, we have

$$\Omega = \{(HH), (HT), (TH), (TT)\}$$

and

$$S = \{\emptyset, \{(HH)\}, \{(HT)\}, \{(TH)\}, \{(TT)\}, \{(HH), (HT)\}, \ldots, \Omega\}.$$

Here, (HH) is a sample point and $\{(HH), (TT)\}$ is an event.
    Using the set operations on events in S, we can get other events in S. For example,

1. $A \cup B$, called union of A and B, represents the event "either A or B or both."
2. $A \cap B$, called intersection of A, B and represents the event "both A and B."
3. A^c, called complement of A, represents the event "not A."

Definition 2.3 (*Equally Likely Outcomes*) The outcomes are said to be equally likely if and only if none of them is expected to occur in preference to the other.

Definition 2.4 (*Mutually Exclusive Events*) Two events are said to be mutually exclusive if the occurrence of one of them rules out the occurrence of the other; i.e., two events A and B are mutually exclusive if the occurrence of A implies B cannot occur and vice versa. We have $A \cap B = \phi$.

Definition 2.5 (*Mutually Exhaustive Events*) Two or more events are said to be mutually exhaustive if there is a certain chance of occurrence of at least one of them when they are all considered together. In that case, we have $\cup A_i = \Omega$.

2.1.1 Definition of Probability

Definition 2.6 (*Classical Definition of Probability*) Let a random experiment results in n mutually exhaustive, mutually exclusive, and equally likely outcomes. If n_A of these outcomes have an attribute event A, then the probability of A is given by

$$P(A) = \frac{n_A}{n}. \tag{2.1}$$

In other words, it is the ratio of the cardinality of the event to the cardinality of the sample space. Note that, the classical definition of probability has a drawback that Ω must be finite. But in real-world problems, Ω may not be finite. Hence, to overcome this, Kolmogorov[1] introduced the axiomatic definition of probability which is stated as follows

Definition 2.7 (*Axiomatic Definition of Probability*) Let (Ω, S) be the sample space. A real-valued function $P(\cdot)$ is defined on S satisfying the following axioms:

1. $P(A) \geq 0 \ \forall \ A \in S$.   (Nonnegative property)
2. $P(\Omega) = 1$.   (Normed property)
3. If $A_1, A_2, \ldots$ is a countable sequence of mutually exclusive events in S, then

$$P\left(\bigcup_{i=1}^{\infty} A_i\right) = \sum_{i=1}^{\infty} P(A_i). \qquad \text{(Countable additivity)}$$

When the above property is satisfied for finite sequences, it is called finite additivity.

Then, P is called the probability function.

The axiomatic definition of probability reduces to the classical definition of probability when Ω is finite, and each possible outcome is equally likely. From the above definition, one can observe that P is a set function which assigns a real number to subsets of Ω. In particular, P is a normalized set function in the sense that $P(\Omega) = 1$. For each subset A of Ω, the number $P(A)$ is called the probability that the outcome of the random experiment is an element of the set A, or the probability of the event A, or the probability measure of the set A. We call (Ω, S, P) a probability space.

An event A is said to be a sure event if $P(A) = 1$. Similarly, an event with probability 0, i.e., $P(A) = 0$ is known as null or impossible event. An event whose probability of occurrence is very small is known as a rare event.

Results:
When (Ω, S, P) is a probability space, the following results hold.

1. $P(A^c) = 1 - P(A), \ \forall A \in S$.

---

[1] Andrey Nikolaevich Kolmogorov (1903–1987) was a twentieth-century Russian mathematician who made significant contributions to the mathematics of probability theory. It was Kolmogorov who axiomatized probability in his fundamental work, Foundations of the Theory of Probability (Berlin), in 1933.

2. $P(\emptyset) = 0$.
3. If $A \subseteq B$, then $P(A) \le P(B)$.

Example 2.1 From the past experience, a stockbroker believes that under the current conditions of the economy, an investor will invest in risk-free bonds with probability 0.6, will invest in risky asset with a probability of 0.3, and will invest in both risk-free bonds and risky asset with a probability of 0.15. Find the probability that an investor will invest

1. in either risk-free bonds or risky asset.
2. in neither risk-free bonds nor risky asset.

Solution:
Let A denote the event that an investor will invest in risk-free bonds and B denote the event that an investor will invest in risky asset. It is given that

$$P(A) = 0.6, \quad P(B) = 0.3, \quad P(A \cap B) = 0.15.$$

1. Probability that the investor will invest in either risk-free bonds or risky asset is given by

$$P(A \cup B) = P(A) + P(B) - P(A \cap B) = 0.6 + 0.3 - 0.15 = 0.75.$$

2. Probability the investor will invest neither in risk-free bonds nor in risky assets is given by
$$P(A^c \cap B^c) = 1 - P(A \cup B) = 1 - 0.75 = 0.25.$$

2.1.2   Conditional Probability

Definition 2.8 (*Conditional Probability*) Let (Ω, S, P) be a probability space. Let $B \in S$ be any event with $P(B) > 0$. For any other event $A \in S$, the conditional probability of A given B, denoted by $P(A/B)$, is defined as

$$P(A/B) = \frac{P(A \cap B)}{P(B)}.$$

If $P(B) = 0$, then the conditional probability is not defined.

Conditional probability provides us a tool to discuss the outcome of an experiment on the basis of partially available information. It can be easily proved that the conditional probabilities $P(A \mid B)$ for a fixed event B is itself a probability function. Hence, one can treat conditional probabilities as probabilities on a reduced sample space, i.e., space obtained by discarding possible outcomes outside B.

Example 2.2 Consider the experiment of tossing an unbiased coin twice. Find the probability of getting a tail in the second toss given that a head has occurred on the first toss.

Solution:
The sample space Ω is $\{\{HH\}, \{HT\}, \{TH\}, \{TT\}\}$. Let the event of getting a head on the first toss and the event of getting a tail in the second toss be A and B, respectively. Then,

$$P(A) = 0.5, \quad P(B) = 0.5, \quad P(A \cap B) = 0.25.$$

Now, according to the definition,

$$P(B/A) = \frac{P(A \cap B)}{P(A)} = \frac{0.25}{0.5} = 0.5.$$

Notice how this is different from the probability of the event of getting a head before a tail in the two tosses.

Example 2.3 An automobile is being filled with petrol. The probability that oil is also need to be changed is 0.25, the probability that a new filter is needed is 0.40, and the probability that both the filter and oil need to be changed is 0.14.

1. Given that oil need to be changed, find the probability that a new oil filter is required?
2. Given that a new oil filter is required, find the probability that the oil need to be changed?

Solution:
Let A be the event that an automobile being filled with petrol will also need an oil change, and B be the event that it will need a new oil filter. Then,

$$P(A) = 0.25, \quad P(B) = 0.40, \quad P(A \cap B) = 0.14.$$

1. Probability that a new oil filter is required given that oil had to be changed is given by

$$P(B/A) = \frac{P(A \cap B)}{P(A)} = \frac{0.14}{0.25} = 0.56.$$

2. Probability that oil has to be changed given that a new filter is needed is given by

$$P(A/B) = \frac{P(A \cap B)}{P(B)} = \frac{0.14}{0.40} = 0.35.$$

Definition 2.9 (*Independent Events*) Two events A and B defined on a probability space (Ω, S, P) are said to be independent if and only if $P(A \cap B) = P(A)P(B)$.

Remark 2.1   1. If $P(A) = 0$, then A is independent of any event $B \in S$.
   2. Any event is always independent of the events Ω and $\emptyset$.
   3. If A and B are independent events and $P(A \cap B) = 0$, then either $P(A) = 0$ or $P(B) = 0$.
   4. If $P(A) > 0$; $P(B) > 0$ and A, B are independent, then they are not mutually exclusive events.

The reader should verify the above remarks using the definition of independence.

Definition 2.10 (*Pairwise Independent Events*) Let U be a collection of events from S. We say that the events in U are pairwise independent if and only if for every pair of distinct events A, $B \in U$, $P(A \cap B) = P(A)P(B)$.

Definition 2.11 (*Mutually Independent Events*) Let U be a collection of events from S. The events in U are mutually independent if and only if for any finite subcollection $A_1, A_2, \ldots, A_k$ of U, we have

$$P(A_1 \cap A_2 \cap \cdots \cap A_k) = \prod_{i=1}^{k} P(A_i) \ \forall \ k.$$

Example 2.4 Suppose that a student can solve 75% of the problems of a mathematics book while another student can solve 70% of the problems of the book. What is the chance that a problem selected at random will be solved when both the students try?

Solution: Let A and B be the events that students can solve a problem, respectively. Then, $P(A) = 0.75$, $P(B) = 0.70$. Since A and B are independent events, we have

$$P(A \cap B) = P(A)P(B) = 0.75 \times 0.70.$$

Hence, the chance that the problem selected at random will be solved when both the students try is obtained as

$$P(A \cup B) = P(A) + P(B) - P(A \cap B) = 0.75 + 0.70 - (0.75 \times 0.70) = 0.925.$$

Example 2.5 Consider a random experiment of selecting a ball from an urn containing four balls numbered 1, 2, 3, 4. Suppose that all the four outcomes are assumed equally likely. Let $A = \{1, 2\}$, $B = \{1, 3\}$, and $C = \{1, 4\}$ be the events. Prove that these events are pairwise independent but not mutually independent.

Solution:
We have, $A \cap B = \{1\} = A \cap C = B \cap C = A \cap B \cap C$. Then,

$$P(A) = \frac{2}{4} = \frac{1}{2}, P(B) = \frac{2}{4} = \frac{1}{2}, P(C) = \frac{2}{4} = \frac{1}{2}.$$

$$P(A \cap B) = \frac{1}{4}, P(A \cap C) = \frac{1}{4}, P(B \cap C) = \frac{1}{4} \text{ and } P(A \cap B \cap C) = \frac{1}{4}.$$

As we know, if $P(A \cap B) = P(A)P(B)$ then A and B are pairwise independent events. We can see A and B, B and C, A and C are pairwise independent events. To be mutually independent events, we have to check $P(A \cap B \cap C) = P(A)P(B)P(C)$. Here,

$$P(A)P(B)P(C) = \frac{1}{8} \neq P(A \cap B \cap C).$$

Hence, A, B, C are pairwise independent events but not mutually independent events.

2.1.3 Total Probability Rule

Definition 2.12 (*Total Probability Rule*) Let B_1, B_2, ..., be countably infinite mutually exclusive events in the probability space (Ω, S, P) such that $\Omega = \bigcup_i B_i$ where $P(B_i) > 0$ for $i = 1, 2, \ldots$. Then, for any $A \in S$, we have

$$P(A) = \sum_i P(A/B_i)P(B_i).$$

Definition 2.13 (*Multiplication Rule*) Let A_1, A_2, ..., A_n be arbitrary events in a given probability space (Ω, S, P) such that $P(A_1 \cap A_2 \cap \cdots \cap A_{n-1}) > 0$, for any $n > 1$, then

$$P(A_1 \cap A_2 \cap \ldots A_n) = P(A_1)P(A_2/A_1)P(A_3/A_1 \cap A_2) \ldots P(A_n/A_1 \cap A_2 \cap \ldots A_{n-1}). \tag{2.2}$$

Example 2.6 Consider an urn which contains ten balls. Let three of the ten balls are red and other balls are blue. A ball is drawn at random at each trial, its color is noted, and it is kept back in the urn. Also, two additional balls of the same color are also added to the urn.

1. What is the probability that a blue ball is selected in the second trial?
2. What is the probability that a red ball is selected in the first three trials?

Solution:

Let R_i be the event that a red ball is selected in the ith trial.

1. We need to find $P(R_2^c)$. By total probability rule,

$$P(R_2^c) = P(R_2^c/R_1)P(R_1) + P(R_2^c/R_1^c)P(R_1^c) = \frac{7}{12} \times \frac{3}{10} + \frac{9}{12} \times \frac{7}{10} = \frac{7}{10}.$$

2. We require to find $P(R_1 \cap R_2 \cap R_3)$. By multiplication rule (Eq. (2.2)), we get the probability that a blue ball is selected in each of the first three trials is

$$P(R_1 \cap R_2 \cap R_3) = \frac{3}{10} \times \frac{5}{12} \times \frac{7}{14} = 0.0625.$$

2.1.4 Bayes' Theorem

We state a very important result in the form of the following theorem in conditional probability which has wide applications.

Theorem 2.1 (Bayes'[2] Rules (or Bayes' Theorem)) *Let $B_1, B_2, \ldots$ be a collection of mutually exclusive events in the probability space (Ω, S, P) such that $\Omega = \bigcup_i B_i$ and $P(B_i) > 0$ for $i = 1, 2, \ldots$. Then, for any $A \in S$ with $P(A) > 0$, we have*

$$P(B_i/A) = \frac{P(A/B_i)P(B_i)}{\sum_j P(A/B_j)P(B_j)}.$$

Proof By the definition of conditional probability, for $i = 1, 2, \ldots$

$$P(B_i/A) = \frac{P(B_i \cap A)}{P(A)}, \quad P(A/B_i) = \frac{P(B_i \cap A)}{P(B_i)}.$$

Combining above two equations, we get

$$P(B_i/A) = \frac{P(A/B_i)P(B_i)}{P(A)}.$$

By total probability rule,

$$P(A) = \sum_j P(A/B_j)P(B_j).$$

Therefore, for $i = 1, 2, \ldots$

$$P(B_i/A) = \frac{P(A/B_i)P(B_i)}{\sum_j P(A/B_j)P(B_j)}.$$

Example 2.7 A company produces and sells three types of products, namely *I*, *II*, and *III*. Based on transactions in the past, the probability that a customer will purchase product *I* is 0.75. Of those who purchase product *I*, 60% also purchase product *III*. But 30% of product *II* buyers purchase product *III*. A randomly

---

[2]Thomas Bayes (1702–1761) was a British mathematician known for having formulated a special case of Bayes' Theorem. Bayes' Theorem (also known as Bayes' rule or Bayes' law) is a result in probability theory, which relates the conditional and marginal probability of events. Bayes' theorem tells how to update or revise beliefs in light of new evidence: a posteriori.

selected buyer purchases two products out of which one product is III. What is the probability that the other product is I?

Solution:

Let A be the event that a customer purchases product I, B be the event that a customer purchases product II, and E be the event that a customer purchases product III. Then,

$$P(A) = 0.75, \quad P(B) = (1 - 0.75) = 0.25, \quad P(E/A) = 0.60 \text{ and } P(E/B) = 0.30.$$

Probability that a customer purchased product I given that he has purchased two products with one product being III is given by

$$P(A/E) = \frac{P(E/A)P(A)}{P(E/A)P(A) + P(E/B)P(B)} = \frac{0.60 \times 0.75}{0.60 \times 0.75 + 0.25 \times 0.30} = \frac{6}{7}.$$

Example 2.8 A box contains ten white and three black balls while another box contains three white and five black balls. Two balls are drawn from the first box and put into the second box, and then a ball is drawn from the second. What is the probability that it is a white ball?

Solution:

Let A be the event that both the transferred balls are white, B be the event that both the transferred balls are black, and C be the event that out of the transferred balls one is black while the other is white. Let W be the event that a white ball is drawn from the second box.

$$P(A) = \frac{15}{26}, \; P(B) = \frac{1}{26}, \; P(C) = \frac{5}{13}.$$

By total probability rule,

$$P(W) = P(W/A)P(A) + P(W/B)P(B) + P(W/C)P(C).$$

If A occurs, box II will have five white and five black balls. If B occurs, box II will have three white seven and black balls. If C occurs, box II will have four white and six black balls.

$$P(W/A) = \frac{5}{10}, \; P(W/B) = \frac{3}{10}, \; P(W/C) = \frac{4}{10}.$$

Thus,

$$P(W) = P(W/A)P(A) + P(W/B)P(B) + P(W/C)P(C)$$
$$= \left(\frac{5}{10} \times \frac{15}{26}\right) + \left(\frac{3}{10} \times \frac{1}{26}\right) + \left(\frac{4}{10} \times \frac{10}{26}\right) = \frac{59}{130}.$$

Example 2.9 There are 2000 autos, 4000 taxis, and 6000 buses in a city. A person can choose any one of these to go from one place to other. The probabilities of an accident involving an auto, taxi, or bus are $0.01, 0.03$, and 0.15, respectively. Given that the person met with an accident, what is the probability that he chose an auto?

Solution:
Let A, B, and C, respectively, be the events that the person hired an auto, a taxi, or a bus; and E be the event that he met with an accident. We have,

$$P(A) = \frac{2000}{12000} = \frac{1}{6}, \quad P(B) = \frac{4000}{12000} = \frac{1}{3}, \quad P(C) = \frac{6000}{12000} = \frac{1}{2}.$$

Given,

$$P(E/A) = 0.01, \quad P(E/B) = 0.03, \quad P(E/C) = 0.15.$$

Thus, the probability that the person who met with an accident hired an auto is

$$P(A/E) = \frac{P(E/A)P(A)}{P(E/A)P(A) + P(E/B)P(B) + P(E/C)P(C)}$$

$$= \frac{\left(0.01 \times \frac{1}{6}\right)}{0.01 \times \frac{1}{6} + 0.03 \times \frac{1}{3} + 0.015 \times \frac{1}{2}} = \frac{2}{23}.$$

2.2 Random Variable and Distribution Function

For mathematical convenience, it is often desirable to associate a real number to every element of the sample space. With this in mind, we define a random variable as follows:

Definition 2.14 (*Random Variable*) A function X, which assigns to each element $w \in \Omega$ a unique real number $X(w) = x$, is called a random variable (RV).

Since X is a real-valued function, the domain of X is the sample space Ω and co-domain is a set of real numbers. The set of all values taken by X, called the image of X or the range of X, denoted by R_X, will be a subset of the set of all real numbers.

Remark 2.2 1. The term "random variable" is actually not an appropriate term, since a random variable X is really a function. When we say that X is a random variable, we mean that X is a function from Ω to the real line; i.e., $(-\infty, \infty)$ and $X(w)$ are the values of the function at the sample point $w \in \Omega$.
2. A random variable partitions the sample space Ω into mutually exclusive and collectively exhaustive set of events. We can write

$$\Omega = \bigcup_{x \in R_X} A_x, \text{ such that } A_x \cap A_y = \emptyset \text{ if } x \neq y$$

where

$$A_x = \{w \in \Omega \mid X(w) = x\}, \quad x \in R_X \tag{2.3}$$

is the collection of the sample points such that $\{X(w) = x\}$ and is an event.

Now, we consider different types of random variables based on R_X, the image of X. We classify them into three cases: the first case in which the random variable assumes at most countable number of values, the second case where the random variable assumes the value of some interval or within any collection of intervals, and the third case in which the random variable assumes both types of values.

Definition 2.15 (*Cumulative Distribution Function*) Let X be a random variable defined on probability space (Ω, S, P). For every real number x, we have

$$P(X \le x) = P\{w \in \Omega : X(w) \le x\}, \quad -\infty < x < \infty.$$

This point function is denoted by the symbol $F(x) = P(X \le x)$. The function $F(x)$ (or $F_X(x)$) is called the cumulative distribution function (CDF) of the random variable X.

$F(x)$ satisfies the following properties:

1. $0 \le F(x) \le 1, -\infty < x < \infty$.
2. $F(x)$ is a nondecreasing function of x. That is, for any $a < b$, we have $F(a) \le F(b)$.
3. $\lim_{x \to -\infty} F(x) = 0$ and $\lim_{x \to +\infty} F(x) = 1$.
4. $F(x)$ is a (right) continuous function of x. That is, for any x, we have $\lim_{h \to 0} F(x + h) = F(x)$.

2.2.1 Discrete Type Random Variable

Definition 2.16 (*Discrete Type Random Variable*) A random variable X is called a discrete type random variable if there exists a countable set $E \subseteq \mathbb{R}$ such that $P(X \in E) = 1$. The points of E that have positive mass are called jump points.

Thus, for a discrete type random variable, we may list down the possible values of X as $x_1, x_2, \ldots, x_n, \ldots$, where the list may terminate for some finite n or it may continue indefinitely. That means if X is a discrete type random variable, then its range space R_X is finite or countably infinite.

Let X be a random variable assuming values $x_1, x_2, \ldots, x_n, \ldots$ with respective probabilities $p(x_1), p(x_2), \ldots, p(x_n), \ldots$. From definition, we have

$$F_X(x) = \sum_{i \in I} p(x_i)$$

where $I = \{i;\ x_i \le x\}$. We note that $F_X(x)$ is right continuous and $P(X = x_i) = p(x_i) = F_X(x_i) - F_X(x_i^-)$. In conclusion, if X is a discrete type random variable, then its probabilities can be obtained from its cumulative distribution function and vice versa.

By the definition of a random variable and the concept of an event space introduced in the earlier section, it is possible to evaluate the probability of the event A_x. We have defined the event A_x, in Eq. (2.3), as the set of all sample points $\{w \in \Omega \mid X(w) = x\}$. Consequently,

$$P(A_x) = P\{X = x\} = P\{w \in \Omega : X(w) = x\} = \sum_{\{w \in \Omega : X(w) = x\}} P\{w\}.$$

This formula provides us with a method of computing $P(X = x)$ for all $x \in \mathbb{R}$.

Definition 2.17 (*Probability Mass Function*) Let X be a discrete type random variable defined on Ω of a random experiment and taking the values $\{x_1, x_2, \ldots, x_i, \ldots\}$. A function with its domain consisting of real numbers and with its range in the closed interval $[0, 1]$ is known as the probability mass function (PMF) of the discrete type random variable X and will be denoted by $p(x)$. It is defined as

$$p(x_i) = P(X = x_i) = \sum_{\{w \in \Omega \mid X(w) = x_i\}} P\{w\}, \quad i = 1, 2, \ldots. \qquad (2.4)$$

Then, the collection of pairs $\{(x_i, p(x_i)), i = 1, 2, \ldots\}$ is called the probability distribution of the discrete type random variable X.

The following properties hold for the PMF:

1. $p(x_i) \ge 0 \ \forall\, i$
2. $\sum_i p(x_i) = 1$.

Clearly, we have the following information associated with a discrete type random variable X as shown in Table 2.1.

The advantage of the probability distribution is that it helps us to evaluate the probability of an event associated with a discrete type random variable X. So far, we have restricted our attention to computing $P(X = x)$, but often we may be interested in computing the probability of the set $\{w \in \Omega \mid X(w) \in B\}$ for some subset B of $\mathbb{R}$ other than a one-point set. It is clear that

$$\{w \in \Omega \mid X(w) \in B\} = \bigcup_{x_i \in B} \{w \in \Omega \mid X(w) = x_i\}.$$

Table 2.1 Probability of the random variable X

$X = x$	x_1	x_2	$\cdots$	x_i	$\cdots$
$p(x)$	$p(x_1)$	$p(x_2)$	$\cdots$	$p(x_i)$	$\cdots$

Usually this event is denoted by $[X \in B]$ and its probability by $P(X \in B)$. Let $I = \{i \in \mathbb{N}^+ \mid x_i \in B\}$. Then, $P(X \in B)$ is evaluated as:

$$P(X \in B) = P(\{w \in \Omega \mid X(w) \in B\})$$
$$= \sum_{i \in I} P(X = x_i) \text{ such that } x_i \in B$$
$$= \sum_{i \in I} p(x_i). \quad \text{(from Eq. (2.4))}$$

Thus, the probability of an event $\{B \subseteq \mathbb{R}\}$ is evaluated as the sum of the probabilities of the individual outcomes consisting the event B. Also, in every finite interval $[a, b]$, there will be at most countable number of possible values of the discrete type random variable X. Therefore

$$P(a \leq X \leq b) = \sum_i P(X = x_i) = \sum_i p(x_i) \text{ such that } x_i \in [a, b].$$

If the interval $[a, b]$ contains none of the possible values x_i, we assign $P(a \leq X \leq b) = 0$. The semi-infinite interval $A = (-\infty, x]$ will be of special interest, and in this case, we denote the event $[X \in A]$ by $[X \leq x]$.

Example 2.10 A fair coin is tossed two times. Let X be the number of heads that appear. Obtain the probability distribution of the random variable X. Also compute $P[0.5 < X \leq 4]$, $P[-1.5 \leq X < 1]$ and $P[X \leq 2]$.

Solution:
In this example, the possible values for X are 0, 1, and 2, respectively. Hence,

$$P_X(0) = P(X = 0) = P((T, T)) = 0.25,$$
$$P_X(1) = P(X = 1) = P((T, H)) + P((H, T)) = 0.25 + 0.25 = 0.5,$$
$$P_X(2) = P(X = 2) = P((H, H)) = 0.25.$$

Now,

$$P(0.5 < X \leq 4) = P(X = 1) + P(X = 2) = 0.75,$$
$$P(-1.5 \leq X < 1) = P(X = 0) = 0.25,$$
$$P(X \leq 2) = P(X = 0) + P(X = 1) + P(X = 2) = 1.$$

The probability mass function of the random variable X is shown in Table 2.2.

Table 2.2 Probability mass function of random variable X (Example 2.10)

$X = x$	0	1	2
$p(x)$	0.25	0.5	0.25

Example 2.11 Two fair dice are tossed and let X be the sum of the numbers on the two faces shown. Find the CDF of X.

Solution:
The possible values of X are $\{2, 3, \ldots, 12\}$. The CDF of X is given by

x	2	3	4	5	6	7	8	9	10	11	12
$F(x)$	$\frac{1}{36}$	$\frac{3}{36}$	$\frac{6}{36}$	$\frac{10}{36}$	$\frac{15}{36}$	$\frac{21}{36}$	$\frac{26}{36}$	$\frac{30}{36}$	$\frac{33}{36}$	$\frac{35}{36}$	1

Example 2.12 Let 30% of the items in a box are defective. Five items are selected from this box. What is the probability distribution of number of defective items out of those five items.

Solution:
The probability that an item is defective is 0.3. Let X denote the number of defective items among the selected 5 items. Then, $X \in \{0, 1, 2, 3, 4, 5\}$. The PMF of X is given by

$$P_X(i) = P(X = i) = \begin{cases} \binom{5}{i} (0.3)^i (1 - 0.3)^{5-i}, & i = 0, 1, \ldots, 5 \\ 0, & \text{otherwise} \end{cases}.$$

Example 2.13 Consider the random variable X that represents the number of people who are hospitalized or died in a single head-on collision on the road in front of a particular spot in a year. The distribution of such random variables is typically obtained from historical data. Without getting into the statistical aspects involved, let us suppose that the CDF of X is as follows:

x	0	1	2	3	4	5	6	7	8	9	10
$F(x)$	0.250	0.546	0.898	0.932	0.955	0.972	0.981	0.989	0.995	0.998	1.000

Find $P(X = 10)$ and $P(X \le 5/X > 2)$.

Solution:
We know that
$$P(X = 10) = P(X \le 10) - P(X < 10).$$

Therefore, $P(X = 10) = 1.000 - 0.998 = 0.002$. Using conditional probability, we have

$$P(X \le 5/X > 2) = \frac{P(2 < X \le 5)}{P(X > 2)} = \frac{P(X \le 5) - P(X \le 2)}{1 - P(X \le 2)}$$
$$= \frac{F(5) - F(2)}{1 - F(2)} = \frac{0.972 - 0.898}{1 - 0.898} = 0.725.$$

2.2.2 Continuous Type Random Variable

In the last section, we studied cases such as tossing of a coin or throwing of a dice in which the total number of possible values of the random variable was at most countable. In such cases, we have studied the probability mass function of random variables. Now, let us consider another experiment say choosing a real number between 0 and 1. Let the random variable be the chosen real number itself. In this case, the possible values of the random variable are uncountable.

In such cases, we cannot claim for $P[X = x]$, but if we say that in choosing a real number between 0 and 1, find the probability $P[a \leq X \leq b]$, then the probability can be calculated for any constants a and b. In such cases, we cannot claim for a nonzero probability at any point, but we can claim for $P[X \leq x]$. From the above experiment, one can consider random variables which assume value in an interval or a collection of intervals. These types of random variables are known as *continuous type random variables*. These random variables generally arise in the experiments like measuring some physical quantity or time.

Unlike in the discrete type random variable, a continuous type random variable assumes uncountably infinite number of values in any specified interval, however small it may be. Thus, it is not realistic to assign nonzero probabilities to values assumed by it. In the continuous type, it can be shown that for any realization x of X,

$$P(X = x) = 0. \qquad (2.5)$$

Hence,

$$P(X \leq x) = P(X < x) \text{ for any } x \in \mathbb{R}. \qquad (2.6)$$

Definition 2.18 (*Probability Density Function*) Let X be a continuous type random variable with CDF $F(x)$. The CDF $F(x)$ is an absolutely continuous function; i.e., there exists a nonnegative function $f(x)$ such that for every real number x, we have

$$F(x) = \int_{-\infty}^{x} f(t)dt.$$

The nonnegative function $f(x)$ is called the probability density function (PDF) of the continuous type random variable X.

Since $F(x)$ is absolutely continuous, it is differentiable at all x except perhaps at a countable number of points. Therefore, using fundamental theorem of integration, we have $f(x) = \frac{dF(x)}{dx}$, for all x, where $F(x)$ is differentiable. $F'(x)$ may not exist on a countable set say $\{a_1, a_2, \ldots, a_i, \ldots\}$ but since the probability of any singleton set is zero, we have

$$P(X \in \{a_1, a_2, \ldots, a_i, \ldots\}) = \sum_i P(X = a_i) = 0.$$

Thus, the set $\{a_1, a_2, \ldots, a_i, \ldots\}$ is not of much consequence and we define

$$\frac{dF(x)}{dx} = 0, \quad for \ x \in \{a_1, a_2, \ldots\}.$$

With this, we can say that

$$f(x) = \frac{dF(x)}{dx} \quad \forall \, x \in \mathbb{R}.$$

For instance, consider a random variable X with CDF $F(x) = \sqrt{2} \sin x$ when $x \in [0, \pi/4]$. Clearly, $F(x)$ has a nonzero derivative in the interval $(0, \pi/4)$. Hence, there exists a function $f(x)$, known as PDF such that

$$f(x) = \begin{cases} \sqrt{2} \cos x, & x \in (0, \pi/4) \\ 0, & \text{otherwise} \end{cases}.$$

We shall use the convention of defining the nonzero values of the PDF $f(x)$ only. Thus, when we write $f(x)$, $a \le x \le b$, it is understood that $f(x)$ is zero for $x \notin [a, b]$. A probability density function satisfies the following properties:

(i) $f(x) \ge 0$ for all possible values of x.

(ii) $\int_{-\infty}^{\infty} f(x)dx = 1$.

Property (i) follows from the fact that $F(x)$ is nondecreasing and hence its derivative $f(x) \ge 0$, while (ii) follows from the property that $\lim_{x \to \infty} F(x) = 1$.

Remark 2.3   1. $f(x)$ does not represent the probability of any event. Only when $f(x)$ is integrated between two limits, it yields the probability. Furthermore, in the small interval Δx, we have

$$P(x < X \le x + \Delta x) \approx f(x)\Delta x. \tag{2.7}$$

2. The CDF $F(x)$ at $x = a$ can be geometrically represented as the area under the probability density curve $y = f(x)$ in the xy–plane, to the left of the abscissa at the point a on the axis. This is illustrated in Fig. 2.1.

3. For any given a, b with $a < b$,

$$P(X \in (a, b)) = F(b) - F(a) = \int_{-\infty}^{b} f(x)dx - \int_{-\infty}^{a} f(x)dx = \int_{a}^{b} f(x)dx.$$

Fig. 2.1 Geometrical
interpretation of $f(x)$ and
$F(a)$

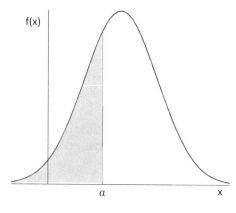

Fig. 2.2 Probability that X
lies between a and b

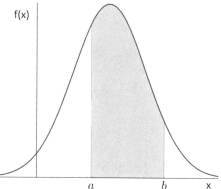

Hence, the area under the curve $y = f(x)$ between the two abscissa at $x = a$
and $x = b$, $a < b$, represents the probability $P(X \in (a, b))$. This is illustrated
in Fig. 2.2.

4. For any continuous type random variable,

$$P(x_1 < X \le x_2) = P(x_1 \le X < x_2) = P(x_1 < X < x_2) = P(x_1 \le X \le x_2)$$

and hence we have,

$$P(X = x) = 0, \ \forall x \ \in \mathbb{R}.$$

5. Every nonnegative real-valued function, that is integrable over $\mathbb{R}$ and satisfies
$\int_{-\infty}^{\infty} f(x)dx = 1$, is a PDF of some continuous type random variable X.

Example 2.14 Consider a random experiment to test the lifetime of a light bulb. The
experiment consists of noting the time when the light bulb is turned on, and the time
when it fails. Let X be the random variable that denotes the lifetime of the light bulb.
Assume that the CDF of X is as given below:

$$F(x) = \begin{cases} 0, & -\infty < x < 0 \\ kx, & 0 \le x < 100 \\ 1, & 100 \le x < \infty. \end{cases}$$

where k is a constant. Find a. What is the probability that the lifetime of light bulb is between 20 and 70 h.

Solution:
Given that $F(x)$ is a cumulative distribution function of the continuous type random variable, it has to be absolutely continuous. Applying continuity at $x = 100$, we get,

$$100k = 1$$
$$\Rightarrow k = \frac{1}{100}.$$

Now,

$$P(20 \le X \le 70) = P(X \le 70) - P(X \le 20) = F(70) - F(20) = 50k = 0.5.$$

Example 2.15 Find the value of k for which the function

$$f(x) = \begin{cases} kx^2, & 0 \le x \le 1 \\ 0, & \text{otherwise} \end{cases}$$

is the PDF of a random variable X and then compute $P\left(\frac{1}{3} < X < \frac{1}{2}\right)$.

Solution:
As the given function is a PDF, its integration over the real line must be unity.

$$k \int_0^1 x^2 dx = 1 \quad \Longrightarrow k = 3.$$

Now,

$$P\left(\frac{1}{3} < X < \frac{1}{2}\right) = \int_{\frac{1}{3}}^{\frac{1}{2}} 3x^2 dx = \frac{19}{216}.$$

2.2.3 Function of a Random Variable

Let X be a RV and assume that its distribution is known. We are interested to find the distribution of $Y = g(X)$, provided that Y is also a RV, where g is a function defined on the real line. We have the following result that gives the distribution of Y.

Table 2.3 Probability mass function of X

X	-2	-1	0	1	2
$P(X = x)$	$\frac{1}{5}$	$\frac{1}{6}$	$\frac{1}{5}$	$\frac{1}{15}$	$\frac{11}{30}$

Result: Let X be a RV defined on the probability space (Ω, S, P). Let g be a Borel measurable function on $\mathbb{R}$. Then, $g(X)$ is also a RV and its distribution is given by

$$P(g(X) < y) = P(X \le g^{-1}(y)) \text{ for all } y \in \mathbb{R}.$$

Note that, every continuous or piecewise continuous functions are Borel measurable functions. For example, $|X|$, $aX + b$ (where $a \neq 0$ and b are constants), X^k (where $k > 0$ is an integer), and $|X|^a$ $(a > 0)$ are all RVs.

Example 2.16 Let X be a discrete type random variable with PMF given in Table 2.3. Define $Y = X^2$. Find PMF of Y.

Solution:
The possible values of Y are $\{0, 1, 4\}$. Now, $P(Y = 0) = P(X = 0)$, $P(Y = 1) = P(X = -1) + P(X = 1)$ and $P(X = 4) = P(X = -2) + P(X = 2)$. Therefore, the PMF of Y is given as

$$P(Y = y) = \begin{cases} \frac{1}{5}, & y = 0 \\ \frac{1}{6} + \frac{1}{15} = \frac{7}{30}, & y = 1 \\ \frac{1}{5} + \frac{11}{30} = \frac{17}{30}, & y = 4 \end{cases}.$$

For continuous type random variable, we have the following result that gives the distribution of a random variable of continuous type.

Result: Let X be a continuous type RV with PDF $f(x)$. Let $y = g(x)$ be a function satisfying following conditions

- $g(x)$ is differentiable for all x,
- $g'(x)$ is continuous and nonzero at all but a finite number of values of x.

Then, for every real number y,

1. There exist a positive integer $n = n(y)$ and real numbers $x_1(y), x_2(y), \ldots, x_n(y)$ such that $g[x_k(y)] = y$ and $g'(x_k(y)) \neq 0, k = 1, 2, \ldots, n(y)$, or
2. There does not exist any x such that $g(x) = y$, $g'(x) \neq 0$, in which case we write $n(y) = 0$.

Then, Y is a continuous type RV with PDF given by

$$h(y) = \begin{cases} \displaystyle\sum_{k=1}^{n(y)} f(x_k(y))|g'(x_k(y))|^{-1}, & n(y) > 0 \\ 0, & \text{otherwise} \end{cases}.$$

Example 2.17 Let X be a continuous type random variable with PDF given as

$$f(x) = \frac{1}{\sqrt{2\pi}} e^{\frac{-x^2}{2}}, \quad -\infty < x < \infty.$$

Let Y be another random variable such that $Y = X^2$. Find the PDF of Y.

Solution:
We know that the PDF of Y is given by

$$f_Y(y) = \sum_{k=1}^{n} f_X(g_k^{-1}(y)) \left| \frac{dg_k^{-1}(y)}{dy} \right|.$$

Here, $g_1^{-1}(y) = \sqrt{y}$ and $g_2^{-1}(y) = -\sqrt{y}$. Therefore, we have, for $0 < y < \infty$

$$f_Y(y) = \frac{1}{2\sqrt{2y\pi}} e^{\frac{-y}{2}} + \frac{1}{2\sqrt{2y\pi}} e^{\frac{-y}{2}} = \frac{1}{\sqrt{2y\pi}} e^{\frac{-y}{2}}. \tag{2.8}$$

We will see later that Y is a chi-square distribution with parameter 1. Thus, the square of a standard normal random variable is a chi-square random variable with parameter 1.

2.3 Moments and Generating Functions

The CDF $F(x)$ completely characterizes the behavior of a random variable X. However, in applications, the study of distribution of a random variable is essentially the study of numerical characteristics associated with the random variable. In this section, we discuss the mean, variance, other higher moments, and generating functions of a random variable X.

2.3.1 Mean

Let X be a random variable defined on $(\Omega,\ S,\ P)$. The expected value or mean of X, denoted by $E(X)$ or μ, is defined as follows:

Definition 2.19 (*Mean of a discrete type RV*) Let X be a discrete type random variable with possible values (or mass points) $x_1, x_2, \ldots, x_i, \ldots$. The *mean* or *average* or *expectation* of the discrete type random variable X, denoted by $E(X)$, is defined by:

$$E(X) = \sum_i x_i p(x_i). \tag{2.9}$$

provided the sum is absolutely convergent, i.e., $\sum_i |x_i| p(x_i) < \infty$ or $E(|X|) < \infty$.

Definition 2.20 (*Mean of a continuous type RV*) If X is a continuous type random variable with PDF $f(x)$, then the expectation of X is defined by:

$$E(X) = \int_{-\infty}^{\infty} x f(x) dx$$

provided the integral converges, i.e., $\int_{-\infty}^{\infty} |x| f(x) dx < \infty$ or $E(|X|) < \infty$.

Example 2.18 Consider the random variable X with PDF (known as Cauchy distribution)

$$f(x) = \tfrac{1}{\pi(1+x^2)}, \quad -\infty < x < \infty.$$

Check whether $E(X)$ exists or not.

Solution:
We know that,

$$\frac{1}{\pi} \int_{-\infty}^{\infty} \frac{x}{1 + x^2} dx = 0 < \infty.$$

However, $E(X)$ does not exist since $E(|X|)$ does not converge as shown below.

$$E(|X|) = \frac{1}{\pi} \int_{-\infty}^{\infty} \frac{|x|}{1 + x^2} dx = \frac{2}{\pi} \int_{0}^{\infty} \frac{x}{1 + x^2} dx = \frac{1}{\pi} \log |1 + x^2| \to \infty \text{ as } x \to \infty.$$

Example 2.19 Let X be a random variable with possible values $\{x_1, x_2, \ldots, x_n\}$ such that $P(X = x_i) = \frac{1}{n} \, \forall \, i$. Find $E(X)$.

Solution:
By definition of expectation, we have

$$E(X) = \sum_{i=1}^{n} x_i P(X = x_i) = \frac{1}{n} \sum_{i=1}^{n} x_i.$$

Thus, $E(X)$ can be interpreted as the weighted average of the distribution of X with weights $p(x_i)$ of x_i. In particular, if $p(x_i) = \frac{1}{n}, i = 1, \ldots, n$, then $E(X)$ is the usual arithmetic mean.

Some Important Results:

1. For any constant a, $E(a) = a$.
2. $E(aX + b) = aE(X) + b$ where a and b are two constants.
3. If $X \geq 0$, then $E(X) \geq 0$.

Remark 2.4   1. It is to be noted that, for a given random variable, expectation is
    always a constant.
  2. Since any random variable X can be written as $X = X^+ - X^-$ and $|X| = X^+ +$
      X^-, $E(|X|) < \infty$ is taking care of removing the cases of infinite mean. Since
      $E(X) = E(X^+) - E(X^-)$, the mean of X exists if $E(X^+)$ and $E(X^-)$ have
      finite values.
  3. $E(X)$ is the center of gravity (centroid) of the unit mass that is determined by
      the probability mass function (if X is a discrete type random variable) or by the
      probability density function (if X is a continuous type random variable). Thus,
      the mean of X is a measure of where the values of the random variable X are
      centered.

2.3.2 Variance

Another quantity of great importance in probability and statistics is called *variance*
and is defined as follows:

Definition 2.21 (*Variance of a RV*) Let X be a discrete type random variable with
possible values $x_1, x_2, \ldots, x_i, \ldots$. The variance of the discrete type random variable
X, denoted by $Var(X)$, is defined by:

$$Var(X) = E\left((X - \mu)^2\right) = \sum_i (x_i - \mu)^2 p(x_i) \qquad (2.10)$$

provided the sum is absolutely convergent. Similarly, the variance of continuous type
random variable is defined as

$$Var(X) = \int_{-\infty}^{\infty} (x - \mu)^2 f(x) dx \qquad (2.11)$$

provided the integral is absolutely convergent.

Note that $Var(X)$ is always a nonnegative number. The positive square root of the
variance is known as the *standard deviation* of the random variable X or *average
squared deviation of X from it's mean*.

Remark 2.5   1. The variance (and so the standard deviation) is a measure of spread
    (or "dispersion" or "scatter") of values of a random variable X about it's mean
    μ. The variance of X is small if the values are concentrated near the mean, and
    variance is large if the values are distributed away from the mean.
  2. Note that if X has certain dimensions or units, such as centimeters (cm), then
      the variance of X has units cm^2 while the standard deviation has the same unit
      as X, i.e., cm.

Some Important Results:

1. $Var(X) \geq 0$. If $Var(X) = 0$, then X is a constant random variable with $P(X = \mu) = 1$.
2. $Var(X) = E(X^2) - [E(X)]^2$.
3. For any constant c,

$$Var(cX) = c^2 Var(X).$$

4. The quantity $E[(X - b)^2]$ is minimum when $b = \mu = E(X)$.

2.3.3 Moment of Order n

Definition 2.22 (*Moment of Order n*) Let X be a random variable and c be any constant. The moment of order n about c of the random variable X is defined by:

$$E((X - c)^n) = \begin{cases} \int_{-\infty}^{\infty} (x - c)^n f(x) dx, & \text{if } X \text{ is continuous type} \\ \sum_i (x_i - c)^n p(x_i), & \text{if } X \text{ is discrete type} \end{cases}$$

where n is a nonnegative integer provided the sum (or integral) is absolutely convergent.

When $c = \mu$, it is called central moment or moment about the mean of order n, denoted by μ_n, and is given by

$$\mu_n = E((X - \mu)^n) = \begin{cases} \int_{-\infty}^{\infty} (x - \mu)^n f(x) dx, & \text{if } X \text{ is continuous type} \\ \sum_i (x_i - \mu)^n p(x_i), & \text{if } X \text{ is discrete type} \end{cases} \quad .(2.12)$$

We have $\mu_0 = 1$, $\mu_1 = E(X)$ and $\mu_2 = Var(X) = \sigma^2$.

When $c = 0$, we define the moment μ_n' of order n about the origin of the random variable X as

$$\mu_n' = E(X^n) = \begin{cases} \int_{-\infty}^{\infty} x^n f(x) dx, & \text{if } X \text{ is continuous type} \\ \sum_i x_i^n p(x_i), & \text{if } X \text{ is discrete type} \end{cases} \quad . \quad (2.13)$$

Note that $\mu_0' = 1$, $\mu_1' = E(X) = \mu$.

Expanding $(X - \mu)^n$ in Eq. (2.12) by binomial theorem and using Eq. (2.13), we get the relationship between the nth central moment and the moments about the origin of order less than or equal to n. we have,

$$\mu_n = E\,(X - E(X))^n = E\left[\sum_{k=0}^{n} \binom{n}{k} X^k (-E(X))^{n-k}\right]$$

$$= \sum_{k=0}^{n} \binom{n}{k} \mu'_k (-\mu'_1)^{n-k}$$

$$= \mu'_n - \binom{n}{1}\mu'_{n-1}\mu + \cdots + (-1)^r \binom{n}{r}\mu'_{n-r}\mu^r + \cdots + (-1)^n \mu^n$$

where $\mu = \mu'_1$. Thus,

$$\begin{aligned}
\mu_2 &= \mu'_2 - (\mu'_1)^2 \\
\mu_3 &= \mu'_3 - 3\mu'_2\mu'_1 + 2(\mu'_1)^3 \\
\mu_4 &= \mu'_4 - 4\mu'_3\mu'_1 + 6\mu'_2(\mu'_2)^2 - 3(\mu'_1)^4.
\end{aligned} \qquad (2.14)$$

Example 2.20 Find the first four moments about the mean for the continuous type random variable whose PDF is given as

$$f(x) = \frac{1}{\sqrt{2\pi}} e^{\frac{-x^2}{2}}, \quad -\infty < x < \infty.$$

Solution: Now,

$$E(X) = \int_{-\infty}^{\infty} x f(x)dx = \int_{-\infty}^{\infty} x \frac{1}{\sqrt{2\pi}} e^{\frac{-x^2}{2}} dx = 0.$$

$$E(X^2) = \int_{-\infty}^{\infty} x^2 f(x)dx = \int_{-\infty}^{\infty} x^2 \frac{1}{\sqrt{2\pi}} e^{\frac{-x^2}{2}} dx = 1.$$

$$E(X^3) = \int_{-\infty}^{\infty} x^3 f(x)dx = \int_{-\infty}^{\infty} x^3 \frac{1}{\sqrt{2\pi}} e^{\frac{-x^2}{2}} dx = 0.$$

$$E(X^4) = \int_{-\infty}^{\infty} x^4 f(x)dx = \int_{-\infty}^{\infty} x^4 \frac{1}{\sqrt{2\pi}} e^{\frac{-x^2}{2}} dx = 3.$$

Using Eq. (2.14), we have

$$\mu_1 = 0, \quad \mu_2 = 1 \quad \mu_3 = 0 \quad \mu_4 = 3.$$

2.3.4   Generating Functions

Now, we discuss the generating functions, namely probability generating function, moment generating function, and characteristic function that generate probabilities or moments of a random variable.

Definition 2.23 (*Probability Generating Function*) Let X be a nonnegative integer-valued random variable. Then, the probability generating function (PGF) of X with probability mass function (pmf) $P(X = k) = p_k$, $k = 0, 1, 2, \ldots$, is defined as

$$G_X(t) = \sum_{k=0}^{\infty} p_k t^k, \quad |t| \le 1.$$

Remark 2.6 1. $G_X(1) = \sum_{k=0}^{\infty} p_k = 1.$

2. $p_k = \frac{1}{k!} \frac{d^k}{dt^k} G_X(t)|_{t=0}, \quad k = 1, 2, \ldots.$

3. $E(X(X-1)\ldots(X-n)) = \frac{d^n}{dt^n} G_X(t)|_{t=1}.$ We say $E(X(X-1)\ldots(X-n))$ as factorial moment of order n.

Example 2.21 Let X be a random variable with PMF given by

$$P(X = k) = \binom{n}{k} p^k (1-p)^{n-k}, \quad k = 0, 1, 2, \ldots, n.$$

Find PGF of X.

Solution:

$$G_X(t) = \sum_{k=0}^{n} t^k \binom{n}{k} p^k (1-p)^{n-k} = \sum_{k=0}^{n} \binom{n}{k} (pt)^k (1-p)^{n-k} = (pt + 1 - p)^n.$$

Definition 2.24 (*Moment Generating Function*) Let X be a random variable. The moment generating function (MGF) of X is defined as

$$M(t) = E(e^{tX}), \quad t \in \mathbb{R}$$

provided the expectation exists in some neighborhood of the origin.

Note that

$$M_X(t) = E(e^{tX}) = E(1 + \frac{tX}{1!} + \frac{t^2 X^2}{2!} + \cdots) = 1 + \frac{t E(X)}{1!} + \frac{t^2 E(X^2)}{2!} + \cdots.$$

Example 2.22 Let X be a continuous type random variable with PDF

$$f(x) = \begin{cases} \frac{1}{3} e^{\frac{-x}{3}}, & x > 0 \\ 0, & \text{otherwise} \end{cases}.$$

Then, find MGF.

Solution:

$$M(t) = \int_{-\infty}^{\infty} e^{tx} f(x)dx = \int_{0}^{\infty} e^{tx} \frac{1}{3} e^{\frac{-x}{3}} dx = \frac{1}{1-3t}, \quad 3t < 1.$$

Remark 2.7 1. The MGF uniquely determines a distribution function and conversely, if the MGF exists, it is unique.

2. The nth order moment of X can be calculated by differentiating the MGF n times and substituting $t = 0$,

$$E(X^n) = \frac{d^n M(t)}{dt^n}\bigg|_{t=0}.$$

Example 2.23 Find MGF of the random variable X with PDF

$$f(x) = \frac{1}{\sigma\sqrt{2\pi}} e^{\frac{-(x-\mu)^2}{2\sigma^2}}, \quad -\infty < x < \infty.$$

Deduce the first four moments about origin.

Solution:

$$M_X(t) = E(e^{tX}) = \int_{-\infty}^{\infty} e^{tx} \frac{1}{\sigma\sqrt{2\pi}} e^{\frac{-(x-\mu)^2}{2\sigma^2}} dx.$$

Let $z = \frac{x-\mu}{\sigma}$, then $x = z\sigma + \mu$ and we have

$$M_X(t) = e^{\mu t} \int_{-\infty}^{\infty} e^{z\sigma t} \frac{1}{\sqrt{2\pi}} e^{\frac{-z^2}{2}} dz = e^{\mu t} e^{\frac{(\sigma t)^2}{2}} = e^{\mu t + \frac{(\sigma t)^2}{2}}.$$

Now, we know that

$$E(X^n) = \frac{d^n M(t)}{dt^n}\bigg|_{t=0}.$$

Therefore, we have

$$E(X) = \mu.$$
$$E(X^2) = \mu^2 + \sigma^2.$$
$$E(X^3) = \mu^3 + 3\mu\sigma^2.$$
$$E(X^4) = \mu^4 + 6\mu^2\sigma^2 + 3\sigma^4.$$

The above results can be verified from Example (2.20) with $\mu = 0$ and $\sigma^2 = 1$.

Definition 2.25 (*Characteristic Function*) Let X be a random variable. The characteristic function of X is defined as,

$$\Psi_X(t) = E(e^{itX}), \ i = \sqrt{-1}, \ t \in \mathbb{R}$$

provided the expectation exists in some neighborhood of the origin.
Note that

$$\Psi_X(t) = E(e^{itX}) = E(1 + \frac{itX}{1!} + \frac{(it)^2 X^2}{2!} + \cdots) = 1 + \frac{itE(X)}{1!} - \frac{t^2 E(X^2)}{2!} + \cdots.$$

For example, the characteristic function for the PDF given in Example 2.23 is $e^{i\mu t - \frac{1}{2}\sigma^2 t^2}$.

Remark 2.8   1. For any random variable X, $E(|e^{itX}|) \leq 1$. Hence, characteristic function of X always exists unlike MGF.
2. $\Psi(0) = 1$.
3. $|\Psi(t)| \leq 1$ for all $t \in \mathbb{R}$.
4. The nth order moment of X can be calculated by differentiating the characteristic function n times and substituting $t = 0$ as follows:

$$E(X^n) = \frac{1}{i^n} \frac{d^n \Psi_X(t)}{dt^n}\Big|_{t=0}.$$

Remark 2.9   1. As we have already seen, the mean of a random variable X provides a measure of central tendency for the values of a distribution. Although the mean is commonly used, two measures of central tendency are also employed. These are the mode and median.
2. The *mode* of a discrete type random variable is that value which occurs most often, or in other words, has the greatest probability of occurring. Sometimes we have two or more values that have the same relatively large probability of occurrence. In such cases, we say that the distribution is bi-model, tri-model, or multi-model. The mode of a continuous type random variable X is where the probability density function has a relative maximum.
3. The *median* is that value x for which $P(X \leq x) \leq \frac{1}{2}$ and $P(X > x) \leq \frac{1}{2}$. In the case of a continuous type random variable, we have $P(X < x) = \frac{1}{2} = P(X > x)$, and the median separates the density curve into two parts having an equal area of $\frac{1}{2}$ each. In the case of a discrete type random variable, a unique median may or may not exist.

2.4   Standard Probability Distributions

In many practical situations, some probability distributions occur frequently. These distributions describe several real-life random phenomena. In this section, we introduce some commonly used probability distributions with various examples.

2.4.1 Standard Discrete and Continuous Distributions

First, we present the standard discrete distributions. Table 2.4 shows some standard discrete distributions along with PMF, mean, variance, and MGF. Note that constants appearing in the probability distribution of random variables are called **parameters** of that probability distribution.

Example 2.24 In a city, suppose that only 60% of the people read a newspaper. What is the probability that four out of a sample of five persons will be reading a newspaper?

Solution:
Let the success be termed as a person reading a newspaper. According to the given data, if p denotes the probability of success, then $p = 0.6$, $q = 1 - p = 0.4$. Here the sample chosen consists of five drivers, $n = 5$.

Let X denote the number of persons reading a newspaper, then the required probability is

$$P(X = 4) = \binom{5}{4} (0.6)^4 (0.4)^1 = 0.259.$$

Example 2.25 If a fair coin is successively tossed, find the probability that the first head appears on the fifth trial.

Solution:
Consider a random experiment of tossing a fair coin. The event "getting head" is termed as success and "p" denotes the probability of success. Then, $p = 0.5$ and $q = 1 - p = 0.5$. We want to find the probability that a head first appears on the fifth trial; that is, first four trials result in a "tail." If X denotes the number of trials until a head appears, then

$$P(X = 4) = q^4 p = \left(\frac{1}{2}\right)^4 \frac{1}{2}.$$

Example 2.26 A market order is a buy or sell order to be executed immediately at current market prices. As long as there are willing sellers and buyers, market orders are filled. Suppose a brokerage firm receives on an average eight market orders per hour and a trader can only process one order at a time. Assume that there are sufficient amount of counter-parties available for each order (that is for every buy/sell order our firm places, there are traders with opposite positions of sell/buy). How many traders should be kept at a time in order that 90% of the orders are met?

Solution:
Let the number of traders required are k and X be the number of orders received by firm in an hour. We have a Poisson distribution in hand with $\lambda = 8$. To obtain k we have,

Table 2.4 Standard discrete distributions

S.No.	Distribution, Notation and X counts	PMF $P(X = k)$	Mean $E(X)$	Variance $Var(X)$	MGF $M(t)$
1	Uniform, $X \sim U(x_1, x_2, \ldots, x_n)$, Outcomes that are equally likely	$\frac{1}{n}, k = x_i, i = 1, 2, \ldots, n$	$\sum_{i=1}^{n} \frac{x_i}{n}$	$\sum_{i=1}^{n} \frac{x_i^2}{n} - (\sum_{i=1}^{n} \frac{x_i}{n})^2$	$\sum_{i=1}^{n} \frac{e^{tx_i}}{n}$
2	Bernoulli, $X \sim b(1, p)$, Success or failure in a trial	$p^k(1-p)^{1-k}, k = 0, 1,$ $0 < p < 1$	p	$p(1-p)$	$(q + pe^t)$
3	Binomial, $X \sim B(n, p)$, Success or failure in n trials	$\binom{n}{k} p^k(1-p)^{n-k},$ $k = 0, 1, 2, \ldots, n,$ $0 < p < 1, n \geq 1$	np	$np(1-p)$	$(q + pe^t)^n$
4	Geometric, $X \sim Geometric(p)$, Number of trials to get first success	$(1-p)^{(k-1)} p, k = 1, \ldots,$ $0 < p < 1$	$\frac{1}{p}$	$\frac{1-p}{p^2}$	$\frac{pe^t}{1-qe^t}$
5	Poisson, $X \sim P(\lambda)$, Number of arrivals in a fixed time interval	$\frac{e^{-\lambda}\lambda^k}{k!}, k = 0, 1, 2, \ldots,$ $\lambda > 0$	λ	λ	$e^{\lambda(e^t-1)}$
6	Negative Binomial, $X \sim NB(r, p)$, Number of trials to get rth success	$x - 1_{C_{r-1}} p^r (1-p)^{x-r},$ $x = r, r+1, r+2, \ldots,$ $0 < p < 1, r \geq 1$	$\frac{r}{p}$	$\frac{r(1-p)}{p^2}$	$(pe^t)^r (1 - qe^t)^{-r}, qe^t < 1$
7	Hypergeometric, $X \sim Hyper(N, M, n)$, Number of marked individuals in a sample of size n drawn from N without replacement	$\frac{M_{C_x} N - M_{C_{n-x}}}{N_{C_n}},$ $\max(0, M + n - N) \leq x \leq$ $\min(M, n)$	$\frac{nN}{M}$	$\frac{nM(N-M)(N-n)}{N^2(N-1)}$	Does not exist

$$P(X = 0) + P(X = 1) + \cdots + P(X = k) = e^{-\lambda}\left(1 + \frac{\lambda}{1} + \cdots + \frac{\lambda^k}{k!}\right) = 0.90.$$

Keep increasing number of terms so that the probability becomes 0.9. Hence, the value of k obtained from the above equation is 12.

Example 2.27 A manufacturer produces electric bulb, 1% of which are defective. Find the probability that there is no defective bulb in a box containing 100 bulbs.

Solution:
Let X denote the number of defective bulbs in the box. Clearly, X follows binomial distribution with $n = 100$ and $p = 0.01$. We need to find the probability that there is no defective bulb, i.e., $P(X = 0)$, which is given by

$$P(X = 0) = \binom{100}{0}(0.01)^0(0.99)^{100} = (0.99)^{100} = 0.366.$$

However, as n is large and p is small, the required probability can be approximated by using the Poisson approximation of binomial distribution as follows. Here, $\lambda = np = 100 \times 0.01 = 1$. Therefore, by the Poisson distribution,

$$P(X = 0) = \frac{e^{-\lambda}\lambda^0}{0!} = e^{-1} = 0.3679.$$

Now, we discuss some standard continuous distributions. Table 2.5 gives the notation, PDF, $E(X)$, Var(X), and MGF for the standard continuous distributions.

Example 2.28 If X is uniform distributed over $(0, 10)$, calculate that (i) $P(X < 3)$ (ii) $P(X > 7)$ (iii) $P(1 < X < 6)$.

Solution:
Since, $X \sim U(0, 10)$, the PDF of X is given by

$$f(x) = \begin{cases} \frac{1}{10}, & 0 \le x \le 10 \\ 0, & \text{otherwise} \end{cases}.$$

(i) $P(X < 3) = \int_0^3 \frac{1}{10}dx = \frac{3}{10}$ (ii) $P(X > 7) = \int_7^{10} \frac{1}{10}dx = \frac{3}{10}$

(iii) $P(1 < X < 6) = \int_1^6 \frac{1}{10}dx = \frac{1}{2}$.

Example 2.29 Lifetimes of VLSI chips manufactured by a semiconductor manufacturer is approximately exponentially distributed with parameter 0.2×10^{-6} h. A computer manufacturer requires that at least 95% of a batch should have a lifetime greater than 2.5×10^5 h. Will the deal be made?

Table 2.5 Standard continuous distributions

S.No.	Distribution and notation	PDF $f(x)$	Mean $E(X)$	Variance $V(X)$	MGF $M(t)$
1	Uniform $X \sim U(a,b)$	$\begin{cases} \frac{1}{(b-a)}, & a<x<b \\ 0, & \text{otherwise} \end{cases}$	$\frac{(a+b)}{2}$	$\frac{(b-a)^2}{12}$	$\frac{e^{tb}-e^{ta}}{t(b-a)}$
2	Exponential $X \sim exp(\lambda)$	$\begin{cases} \lambda e^{-\lambda x}, & x>0 \\ 0, & \text{otherwise} \end{cases}, \lambda>0$	$\frac{1}{\lambda}$	$\frac{1}{\lambda^2}$	$\frac{\lambda}{\lambda-t}, t<\lambda$
3	Gamma $X \sim G(r,\lambda)$	$\begin{cases} \frac{x^{r-1}\lambda^r e^{-\frac{x}{\lambda}}}{\Gamma(r)}, & x\geq 0 \\ 0, & \text{otherwise} \end{cases}$ $r>0, \lambda>0$	$\frac{r}{\lambda}$	$\frac{r}{\lambda^2}$	$\left(1-\frac{t}{\lambda}\right)^{-r}, t<\lambda$
4	Beta $X \sim B(\alpha,\beta)$	$\begin{cases} \frac{x^{\alpha-1}(1-x)^{\beta-1}}{B(\alpha,\beta)}, & 0<x<1 \\ 0, & \text{otherwise} \end{cases}$ $\alpha>0, \beta>0$	$\frac{\alpha}{\alpha+\beta}$	$\frac{\alpha\beta}{(\alpha+\beta)^2(\alpha+\beta+1)}$	$\sum_{j=0}^{\infty}\frac{t^j}{\Gamma(j+1)}\frac{\Gamma(\alpha+j)\Gamma(\alpha+\beta)}{\Gamma(\alpha+\beta+j)\Gamma(\alpha)}$
5	Lognormal $X \sim Lognormal(\mu,\sigma^2)$	$\frac{1}{x\sigma\sqrt{2\pi}}e^{-\frac{1}{2}\left(\frac{\log_e x-\mu}{\sigma}\right)^2}, x>0,$ $\mu\in\mathbb{R}, \sigma>0$	$e^{\mu+\frac{1}{2}\sigma^2}$	$(e^{\sigma^2}-1)e^{2\mu+\sigma^2}$	Does not exist
6	Cauchy $X \sim C(\mu,\theta)$	$\frac{1}{\pi}\frac{\mu}{\mu^2+(x-\theta)^2}, -\infty<x<\infty, \mu>0, \theta\in\mathbb{R}$	Does not exist	Does not exist	Does not exist
7	Weibull $X \sim Weibull(\alpha,\beta)$	$\begin{cases} \frac{\alpha}{\beta}x^{\alpha-1}e^{-\frac{x^\alpha}{\beta}}, & x\geq 0 \\ 0, & \text{otherwise} \end{cases}$ $\alpha>0, \beta>0$	$\beta^{\frac{1}{\alpha}}\Gamma(1+\frac{1}{\alpha})$	$\beta^{\frac{2}{\alpha}}\left[\Gamma(2+\frac{1}{\alpha})-\Gamma^2(1+\frac{1}{\alpha})\right]$	$\sum_{n=0}^{\infty}\frac{t^n\beta^n}{n!}\Gamma(1+\frac{1}{\alpha}), \alpha\geq 1$

Solution:
Let X be the random variable that denotes the lifetime of VLSI chips. Then,

$$P(X \geq 2.5 \times 10^5) = \int_{2.5 \times 10^5}^{\infty} 0.2 \times 10^{-6} e^{-0.2 \times 10^{-6} x} dx$$

$$= e^{-0.2 \times 10^{-6} \times 2.5 \times 10^5} = 0.9512 > 0.95.$$

Therefore, the deal will be made.

2.4.2 The Normal Distribution

Normal distribution or the Gaussian[3] distribution is the most frequently used probability distribution since it serves as a realistic model in many practical scenarios. Also due to an important result called central limit theorem, normal distribution can be used to approximate a large family of probability distributions. Normal distribution is the underlying assumption in many of the commonly used statistical procedures. Even if the underlying distribution deviates slightly from normal distribution assumption, still these procedures are useful to draw inferences about the underlying variable. The significance of normal random variable in statistics will be discussed in the remaining sections.

A random variable X is said to have a **normal distribution** if its PDF is given by

$$f(x) = \frac{1}{\sigma\sqrt{2\pi}} e^{-\frac{1}{2}\left(\frac{x-\mu}{\sigma}\right)^2}, \quad -\infty < x < \infty. \tag{2.15}$$

The constants $-\infty < \mu < \infty$ and $\sigma > 0$ are the parameters of the distribution. When X follows a normal distribution defined by Eq. (2.15), we write $X \sim N(\mu, \sigma^2)$.

One can easily verify that the function $f(x)$ in Eq. (2.15) represents a PDF. Firstly, $f(x) > 0$ for every real x, since e^x is positive for $x \in \mathbb{R}$. Further by changing the variable $z = (x - \mu)/\sigma$, we have

$$\int_{-\infty}^{\infty} f(x)dx = \int_{-\infty}^{\infty} \frac{1}{\sqrt{2\pi}} e^{-\frac{1}{2}z^2} dz = 2\int_{0}^{\infty} \frac{1}{\sqrt{2\pi}} e^{-\frac{1}{2}z^2} dz.$$

---

[3] Johann Carl Friedrich Gauss (30 April 1777–23 February 1855) was a German mathematician who contributed significantly to many fields, including number theory, algebra, statistics. Sometimes referred to as "greatest mathematician since antiquity," Gauss had exceptional influence in many fields of mathematics and science and is ranked as one of history's most influential mathematicians. He discovered the normal distribution in 1809 as a way to rationalize the method of least squares.

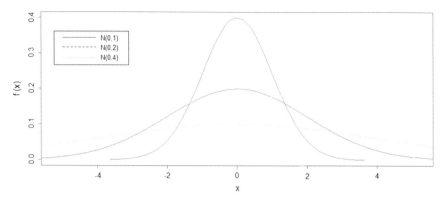

Fig. 2.3 Form of PDF of normal distribution for different values of the σ^2

By changing the variable $t = z^2/2$, we have

$$= \frac{\sqrt{2}}{\sqrt{\pi}} \int_0^\infty e^{-t} \frac{t^{-\frac{1}{2}}}{\sqrt{2}} dt = \frac{1}{\sqrt{\pi}} \int_0^\infty e^{-t} t^{-1/2} dt = \frac{1}{\sqrt{\pi}} \dot{\Gamma}(1/2) = \frac{1}{\sqrt{\pi}} \sqrt{\pi} = 1.$$
(2.16)

Thus, $f(x)$ in Eq. (2.15) represents a valid PDF.

The normal distribution with parameters $\mu = 0$ and $\sigma = 1$ is referred to as the **standard normal distribution** and is denoted by Z. If a random variable follows standard normal distribution, we write $Z \sim N(0, 1)$. The PDF of Z is

$$f(z) = \frac{1}{\sqrt{2\pi}} e^{-\frac{1}{2}z^2}, \quad -\infty < z < \infty.$$
(2.17)

Parameters do play a very important role in the making of a distribution. It decides upon the nature of the distribution and its graph. For example, normal distribution is a distribution with two parameters μ and σ. The form of the normal curve $y = f(x)$ defined in Eq. (2.15) is graphed in Fig. 2.3 for different values of σ.

Note that the curve is symmetrical about the point $x = \mu$. The symmetry about point μ is due to fact that

$$f(\mu + t) = \frac{1}{\sigma\sqrt{2\pi}} e^{-\frac{t^2}{2\sigma^2}} = f(\mu - t) \text{ for any } t,$$

Using the calculus methods, we observe that the PDF $f(x)$ attains its maximum at $x = \mu$. Also, the PDF is symmetric about the mean μ. Similarly, one can show that the PDF has two points of inflection at $x = \mu \pm \sigma$ and x-axis is an asymptote to the PDF. These properties of the PDF make the PDF curve look like a bell and so it is called a bell-shaped curve.

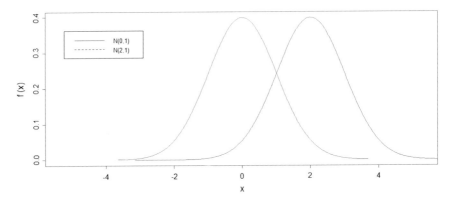

Fig. 2.4 Form of normal PDF for different values of the μ

Observe from Fig. 2.4 that for a fixed value of σ, the PDF curve shifts its point of symmetry to the new value when the value of μ changes without affecting the shape of the curve. Thus, the parameter μ determines the location of the point of symmetry of the curve. Therefore, the parameter μ is known as the location parameter of the normal distribution.

By symmetry, the mean, the median, and the mode of the normal distribution each equal μ. We can explicitly find the mean using definition as follows

$$E(X) = \int_{-\infty}^{\infty} x \frac{1}{\sigma\sqrt{2\pi}} e^{-\frac{1}{2}(\frac{x-\mu}{\sigma})^2} = \int_{-\infty}^{\infty} (\mu + \sigma z) \frac{1}{\sqrt{2\pi}} e^{-z^2/2} dz, \quad (\text{substituting } z = (x-\mu)/\sigma)$$

$$= \mu \int_{-\infty}^{\infty} \frac{1}{\sqrt{2\pi}} e^{-z^2/2} dz + \sigma \int_{-\infty}^{\infty} z e^{-z^2/2} dz = \mu.$$

where we have used the fact that the integrand in the first integral is PDF of standard normal random variable and hence integral is 1. Since the integral is an odd function, the second integral vanishes.

Observe in Fig. 2.3 that for a fixed value of μ, the spread of the distribution changes with change in the value of σ. Greater the value of σ, greater is the spread. If the value tends to be concentrated near mean, the variance is small while the values tend to be distributed far from the mean if the variance is large. Therefore, the parameter σ is known as the scale parameter of the distribution and this behavior is because of the fact that in case of normal distribution $Var(X) = \sigma^2$. To prove it, we proceed as follows:

$$Var(X) = \int_{-\infty}^{\infty} (x-\mu)^2 f(x) dx$$

where $f(x)$, PDF of the normal distribution, is defined in Eq. (2.15). Substituting $z = (x-\mu)/\sigma$ in the above equation, we get

$$Var(X) = \sigma^2 \int_{-\infty}^{\infty} \frac{1}{\sigma\sqrt{2\pi}} - z^2 e^{-z^2/2} \sigma \, dz = \sigma^2 \frac{1}{\sqrt{2\pi}} 2 \int_0^{\infty} z^2 e^{-z^2/2} \, dz.$$

Put $\frac{z^2}{2} = t$, we get

$$Var(X) = \sigma^2 \frac{\sqrt{2}}{\sqrt{\pi}} \int_0^{\infty} 2t e^{-t} \frac{dt}{\sqrt{2t}} = \sigma^2 \frac{2}{\sqrt{\pi}} \int_0^{\infty} t^{\frac{1}{2}} e^{-t} \, dt = \sigma^2.$$

since $\int_0^{\infty} t^{\frac{1}{2}} e^{-t} \, dt = \Gamma(\frac{3}{2})$ and $\Gamma(\frac{3}{2}) = \frac{1}{2}\Gamma(\frac{1}{2})$.

The first four moments of the normal distribution are given by (refer Example 2.20)

$$E(X) = \mu.$$
$$E(X^2) = \mu^2 + \sigma^2.$$
$$E(X^3) = \mu^3 + 3\mu\sigma^2.$$
$$E(X^4) = \mu^4 + 6\mu^2\sigma^2 + 3\sigma^4.$$

The CDF $F(x)$ of a normal distribution is

$$F(x) = P(X \le x) = \int_{-\infty}^{x} \frac{1}{\sigma\sqrt{2\pi}} e^{-\frac{1}{2}\left(\frac{t-\mu}{\sigma}\right)^2} \, dt. \tag{2.18}$$

The integral on the right can not be explicitly expressed as a function of x. Using the transformation $z = (t - \mu)/\sigma$ in Eq. (2.18), we can write $F(x)$ in terms of a function $\Phi(z)$ as follows

$$F(x) = \Phi\left(\frac{x - \mu}{\sigma}\right)$$

where

$$\Phi(z) = \frac{1}{\sqrt{2\pi}} \int_{-\infty}^{z} e^{-\frac{s^2}{2}} \, ds. \tag{2.19}$$

Here, $\Phi(z)$ is the CDF of the standard normal random variable also called the **standard normal probability integral**. Thus, in order to find $F(x)$ for a normal random variable, i.e., $N(\mu, \sigma^2)$, we need to find $\Phi(z)$ is known where $z = (x - \mu)/\sigma$. The value of CDF $\Phi(z)$ of a standard normal random variable is extensively tabulated and is given in Tables A.6 and A.7 in Appendix at the end of this book. Hence $F(x)$, for any general normal distribution for a given x, may be computed using the table.

One can observe by the symmetry of the standard normal random variable that $\Phi(-z) = 1 - \Phi(z)$ for any real z. The relation is established by noting that because of symmetry the two unshaded areas under the standard normal curve in Fig. 2.5 are equal. The two unshaded areas represent $P(Z \le z) = \Phi(-z)$ and $P(Z \ge z) =$

Fig. 2.5 Area under PDF of
standard normal distribution

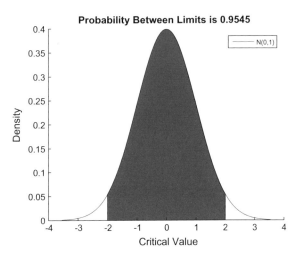

$1 - \Phi(z)$, respectively. Because of this result, it is only necessary to tabulate $\phi(z)$ for positive vales of z and Tables A.6 and A.7 in Appendix give $\Phi(z)$ for nonnegative values of z.

$$P(\mu - \sigma < X < \mu + \sigma) = P(-1 < Z < 1) = \Phi(1) - \Phi(-1) \approx 0.68$$
$$P(\mu - 2\sigma < X < \mu + 2\sigma) = P(-2 < Z < 2) = \Phi(2) - \Phi(-2) \approx 0.95$$
$$P(\mu - 3\sigma < X < \mu + 3\sigma) = P(-3 < Z < 3) = \Phi(3) - \Phi(-3) \approx 0.99.$$

Hence, the following empirical rule holds as shown in Fig. 2.6.

1. 68% of all the observations on X will fall within the interval $X - \mu \pm \sigma$.
2. 95% of all the observations on X will fall within the interval $X - \mu \pm 2\sigma$.
3. 99.7% of all the observations on X will fall within the interval $X - \mu \pm 3\sigma$.

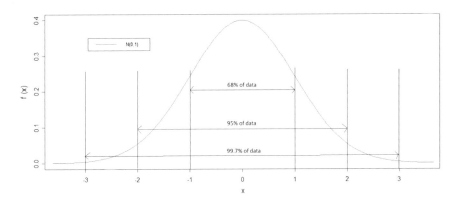

Fig. 2.6 Area under PDF of standard normal distribution

Example 2.30 Assume that the height of students in a class is normal distributed with mean 5.6 feet and variance 0.04 feet. Assume further that the in order to participate in an athletic tournament, height of students must be greater than 5.7 feet. Assume that a student is selected, what is the probability that his height is more than 5.8 feet?

Solution:
Let H denote the height of a student in the class. It is given that $H \sim N(5.6, 0.04)$, and the height of the selected student, like the height of any student to be selected, is greater than 5.7. Hence the required probability is $P(H \geq 5.8/H \geq 5.7)$. Using the definition of the conditional probability, the probability equals

$$\frac{P(H \geq 5.8, H \geq 5.7)}{P(H \geq 5.7)} = \frac{P(H \geq 5.8)}{P(H \geq 5.7)} = \frac{1 - F(5.8)}{1 - F(5.7)} = \frac{1 - \Phi(1.0)}{1 - \Phi(0.5)} = \frac{0.1587}{0.3085} = 0.5144$$

where $\Phi(1.0)$ and $\Phi(0.5)$ are obtained from Appendix in Table A.7.

Normal Probability Points

The table for $\Phi(z)$ can be used to find by interpolation, the probability points of normal distribution $N(0, 1)$. In other words, given α, one can find z_α such that $\Phi(z_\alpha) = \alpha$. For a given probability $0 < \alpha < 1$, the point $z = z_\alpha$ is called the lower 100α percent normal probability point. It may also be referred to as the upper $100(1 - \alpha)$ percent normal probability point. Values of z_α given below are obtained from Appendix. Table A.7

α	0.25	0.50	0.90	0.95	0.975	0.99
z_α	0.675	0.00	1.28	1.64	1.96	2.33

Sum of n Independent Normal random variables: Let $X_1, \ldots, X_n$ be n independent normal distributed random variables with mean and variance of X_i as μ_i and σ_i^2. The product of moment generating function of the sum $S = X_1 + \cdots + X_n$ is the product of the MGF's of $X_1, \ldots, X_n$. Hence, we have MGF of sum is

$$M_S(t) = \Pi_{i=1}^n M_{X_i}(t) = \exp\left(\mu t + \frac{\sigma^2 t^2}{2}\right)$$

where $\mu = \sum_{i=1}^n \mu_i$ and $\sigma^2 = \sum_{i=1}^n \sigma_i^2$ so that the sum is exactly normal distributed with the mean equal to the sum of the means and the variance equal to sum of the variances. The converse of above result is also true which states that if the sum of n independent variables is exactly normal distributed, then each random variable is normal distributed.

Remark 2.10   1. Let $X \sim N(\mu, \sigma^2)$, then distribution of $Y = X^2$ is known as Rayleigh distribution and its PDF is given by

$$f(y) = \frac{1}{\sigma\sqrt{2\pi y}} \exp\left(-\frac{y + \mu^2}{2\sigma^2}\right) \cosh\frac{\mu\sqrt{y}}{\sigma^2}, \quad 0 < y < \infty.$$

2. A positive random variable X is said to have a lognormal distribution if $Y = \log_e X$ has a normal distribution, i.e.,

$$Y = \log_e X \sim N(\mu, \sigma^2).$$

The PDF of a lognormal distribution is given by

$$f(x) = \frac{1}{x\sigma\sqrt{2\pi}} e^{-\frac{1}{2}\left(\frac{\log_e x - \mu}{\sigma}\right)^2}, \quad 0 < x < \infty \qquad (2.20)$$

where $\sigma > 0$ and $-\infty < \mu < \infty$. Further, $E(X) = e^{\mu + \frac{1}{2}\sigma^2}$ and $Var(X) = (e^{\sigma^2} - 1)e^{2\mu + \sigma^2}$.

Example 2.31 A manufacturer wishes to give a safe guarantee for his product against manufacturing defects. He proposes to replace a product if it fails to work any time within the period of guarantee. He considers that a guarantee period is safe if he is required to replace not more than 6% of his products that fail within the period of guarantee. If the lifetime of his product is normal distributed with mean life 2 years and standard deviation 4 months, then what should be the maximum period of guarantee, in terms of whole months, so that the guarantee is safe for him.

Solution:
Let X be the lifetime of the product and t be the required guarantee time. According to the problem $X \sim N(2, 1/9)$. It is required to find such that

$$P(X \le t) = F(t) \le 0.06$$
$$\Phi\left(\frac{t-2}{1/3}\right) \le 0.06 \simeq \Phi(-1.55)$$
$$3(t-2) = -1.55$$
$$t = 1.49 \text{ yrs } = 17.88 \text{ months.}$$

Hence, the manufacturer can give a guarantee of 17 months safely.

Definition 2.26 If the sum of independent random variables following some distribution has the same family of distribution as the individual random variables, then we say that this family of distribution has reproductive property.

Some of the distributions that have reproductive property are binomial, geometric, Poisson, normal distribution while exponential distribution and uniform distribution does not have reproductive property.

2.5 Central Limit Theorem

The central limit theorem argues that no matter how the distribution of original variable looks, the distribution of sample mean can be approximated by a normal distribution provided the sample size is sufficiently large (usually at least 30) and size of every sample is same. This gives us the ability to measure how the sample means of different samples will behave, without comparing them to mean of other samples. As a consequence, samples can be used to answers many questions regarding the population.

Moreover, CLT does not apply only to the sample mean but to some other functions of the sample as well, for instance, sample proportion. As the knowledge about the normal distribution is vast, many applications and analysis become easier by using the central limit theorem, in particular, the two areas called hypothesis testing and confidence intervals.

Now we present the central limit theorem as follows.

Theorem 2.2 *Let $X_1, X_2, \ldots, X_n$ be independent random variables that are identically distributed and have finite mean μ and variance σ^2. Then, if $S_n = X_1 + \cdots + X_n$ ($n = 1, 2, \ldots$), we have*

$$\lim_{n \to \infty} P \left\{ a \leq \frac{S_n - n\mu}{\sigma \sqrt{n}} \leq b \right\} = \frac{1}{\sqrt{2\pi}} \int_a^b e^{-\frac{u^2}{2}} du.$$

i.e., the random variable $\frac{S_n - n\mu}{\sigma \sqrt{n}}$ is asymptotically normal distributed.

This theorem is also true under more general conditions. For example, it holds when $X_1, X_2, \ldots, X_n$ are independent random variables with the same mean and variance but not necessarily identically distributed.

Example 2.32 A random variable having a Poisson distribution with mean 100 can be thought of as the sum Y of the observations of a random sample of size 100 from a Poisson distribution with mean 1. Thus, $W = \frac{Y - 100}{\sqrt{100}}$ has asymptotically normal distribution.

$$P(75 \leq Y \leq 125) = P \left(\frac{75 - 100}{\sqrt{100}} \leq \frac{Y - 100}{\sqrt{100}} \leq \frac{125 - 100}{\sqrt{100}} \right).$$

In general, if Y has a Poisson distribution with mean λ, then the distribution of $W = \frac{Y - \lambda}{\sqrt{\lambda}}$ is asymptotically $N(0, 1)$ when λ is sufficiently large.

Example 2.33 Let $X \sim \chi_{31}^2$. Approximate $P(\chi_{31}^2 \leq 38.307)$ using CLT and compare with the tabulated value.

Solution:
We know that

$$E(X) = 31, \, Var(X) = 62.$$

Using CLT, we have

$$P(X \leq 38.307) = P\left(\frac{X-31}{\sqrt{62}} \leq \frac{38.307-31}{\sqrt{62}}\right) \approx P(Z \leq 0.92799) = 0.8233.$$

Using tabulated values, we have

$$P(\chi^2 \leq 38.307) = 0.8282.$$

The interested readers to know more about probability may refer to Castaneda et al. (2012), Feller (1968), Rohatgi and Saleh (2015) and Ross (1998).

Problems

2.1 Determine the sample space for each of the following random experiments.

1. A student is selected at random from a probability and statistics lecture class, and the student's total marks are determined.
2. A coin is tossed three times, and the sequence of heads and tails is observed.

2.2 One urn contains three red balls, two white balls, and one blue ball. A second urn contains one red ball, two white balls, and three blue balls:

1. One ball is selected at random from each urn. Describe the sample space.
2. If the balls in two urns are mixed in a single urn and then a sample of three is drawn, find the probability that all three colors are represented when sampling is drawn (i) with replacement (ii) without replacement.

2.3 A fair coin is continuously flipped. What is the probability that the first five flips are (i) H, T, H, T, T (ii) T, H, H, T, H.

2.4 The first generation of a particle is the number of offsprings of a given particle. The next generation is formed by the offsprings of these members. If the probability that a particle has k offsprings (split into k parts) is p_k where $p_0 = 0.4$, $p_1 = 0.3$, $p_2 = 0.3$, find the probability that there is no particle in the second generation. Assume that the particles act independently and identically irrespective of the generation.

2.5 A fair die is tossed once. Let A be the event that face 1, 3, or 5 comes up, B be the event that it is 2, 4, or 6, and C be the event that it is 1 or 6. Show that A and C are independent. Find $P(A, B, \text{or } C \text{ occurs})$.

2.6 An urn contains four tickets marked with numbers 112, 121, 211, 222, and one ticket is drawn at random. Let A_i ($i = 1, 2, 3$) be the event that ith digit of the number of the ticket drawn is 1. Discuss the independence of the events A_1, A_2 and A_3.

2.7 There are two identical boxes containing, respectively, four white and three red balls; three white and seven red balls. A box is chosen at random, and a ball is drawn from it. Find the probability that the ball is white. If the ball is white, what is the probability that it is from the first box?

2.8 Let A and B are two independent events. Show that A^c and B^c are also independent events.

2.9 Five percent of patients suffering from a certain disease are selected to undergo a new treatment that is believed to increase the recovery rate from 30 to 50%. A person is randomly selected from these patients after the completion of the treatment and is found to have recovered. What is the probability that the patient received the new treatment?

2.10 Four records lead away from the country jail. A prisoner has escaped from the jail. The probability of escaping is $1/6$, if road 2 selected, the probability of success is $1/6$, if road 3 is selected, the probability of escaping is $1/4$, and if road 4 is selected, the probability of escaping is $9/10$.

1. What is the probability that the prisoner will succeed in escaping?
2. If the prisoner succeeds, what is the probability that the prisoner escaped by using road 4 and by using road 1?

2.11 The probability that an airplane accident which is due to structure failure is identified correctly is 0.85, and the probability that an airplane accident which is not due to structure failure is identified as due to structure failure is 0.15. If 30% of all airplane accidents are due to structure failure, find the probability that an airplane accident is due to structure failure given that it has been identified to be caused by structure failure.

2.12 The numbers $1, 2, 3, \ldots, n$ are arranged in random order. Find the probability that the digits $1, 2, \ldots, k$ $(k < n)$ appear as neighbors in that order.

2.13 In a town of $(n + 1)$ inhabitants, a person tells a rumor to a second person, who in turn, repeats it to a third person, etc. At each step, the recipient of the rumor is chosen at random from the n people available. Find the probability that the rumor will be told r times without returning to the originator.

2.14 A secretary has to send n letters. She writes addresses on n envelopes and absentmindedly places letters one in each envelope. Find the probability that at least one letter reaches the correct destination.

2.15 A pond contains red and golden fish. There are 3000 red and 7000 golden fish, of which 200 and 500, respectively, are tagged. Find the probability that a random sample of 100 red and 200 golden fish will show 15 and 20 tagged fish, respectively.

2.16 A coin is tossed four times. Let X denote the number of times a head is followed immediately by a tail. Find the distribution, mean, and variance of X.

2.17 In a bombing attack, there is 50% chance that a bomb can strike the target. Two hits are required to destroy the target completely. How many bombs must be dropped to give a 99completely destroying the target?

2.18 For what values of α and p does the following function represent a PMF $p_X(x) = \alpha p^x$, $x = 0, 1, 2, \ldots$.

2.19 Let the probability density function of X be given by

$$f(x) = \begin{cases} c(4x - 2x^2), & 0 < x < 2 \\ 0, & \text{otherwise} \end{cases}.$$

1. What is the value of c?
2. What is the distribution of X?
3. $P\left(\frac{1}{2} < X < \frac{3}{2}\right)$?

2.20 A bombing plane flies directly above a railroad track. Assume that if a larger(small) bomb falls within 40(15) feet of the track, the track will be sufficiently damaged so that traffic will be disrupted. Let X denote the perpendicular distance from the track that a bomb falls. Assume that

$$f_X(x) = \begin{cases} \frac{100-x}{5000}, & if\ x \in 0 < x < 100 \\ 0, & \text{otherwise} \end{cases}.$$

1. Find the probability that a larger bomb will disrupt traffic.
2. If the plane can carry three large(eight small) bombs and uses all three(eight), what is the probability that traffic will be disrupted?

2.21 A random variable X has the following PMF

$X = x$	0	1	2	3	4	5	6	7	8
$P(X = x)$	k	$3k$	$5k$	$7k$	$9k$	$11k$	$13k$	$15k$	$17k$

1. Determine the value of k.
2. Find $P(X < 4)$, $P(X \geq 5)$, $P(0 < X < 4)$.
3. Find the CDF of X.
4. Find the smallest value of x for which $P(X \leq x) = 1/2$.

2.22 An urn contains n cards numbered $1, 2, \ldots, n$. Let X be the least number on the card obtained when m cards are drawn without replacement from the urn. Find the probability distribution of random variable X. Compute $P(X \geq 3/2)$.

2.23 Let X be binomial distributed with $n = 25$ and $p = 0.2$. Find expectation, variance, and $P(X < E(X) - 2\sqrt{Var(X)})$.

2.24 Let X be a Poisson distributed random variable such that $P[X = 0] = 0.5$. Find the mean of X.

2.25 In a uniform distribution, the mean and variance are given by 0.5 and $\frac{25}{12}$, respectively. Find the interval on which the probability is uniform distributed.

2.26 Let

$$F_X(x) = \begin{cases} 0, & x < 0 \\ 1 - 2e^{-x} + e^{-2x}, & x \geq 0 \end{cases}.$$

Is F_X a distribution function? What type of random variable is X? Find the PMF/PDF of X?

2.27 Let X be a continuous type random variable with PDF

$$f_X(x) = \begin{cases} a + bx^2, & 0 < x < 1 \\ 0 & \text{otherwise} \end{cases}$$

If $E(X) = \frac{3}{5}$, find the value of a and b.

2.28 Let X be a random variable with mean μ and variance σ^2. Show that $E[(aX - b)^2]$, as a function of b, is minimized when $b = \mu$.

2.29 Let X and Y be two random variables such that their MGFs exist. Then, prove the following:

1. If $M_X(t) = M_Y(t)$, $\forall t$, then X and Y have same distribution.
2. If $\Psi_X(t) = \Psi_Y(t)$, $\forall t$, then X and Y have same distribution.

2.30 Let $\Omega = [0, 1]$. Define $X : \Omega \to \mathbb{R}$ by

$$X(w) = \begin{cases} w, & 0 \leq w \leq 1/2 \\ w - 1/2, & 1/2 \leq w \leq 1 \end{cases}.$$

For any interval $I \subseteq [0, 1]$, let $P(I) = \int_I 2x\,dx$. Determine the distribution function of X and use this to find $P(X > 1/2)$, $P(1/4 < X < 1/2)$, $P(X < 1/2/X > 1/4)$.

2.31 A random number is chosen from the interval $[0, 1]$ by a random mechanism. What is the probability that (i) its first decimal will be 3 (ii) its second decimal will be 3 (iii) its first two decimal will be 3's?

2.32 Prove that, the random variable X has exponential distribution and satisfies a memoryless property or Markov property which is given as

$$P(X > x + s / X > s) = P(X > x) \quad x, s \in \mathbb{R}^+. \tag{2.21}$$

2.33 Suppose that diameters of a shaft s manufactured by a certain machine are normal random variables with mean 10 and s.d. 0.1. If for a given application the shaft must meet the requirement that its diameter falls between 9.9 and 10.2 cm. What proportion of shafts made by this machine will meet the requirement?

2.34 A machine automatically packs a chemical fertilizer in polythene packets. It is observed that 10% of the packets weigh less than 2.42 kg while 15% of the packets weigh more than 2.50 kg. Assuming that the weight of the packet is normal distributed, find the mean and variance of the packet.

2.35 Show that the PGF's of the geometric, negative binomial and Poisson distribution exists and hence calculate them.

2.36 Verify that the normal distribution, geometric distribution, and Poisson distribution have reproductive property, but the uniform distribution and exponential distributions do not.

2.37 Let $Y \sim N(\mu, \sigma^2)$ where $\mu \in \mathbb{R}$ and $\sigma^2 < \infty$. Let X be another random variable such that $X = e^Y$. Find the distribution function of X. Also, verify that $E(\log(X)) = \mu$ and $Var(\log(X)) = \sigma^2$.

2.38 Let $X \sim B(n, p)$. Use the CLT to find n such that: $P[X > n/2] \leq 1 - \alpha$. Calculate the value of n when $\alpha = 0.90$ and $p = 0.45$.

2.39 Suppose that the number of customers who visit SBI, IIT Delhi on a Saturday is a random variable with $\mu = 75$ and $\sigma = 5$. Find the lower bound for the probability that there will be more than 50 but fewer than 100 customers in the bank?

2.40 Does the random variable X exist for which

$$P[\mu - 2\sigma \leq X \leq \mu + 2\sigma] = 0.6.$$

2.41 Suppose that the life length of an item is exponentially distributed with parameter 0.5. Assume that ten such items are installed successively so that the ith item is installed immediately after the $(i - 1)$th item has failed. Let T_i be the time to failure of the ith item $i = 1, 2, \ldots, 10$ and is always measured from the time of installation. Let S denote the total time of functioning of the 10 items. Assuming that $T_i's$ are independent, evaluate $P(S \geq 15.5)$.

2.42 A certain industrial process yields a large number of steel cylinders whose lengths are distributed normal with mean 3.25 inches and standard deviation 0.05 inches. If two such cylinders are chosen at random and placed end to end what is the probability that their combined length is less than 6.60 inches?

2.43 A complex system is made of 100 components functioning independently. The probability that any one component will fail during the period of operation is equal to 0.10. For the entire system to function at least 85 of the components must be working. Compute the approximate probability of this.

2.44 Suppose that $X_i, i = 1, 2, \ldots, 450$ are independent random variables, each having a distribution $N(0, 1)$. Evaluate $P(X_1^2 + X_2^2 + \cdots + X_{450}^2 > 495)$ approximately.

2.45 Suppose that $X_i, i = 1, 2, \ldots, 20$ are independent random variables, each having a geometric distribution with parameter 0.8. Let $S = X_1 + \cdots + X_{20}$. Use the central limit theorem $P(X \geq 18)$.

2.46 A computer is adding number, rounds each number off to the nearest integer. Suppose that all rounding errors are independent and uniform distributed over $(-0.5, 0.5)$.
(a) If 1500 numbers are added, what is the probability that the magnitude of the total error exceeds 15?
(b) How many numbers may be added together so that the magnitude of the total error is less than 10 with probability 0.90?

2.47 Let $X \sim B(n, p)$. Use CLT to find n such that $P[X > n/2] \geq 1 - \alpha$. Calculate the value of n, when $\alpha = 0.90$ and $p = 0.45$.

2.48 A box contains a collection of IBM cards corresponding to the workers from some branch of industry. Of the workers, 20% are minors and 30% adults. We select an IBM card in a random way and mark the age given on this card. Before choosing the next card, we return the first one to the box. We observe n cards in this manner. What value should n have so that the probability that the frequency of cards corresponding to minors lies between 0.10 and 0.22 is 0.95?.

2.49 Items are produced in such a manner that the probability of an item being defective is p (assume unknown). A large number of items say n are classified as defective or nondefective. How large should n be so that we may be 99% sure that the relative frequency of defective differs from p by less than 0.05?

2.50 A person puts some rupee coins into a piggybank each day. The number of coins added on any given day is equally likely to be 1, 2, 3, 4, 5 or 6 and is independent from day to day. Find an approximate probability that it takes at least 80 days to collect 300 rupees?

2.51 Suppose that 30 electronic devices say $D_1, D_2, \ldots, D_{30}$ are used in the following manner. As soon as D_1 fails, D_2 becomes operative. When D_2 fails, D_3 becomes operative, etc. Assume that the time to failure of D_i is an exponentially distributed random variable with parameter $= 0.1$ (h)$^{-1}$. Let T be the total time of operation of the 30 devices. What is the probability that T exceeds 350 h?

2.52 Suppose that X_i, $i = 1, 2, \ldots, 30$ are independent random variables each having a Poisson distribution with parameter 0.01. Let $S = X_1 + X_2 + \cdots + X_{30}$.
(a) Using central limit theorem evaluate $P(S \geq 3)$.
(b) Compare the answer in (a) with exact value of this probability.

2.53 Use CLT to show that

$$\lim_{n \to \infty} e^{-nt} \sum_{k=0}^{n} \frac{(nt)^k}{k!} = 1 = \begin{cases} 1, & 0 < t < 1 \\ 0.5, & t = 1 \\ 0, & t > 1 \end{cases}.$$

2.54 Consider polling of n voters and record the fraction S_n of those polled who are in favor of a particular candidate. If p is the fraction of the entire voter population that supports this candidate, then $S_n = \frac{X_1+X_2+\cdots+X_n}{n}$, where X_i are independent Bernoulli distributed random variables with parameter p. How many voters should be sampled so that we wish our estimate S_n to be within 0.01 of p with probability at least 0.95?

2.55 Let $X_1, X_2, \ldots$ be a sequence of independent and identically distributed random variables with mean 1 and variance 1600, and assume that these variables are nonnegative. Let $Y = \sum_{k=1}^{100} X_k$. Use the central limit theorem to approximate the probability $P(Y \geq 900)$.

2.56 If you wish to estimate the proportion of engineers and scientists who have studied probability theory and you wish your estimate to be correct within 2% with probability 0.95, how large a sample should you take when you feel confident that the true proportion is less than 0.2?

References

Castaneda LB, Arunachalam V, Dharmaraja S (2012) Introduction to probability and stochastic processes with applications. Wiley, New York

Feller W (1968) An introduction to probability theory and its applications: volume I. Wiley, New York

Rohatgi VK, Md. Ehsanes Saleh AK (2015) An introduction to probability and statistics. Wiley, New York

Ross SM (1998) A first course in probability. Prentice Hall, Englewood Cliffs

Chapter 3
Descriptive Statistics

3.1 Introduction

Statistics is an art of learning from data. One of the tasks to be performed after collecting data from any observed situation, phenomena, or interested variable is to analyze that data to extract some useful information. Statistical analysis is one of the most applied tools in the industry, decision making, planning, public policy, etc. Many practical applications start from analyzing data, which is the main information source. Given this data, the analyst should be able to use this data to have an idea of what the collected data have to say, either by providing a report of his/her findings or making decisions.

The first step after having data is to make it useful, which is possible if we can extract some useful information from that data. The obtained information is then used to *describe* the observed phenomena and its behavior. After having described the phenomena, the analyst can *infer* some characteristics, using appropriate tools. And finally, after inferring the main characteristics of the phenomena, it is possible to *model* observed trends, behaviors, unusual observations, etc.

For example, the government may want to get some idea about the income of its population to make economic decisions. The first step will be to collect as much data as possible across different classes and age groups. Now, this data will be processed to get meaningful information, e.g., mean, standard deviation, etc. After calculating different quantities, government can make inferences, e.g., the average income of 30–40 years age group is more than 10–20 years age group. Also, the government can use this data to model the income of middle-class population or classify a person as middle-class depending on other factors.

This chapter is intended to provide some tools for performing basic descriptive statistical analysis of data. Furthermore, some tools and examples are designed to introduce the reader to advanced statistical concepts, such as inference and modeling.

© Springer Nature Singapore Pte Ltd. 2018
D. Selvamuthu and D. Das, *Introduction to Statistical Methods,
Design of Experiments and Statistical Quality Control*,
https://doi.org/10.1007/978-981-13-1736-1_3

3.2   Data, Information, and Description

Data are a collection of unrefined facts and figures that do not have any added interpretation or analysis, whereas information is processed data that is meaningful to the user. Statistics is an art of learning from the data. Applicability and usefulness of statistics are almost unlimited, as most of the natural phenomena have a random behavior. This way, the analysis, description, and inference about data become a very important part of modern life, even in minor aspects of daily life such as buying stuff or going to work.

Now, an *observed data* are a measurement of a variable, in a given state. For instance, the height of a person (at a particular point of his life), the weight of a package, the pressure of a riveting machine, etc., are the observed data. Note that if we take similar *observational units* (a.k.a individuals) coming from a homogeneous group, there is a chance of having different measurements and we are not really sure of the exact reasons for those differences between measurements. These differences generally account for variation within a population or variations due to time.

It is important to introduce the concept of *samples* and *populations*. A population is the set of all observational units that we can analyze, and a sample is a representative subset of a population (which should be easy to access and measure). Thus, the size of the population, called the population size, denoted by N can be finite or infinite. Similarly, the number of units in a sample is called the sample size. It is usually denoted by n and is generally considered to be a finite number. The domain of the measured data is denoted by Ω so that the measurements collected over the population should be defined over Ω as well. In the next example, we explain these concepts.

For example, suppose we want to draw some conclusion about the weights of 8,000 students (the population) by examining only 200 students selected from this population. Here, $N = 8,000$ and $n = 200$. Another example is to conclude the whether a particular coin is fair or not by repeatedly tossing it. All the possible outcomes of the tosses of the coin form the population. A sample can be obtained by examining, say, the first 100 tosses of the coin and notify the percentage of heads and tails. Here N is infinite and $n = 100$.

Example 3.1 The national army of a country is composed of 1,500 officers, in different confinements and bases. The commanders of the army have selected a sample of 20 officers for whom four variables were measured: height (in m), marital status ($0 =$ single, $1 =$ married), education level ($0 =$ high school, $1 =$ technician, $2 =$ graduate, $3 =$ postgraduate, $4 =$ Ph.D.), and weight (in kg). The observed data are given in Table 3.1. The commanders want to infer certain information from the observed data to plan future supplies for their officers and duties to be assigned based on the skills of their officers. To do this, the provided data should be analyzed.

In this case, the addressed population size is $N = 1500$ officers, the sample size is $n = 20$ officers (observational units), and the domains for all the variables are: $\Omega_1 = [a, b]$, $a < b, a, b \in \mathbb{R}^+$, $\Omega_2 = \{0, 1\}$, $\Omega_3 = \{0, 1, 2, 3, 4\}$, and $\Omega_4 = [a, b]$,

Table 3.1 Observed data for
Example 3.1

Officer	Height (m)	M.S.	E.L.	Weight (kg)
1	1.76	0	2	83.1
2	1.83	1	2	91.8
3	1.79	1	3	91.3
4	1.71	1	2	85.6
5	1.81	0	0	88.0
6	1.71	0	2	89.2
7	1.96	1	2	92.8
8	1.80	0	2	89.1
9	2.10	0	3	90.8
10	1.89	0	1	87.0
11	2.13	0	3	90.2
12	1.82	0	3	85.9
13	2.07	1	3	93.2
14	1.73	0	2	89.6
15	1.72	1	2	89.1
16	1.86	0	4	90.5
17	1.82	1	2	87.1
18	1.94	1	3	88.5
19	1.74	1	4	89.9
20	1.99	1	2	88.3

$a < b, a, b \in \mathbb{R}^+$. Note that there are some differences among variables, since variables 1 and 4 have continuous domains, unlike variables 2 and 3 whose domains are discrete.

3.2.1 Types of Data

In statistical analysis, there are different kinds of data, whose values are closely related to the nature of the variables. There are two main types of data that are mostly observed in practical applications which further are of different types:

Categorical Data: It is also described as qualitative data. This data arise when the observations fall into separate distinct categories. Such data are inherently discrete, i.e., there are finite number of possible categories into which each observation may fall. Categorical data can further be classified as

Nominal Data: It is a variable whose measurement indicates a category or characteristic, more than an exact mathematical measure. In nominal variables, there is not a clear order among categories, and so a nominal variable is just a label of a characteristic of the observational unit without a rating scale (order). For example gender, eye color, religion, brand.

Ordinal Data: It is a variable whose measurement indicates a clear ordered category or characteristic. In ordinal variables, there is a clear order among categories. So an ordinal variable points out a characteristic of an observational unit that can be ranked regarding a rating scale. For example, a student's grades such as (A, B, C), clothing size (small, medium, large).

Numerical Data: This kind of data, also known as quantitative data, arise when the observations are counts or measurements. For example, the quantities such as number of students in the class, weight of an individual, temperature at a particular place, etc. The numerical data can further be of two types.

Discrete Data: The domain of discrete data is integers. For example, number of houses in a society, number of chapters in a book, etc.

Continuous Data: The domain of a continuous variable is $\Omega \in (-\infty, \infty)$ or $[a, b]$ or some interval on real line. Continuous domains are lattices and are clearly well ordered.

As seen in Example 3.1, we have the variables 1 and 4 are numerical data in particular continuous data. Variable 2 can be considered as an ordinal data since there is a clear consensus that single is better and married is worse. Variable 3 is an ordinal data since it is clear that postgraduate skills are better than high school skills, so category 1 implies to be better than 0, 2 is better than 1, and so on. In conclusion, the above classification of data is shown in Fig. 3.1.

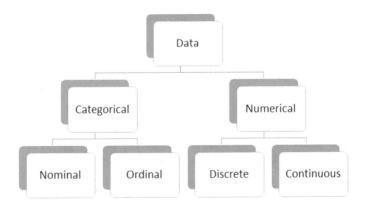

Fig. 3.1 Types of data

3.2.2 Data, Information, and Statistic

Based on the knowledge of nature of variables, we can establish some concepts before analyzing data. These concepts are *Data, Information*, and *Statistic*.

Data: These are the observed values of a variable, at a specific observational unit.

Information: Information, retrieved from the data, is a logical interpretation and/or statement which can be observed from the data.

Statistic: It is a fully computable quantity which can be obtained from the data (usually defined by a mathematical function). A statistic is based on logical reasoning, e.g., mean gives the average, and standard deviation gives a measure of variation in the sample.

This leads us to think that **data** are just measurements collected over **observational units** which can provide **information** to the analyst through **statistics**. More formally:

In simple words, a statistic is a function which takes values observed in a sample as input and gives a number with some meaning/information about the data as output. For example, mean is a statistic which takes all values of the sample as input and gives a number (mean of the sample) as output. This number gives an estimate of the population mean.

Example 3.2 (*Example* 3.1 *Continued*) Let us go back to Example 3.1. Denote every variable as X_i and the observed data as x_{ij} where i means the observational unit, and j means the measured variable. Let be $T_j = \max_i(x_{ij}) \; \forall \; j \in \{1, 2, 3, 4\}$, the question is: Is T_j a statistic? To do so, we compute T_j, as follows:

$$T_1 = 2.13 \quad T_2 = 1 \quad T_3 = 4 \quad T_4 = 93.24.$$

The answer is Yes, T_j is a statistic of all X_j variables.

Now consider $T_j = \sum_i (ln(x_{ij})) \; \forall \; j \in \{1, 2, 3, 4\}$, is T_j a statistic?

The answer is No, since $ln(x_{ij})$ is not computable for variables 2 and 3.

3.2.3 Frequency Tables

One aim of analyzing data is to categorize it into *classes* and then visualize its behavior. The idea is to divide the sample into well-defined subsets (classes) as a function of its domain and see how many data items are in each class. This analysis is called *frequency table approach*. A frequency table records how often each value (or a set of values) of the variable in question occurs, no matter if some events are not present in the sample, since the main idea is to *aggregate* data samples.

A frequency table is used to summarize nominal, ordinal, and continuous data, once the data set has been divided up into sensible groups. In nominal and ordinal data, classes are defined by their own labels, and in continuous data, classes should be defined by the observed domain of the data set, called its *Range*, where the range has to be divided into subsets to classify the data. The range of a continuous variable X_j namely R_j is defined as follows:

$$R_j = \max_i(x_{ij}) - \min_i(x_{ij}).$$

R_j is divided into an appropriate number (k) of classes $C_{j,k}$. Then the *Frequency* $f_{j,k}$ of each class is calculated by counting the data included in each class. An easy way to obtain classes and their widths are given below:

1. Compute the Range R_j.
2. Select the desired number of classes desired denoted by k. Usually, k is selected between 5 and 20.
3. Compute the class width w_k by dividing R_j by the number of classes and rounding up (not rounding off). If it is an integer, then you have two options: either increase the number of classes or the class width by one.
4. Select a starting point less than or equal to $\min_i(x_{ij})$. It is mandatory to cover all the range, so we have that $k \times w_k > R_j$ which leads to cover one more value than the range. Now set the starting point as the lower limit of the first class. Continue to add the class width to this lower limit to get the rest of the lower limits. These are called the *Limits* of the class.
5. The upper limit of the first class is located by subtracting one unit from the lower limit of the second class. Continue adding w_k to this upper limit to find all upper limits.
6. Define the *Real limits* of a class by subtracting 0.5 units and adding 0.5 units from the lower limits and to the upper limits, respectively.
7. Find the frequencies $f_{j,k}$.

Remark 3.1 When we have more than one categorical variable in a sample, a frequency table is often called a contingency table since there is a row data dependence upon column data.

Contingency table is a display format used to analyze and record the relationship between two or more categorical variables. These are constructed by noting all levels of one variable as rows and levels of another variable as a column and finding a joint frequency.

Consider an example as shown in Table 3.2. let Male $= M = 0$ and Female $= F = 1$. Let diseased Yes $= Y = 0$ and No $= N = 1$. The data observed for a sample of three people. The corresponding contingency table is shown in Table 3.3.

Another interesting statistic is called the **Relative frequency** of $C_{j,k}$, namely $Rf_{j,k}$, which is simply the ratio of number of data points present in $C_{j,k}$ to the total number of observations n which is defined as follows:

$$Rf_{j,k} = \frac{f_{j,k}}{n}.$$

Table 3.2 Observed data

	Diseased	Gender
1st person	0	1
2nd person	1	1
3rd person	0	1

Table 3.3 Contingency table

Disease	Gender		
	Male	Female	Marginal total
Yes	0	2	2
No	1	0	1
Marginal total	1	2	Sum total 3

The information provided by the frequency and relative frequency of a class indicates as to how often a subgroup of data is present in a sample. In most cases, $Rf_{j,k}$ shows the most and least frequent classes which reveal the most and least probable values of the sample (and the values within).

Cumulative Relative Frequency $Cf_{j,k}$ is another statistic which can be found by summing up the frequencies of the $k < k'$ classes, where k' is the desired class. It can be represented as follows:

$$Cf_{j,k'} = \sum_{k=1}^{k'} Rf_{j,k}.$$

The cumulative frequency table provides information about how many data items belong to a particular class, which is useful to determine how the sample is spread and where the major amount of data is located.

Example 3.3 (*Example 3.1 continued*) Recalling the information of the variable 4 in Example 3.1, we can see that $R_j = \max_i(x_{ij}) - \min_i(x_{ij}) \rightarrow 93.2 - 83.1 = 10.1$. The obtained results are shown in Table 3.4.

Table 3.4 Frequency table for Example 3.1

Class	Limits	Actual limits	$f_{j,k}$	$Rf_{j,k}$	$Cf_{j,k}$
1	(83.1–85.11)	(83.095–85.115)	1	0.05	0.05
2	(85.12–87.13)	(85.115–87.135)	4	0.2	0.25
3	(87.14–89.15)	(87.135–89.155)	5	0.25	0.5
4	(89.16–91.17)	(89.155–91.175)	6	0.3	0.8
5	(91.18–93.2)	(91.175–93.205)	4	0.2	1
Total			20	1	

Table 3.4 shows all 5 classes, their limits and real limits, their frequencies $f_{j,k}$, relative frequencies $Rf_{j,k}$, and cumulative frequencies as well. Note that the fourth class is the most frequent class (89.16–91.17), while first class is the least frequent. This means that most of the officers weigh around (89.16–91.17) kg and the less popular weight is around (83.1–85.11) kg. Also note that at least 80% of the officers weigh less than 91.175 kg, and at least 75% of the officers weigh more than 87.135 kg $(1-Cf_{4,2})$.

3.2.4   Graphical Representations of Data

Frequency tables are indeed useful tool to represent the data; however, they have their own limitations. Thus, in situations where we have large amounts of data, representing the data using graphical methods is often clearer to understand and is a comprehensive way to understand and visualize the behavior of the data. Most of the analysts request for graphs to understand some of the statistics computed from samples and so there is a need for presenting results in a clear way. In next subsections, some useful graphical representations are introduced.

Bar Chart

Bar charts provide a clear and commonly used method of presenting un-grouped discrete frequency observations or any categorical data. Bar charts provide a simple method of quickly spotting simple patterns of popularity within a discrete data set. In bar chart, rectangular blocks of equal width are plotted on the x-axis, each representing independent variable placed at equal distance from each other. The height of each block represents the frequency of the categories and is proportional to the number of percentage in each category. Because each bar represents a completely separate category, the bars must not touch each other, i.e., always leave a gap between the categories. The bar diagrams are drawn through columns of equal width. The bar diagram is also called a columnar diagram. The steps to create a bar chart are the following:

1. Put the data into a frequency table.
2. Decide the data to be represented on each of the axes of the chart. Conventionally, the frequency is represented on the y-axis (i.e., vertical axis) and the data which is being measured is on the x-axis (i.e., horizontal axis).
3. Decide the scale for the y-axis or the frequency axis which represent the frequency in each category on the x-axis by its height. Label this axis with suitable number scale.
5. Draw both the axes and appropriately label them.
6. Draw a bar for each category. While drawing the bars, one must ensure that each bar is of same width and there are equally sized gaps in between the bars.

For example, variable 3 (education level) in our example can be represented by bar chart as shown in Fig. 3.2.

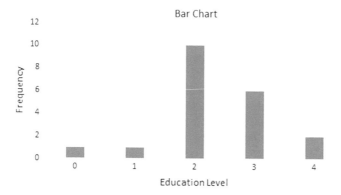

Fig. 3.2 Bar chart

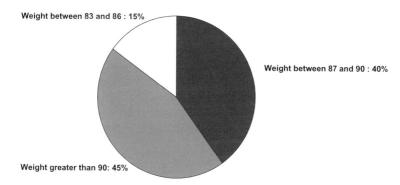

Fig. 3.3 Pie chart graph

Pie Chart

A pie chart is used to represent the relative frequencies for a non-numerical data i.e., it is only applicable for categorical data or grouped data. A pie chart presents the categories of data as parts of a circle or "slices of a pie." They are used to show the proportions of a whole and are useful when there are small number of categories to display. The underlying concept is to construct a circle and then slice it into different sectors, one for each category. The area of each sector being equal to the product of total area of circle with the relative frequency of the category that sector is representing, i.e.,

$$\text{angle} = \frac{\text{Frequency of the category}}{\text{Total number in sample}} \times 100.$$

For example, Fig. 3.3 shows the pie chart for the data of Table 3.1.

Frequency Histogram and Dot-Plot

In case of continuous data grouped into classes, a histogram is constructed for graphical representation of data. A histogram is a vertical bar chart drawn over a set of class intervals that cover the range of observed data. Before drawing a histogram, organize the data into a frequency distribution table, as we do in case of grouped data. The class intervals should be formed with class boundaries since histograms are prepared for continuous data. Thus, the upper boundary of a class interval will be the same with the lower boundary of the subsequent class interval. Then, for each class, a rectangle is constructed with a base length equal to its class width and height is equal to the observed frequency in the class. Usually, the rectangles do not have an equal length.

It is often used in descriptive data analysis to visualize the major features of the distribution of the data in a convenient form. A histogram can give the following information about the data (a) the shape of the distribution, (b) the typical values of the distribution, the spread of the distribution, (c) and the percentage of distribution falling within a specified range of values.

A histogram can also help detect any unusual observations *(outliers)*, most and least frequent classes, or any gaps in the sample. When having nongrouped discrete data, the histogram cannot be constructed. In that case, the most recommended graph is called **Dot-plot** graph, which draws the frequency of each category of the sample using points and lines instead of classes and rectangles. A dot-plot graph of the data can be presented in a frequency table.

Most of the information provided by a histogram or dot-plot regards to the most and least frequent values, the range of the sample and possible unusual behaviors in samples. Next example illustrates the main idea of drawing a histogram and a dot-plot.

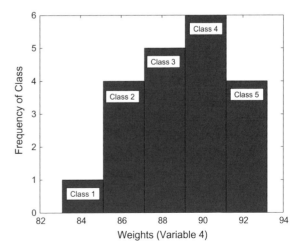

Fig. 3.4 Frequency histogram for Variable 4 (weight)

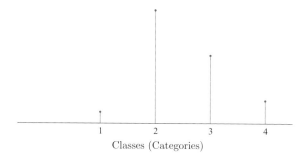

Fig. 3.5 Dot-plot graph for Variable 3 (education level)

Example 3.4 (*Example* 3.1 *continued*) In our example, we want to draw the histogram of the variable 4, for which we computed the frequencies. The results are shown in Fig. 3.4.

Figure 3.4 shows that the most frequent class is fourth class (89.16–91.17) kg. This means that the commanders of the army should plan the supplies for their officers keeping in mind that most of them weigh (87.14–89.15) kg and only a few of them weigh (83.1–85.11) kg.

Now, the variable 3 (education level) can be plotted using a dot-plot to display its frequencies, shown as follows.

Figure 3.5 shows a dot-plot where we can see that the category 2 (graduate) is the most prevalent among officers, and only a few of them have only high school education level. Also note that most of the officers have at least graduate degree, so the commanders have a well-trained body of officers.

Many of the large data sets observed in practice have histograms that are similar in shape which often reach their peaks at the sample mean and then decrease on both sides of this peak in a bell-shaped symmetric fashion. Such histograms are known as normal histograms, and such data sets are called normal. Figure 3.6 shows normal histograms.

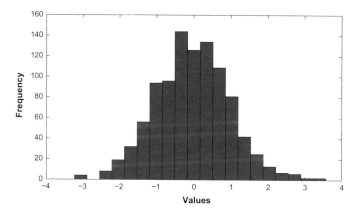

Fig. 3.6 Normal histograms

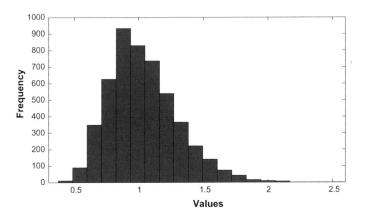

Fig. 3.7 Skewed histograms

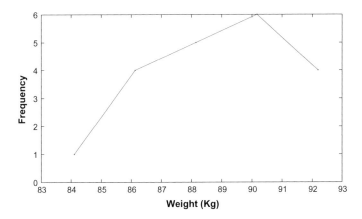

Fig. 3.8 Frequency polygon

The data set for which histogram is not even approximately symmetric about its sample mean is said to be skewed data. It is called "skewed to the right" if its right tail is longer as compared to its left tail and "skewed to the left" if it has a longer left tail. Figure 3.7 shows skewed to the right.

Frequency Polygon

Frequency polygon is another way to represent the data in a frequency table. Frequency polygon plots the frequencies on the y-axis (i.e., vertical axis) against different data values on the horizontal axis and then connects the plotted points with straight lines. Figure 3.8 represents a frequency polygon for the data of Table 3.1.

Box Plot

This is the most commonly used graphical statistics used to estimate the distribution of quantitative data of a population. Box plot is defined as a graphical statistic that can

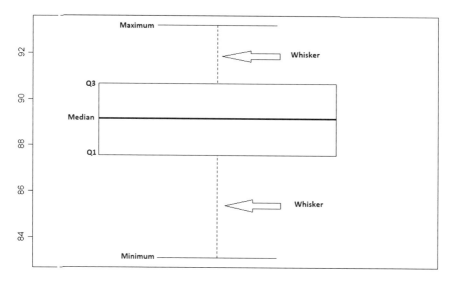

Fig. 3.9 Box plot for Variable 4 (Weight)

be used to summarize the observed sample data and can provide useful information on the shape and the tail of the distribution. This is also called as box and whisker plot.

A box plot is based on five number summary associated with the data: (a) the minimum value, (b) the maximum value, (c) the first quartile (Q1) value, (d) the second quartile (Q2) value or the median, and (e) the third quartile (Q3) value. From the box plot, one can get an idea of interquartile range from the length of the box. The larger the size of the box, greater is the range. The length of line from Q1 to the lowest value and from Q3 to the highest value gives an idea of the spread of the data beyond Q3 and below Q1, respectively. Also, one can infer one can infer about the tails of the distribution or the extreme values, the typical values in the distribution (minimum, maximum, median).

For example, we want to draw the box plot of the variable 4. The results are shown in Fig. 3.9. We can observe the long tail or whisker to the below of Q1 that indicates the spread of data is more below Q1. Similarly, median is around 91 kg, Q1 is around 87.5 and Q3 is around 90.5.

Cumulative Frequency Graph

A *Cumulative frequency graph* displays the behavior of the cumulative frequencies of a sample in an increasing fashion. This graph starts from zero (0) and ends at one (1), since the cumulative sum of all frequencies shall not be more than one (which is equivalent to the 100% of data).

Ogives

The partition values, namely quartiles, deciles, and percentiles can be conveniently located with the help of a curve called the "cumulative frequency graph" or "Ogive."

Fig. 3.10 Cumulative
frequency graph for
Variable 4 (weight)

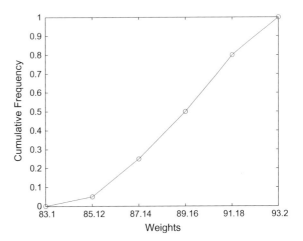

Example 3.5 (*Example* 3.1 *Continued*) Now, the commanders give special attention
to the weight of their officers so that they can see how the sample's frequency is
increasing. To do so, Fig. 3.10 shows the cumulative frequency graph for the weight
of the officers. This curve is called less than cumulative frequency curve or less than
Ogive.

Note that the behavior of the frequency is increasing from 83.1 to 93.2 kg, and
its behavior is very well distributed (which means that there are no sudden changes
among classes).

The commanders have realized that 80% of their officers weigh less than 91.18 kg,
and just a 25% of them weigh less than 87.14 kg. This is equivalent to saying that the
commanders used the first quartile to determine where the 25% of the officers are,
and the eighth quantile to determine where the 80% of the officers are. Figure 3.11
shows where $q_{0.25}$ and $q_{0.8}$ are.

Fig. 3.11 Quantiles for
Variable 4 (weight)

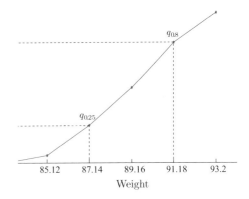

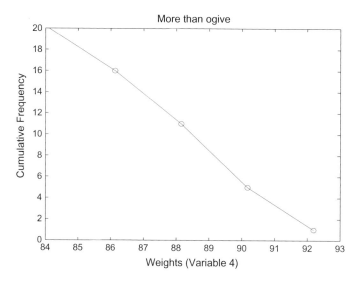

Fig. 3.12   More than ogive for Variable 4 (weight)

Other interesting analysis can be performed: 55% of the officers weigh between 87.14 and 91.18 kg. This means that more than half of the officers weigh between 87.14 and 91.18 kg, which would be a reference weight for planning supplies.

There are two types of Ogives: more than type ogive and less than type ogive. Figure 3.12 shows more than type ogive of weights. Similarly, less than ogive for weights is shown in Fig. 3.13.

The x-coordinate of the point of intersection of less than ogive and more than ogive gives the median of data. This is shown in Fig. 3.14.

We often come across the data sets having paired variables, related to each other in some way. A useful way of representing a data with paired values (x_i, y_i), $i = 1, 2, \ldots, n$ is to plot the data on a two-dimensional graph with x values being representing on the x-axis and the y values on the y-axis. Such a plot is known as a scatter diagram. For example, Fig. 3.15 shows the scatter diagram for the data in Example 3.1.

3.3   Descriptive Measures

Sometimes, the analyst requires a single measure to make a decision or has an idea about the behavior of the sample. A natural thought is how to answer the following questions: What is the average of the sample? how much variation is there in the sample?

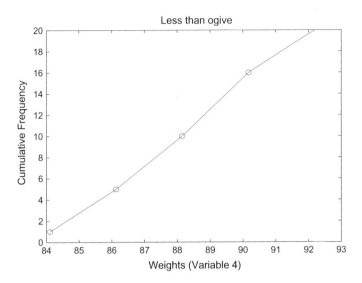

Fig. 3.13 Less than ogive for Variable 4 (weight)

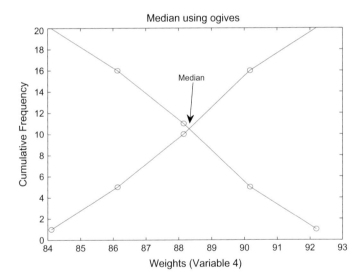

Fig. 3.14 Median using Ogives

In the theory of statistics, these questions are answered by the measures of *Central tendency* and *Variability* of a sample. The most useful statistics to answer these questions are the *mean* and *variance*, defined in the next section. The measures that we are defining in the following subsections are for sample.

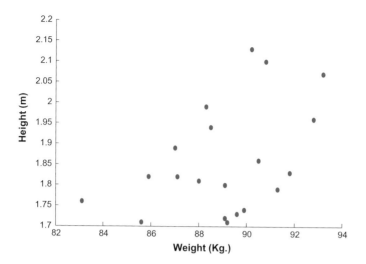

Fig. 3.15 Scatter diagram

3.3.1 *Central Tendency Measures*

Central tendency is a concept closely related to what is expected in a sample, its most frequent value, or their average behavior. A central tendency measure give us an idea of how a data sample has been grouped around a value. Some important central tendency measures are the *Mean*, *Median*, *Mode*, and the *Geometric mean*, which are defined as follows:

Definition 3.1 *(Mean)* The **mean** of a sample composed of observations $x_1, x_2, \ldots, x_n$ is its arithmetic mean, denoted by $\overline{x}$

$$\overline{x} = \frac{\sum\limits_{i=1}^{n} x_i}{n}. \tag{3.1}$$

The mean is an appropriate measure in most cases. However, as it involves all data samples, extreme values (a.k.a as outliers) can affect its value. Note that the mean of any variable *has the same units* as the variable. Sum of deviation of a set of n observations $x_1, x_2, \ldots, x_n$ from their mean $\overline{x}$ is zero, i.e.,

$$\sum_{i=1}^{n} (x_i - \overline{x}) = 0.$$

For a discrete data, if the observations $x_1, x_2, \ldots, x_n$ occur with the frequencies $f_1, f_2, \ldots, f_n$, respectively, and $d_i = x_i - A$, the deviation from a number A, then

Variable	Height (m)	Weight (kg)
Mean	1.859	89.05

Table 3.5 Sample means for Example 3.1

$$\bar{x} = A + \frac{1}{N} \sum_{i=1}^{n} f_i d_i$$

where $\sum_{i=1}^{n} f_i = N$. For a continuous data, if all the k class intervals had same width h and $d_i = x_i - A$, where x_i is the class mid point of the ith class interval, then define Now,

$$\bar{x} = A + \frac{h}{N} \sum_{i=1}^{n} f_i u_i$$

where $\sum_{i=1}^{n} f_i = N$.

Example 3.6 (*Example* 3.1 *Continued*) The commanders of the army want to know the average value of the data samples provided, to have an idea about the main characteristics of their officers. For instance, the mean height of the officers is computed as follows

$$\bar{x}_{height} = (1.76 + 1.83 + \cdots + 1.74 + 1.99)\,\text{m}/20 = 1.859\,\text{m}.$$

And the mean of variables 1 and 4 is shown in Table 3.5.

Now, the commanders can see that the average height of their officers is 1.859 m and their average weight is 89.05 kg. These are useful information for planning supplies in the future, assuming the average values should be supplied more efficiently than extreme values.

Definition 3.2 (*Median*) The **median** of an ordered sample $x_1 \leqslant x_2 \leqslant \cdots \leqslant x_n$ is the value, namely (med(x)) for which a half of the observations are less than this value, and the other half are greater than this value.

If the sample size n is odd, then the value in the middle of the ordered sample is its median. If n is even, then the arithmetic mean of the two central values is its median.

For a grouped data, median is given by

$$med(x) = L_m + h \left(\frac{N/2 - C}{f_m} \right)$$

Table 3.6 Sample medians for Example 3.1

Variable	Height (m)	Weight (kg)
Median	1.82	89.15

where

$$N = \text{total frequency} = \sum_{i=1}^{n} f_i$$

n = number of classes

f_m = frequency of the class where the median lies

L_m = lower class boundary of the class where median lies

C = cumulative frequency of class below the class where the median lies

h = width of the class interval.

Example 3.7 (Example 3.1 Continued) The commanders wish to know the median height of their officers to see if the officer's height show any kind of symmetry.

For instance, the median of the variable height is computed from the ordered sample, as follows:

$$(1.71, 1.71, \ldots, 1.82, 1.82, \ldots, 2.1, 2.13)$$
$$\rightarrow \text{med}(x_{height}) = (1.82 + 1.82)/2 = 1.82.$$

The median of variables 1 and 4 is summarized in Table 3.6.

As the medians are close to the means of those variables, the commanders can see that the officers have a symmetric behavior, since half of their officers are less than its average in those variables.

Definition 3.3 (*Quantiles, Percentiles, and Quartiles*)

Quantiles

Particularly in a frequency graph, it is possible to locate interesting points known as *Quantiles*. A Quantile q is a value below which a specific proportion of the sample (or population) is located. Quantiles are regular intervals taken from the cumulative frequency graph of a random variable. It can be obtained by dividing ordered data into k essentially equal-sized proportions. Quantiles are the measures that divide the observations into equal parts provided that the data values are arranged in an ascending order).

Percentile: The Kth percentile of an ordered sample $x_1 \leq x_2 \leq \cdots \leq x_n$ is a quantity such that at least $K\%$ of the data values are less than or equal to and at least $(100\text{-}K)$ % of the data values are greater than or equal to it.

Quartiles: The quartiles of an ordered sample $x_1 \leq x_2 \leq \cdots \leq x_n$ divide the data into four equal parts. The lower or first quartile Q_1 is the quantity such that at least 25% of the observations are less than or equal to it, and 75% are greater than or equal to it. The upper or third quartile is the quantity such that 75% of the observations are less than or equal to it, and 25% are greater than or equal to it. The middle or second quartile Q_2 is nothing but the median, which divides data into two equal parts.

Summarizing these quantities, we have **Median**: Divide the data into two equal parts,

Quartiles: Divide the data into four equal parts,

Deciles: Divide the data into 10 equal parts,

Percentiles: Divide the data into 100 equal parts.

For a given un-grouped data, we compute three quartiles q_1, q_2, and q_3, that will divide the observations into four equal parts. We have q_1 is located at $\frac{n+1}{4}$ and 25% of all observations lie below q_1. Similarly, q_2 is located at $\frac{n+1}{2}$ and 50% of all observations lie below q_2. q_3 is located at $\frac{3(n+1)}{4}$ and 75% of all observations lie below q_3 provided that the data have been arranged in ascending order.

The range between the first and third quartiles is known as interquartile range (IQR) and is given by

$$IQR = Q_3 - Q_1.$$

Interquartile range helps us to locate the middle 50% of the total observations. Similarly, the quartile deviation (q.d.) is given by

$$q.d. = \frac{IQR}{2} = \frac{(q_3 - q_1)}{2}$$

and it can be used as a measure of dispersion in the middle half of distribution. Quartile deviation is useful when the data contain outliers.

Similarly, there are nine quantities known as deciles $d_1, d_2, \ldots, d_9$ that divide the data into ten equal parts. Thus, 10% of the data lies below d_1, 20% of the data lies below d_2, and so on provided the data are arrange in ascending order. The deciles can be calculated as follows for an un-grouped data.

$$d_i \text{ is located at } \frac{i(n+1)}{10}, \quad i = 1, \ldots, 9.$$

For instance, after arranging data in ascending order, d_4 occupies the position $\frac{4}{10}(n + 1)$. Finally, there are 99 quantities called percentiles $p_1, p_2, \ldots, p_{99}$ that divide the data into 100 equal parts. Here, 1% of the data lies below p_1, 2% of the data lies below p_2, and so on provided the data are arranged in ascending order. Percentiles can be calculated as follows for an un-grouped data.

$$p_i \text{ is located at } \frac{i(n+1)}{100}, \quad i = 1, \ldots, 99,$$

with, say, p_{30} found at $\frac{30}{100}(n + 1)$.

Example 3.8 Consider the height of the officers given in Table 2.1 Upon arranging them in increasing order, we have 1.71, 1.71, 1.72, 1.73, 1.74, 1.76, 1.79, 1.80, 1.81, 1.82, 1.83, 1.86, 1.89, 1.9, 1.94, 1.96, 1.99, 2.07, 2.10, 2.13, we can easily determine that:

1. q_1 is located at the position $\frac{20+1}{4} = \frac{21}{4} = 5.25$, i.e., q_1 is $\frac{1}{4}$ of the distance between the fifth and sixth data points. Hence, $q_1 = 1.745$.
2. Similarly, d_2 is located at the position $\frac{2(20+1)}{10} = 4.2$ or d_2, i.e., d_2 is $\frac{1}{5}$ of the distance between the fourth and fifth data points. Hence, $d_2 = 1.732$.
3. Also, p_{40} is located at the position $\frac{40(20+1)}{100} = 8.4$ or p_{40}, i.e., p_{40} is $\frac{2}{5}$ of the distance between the eighth and ninth data points. Hence, $p_{40} = 1.804$.

Similar to the formula for median, for a grouped data, the ith percentile for a grouped data is given by

$$p_i = L_m + h \left(\frac{iN/100 - C}{f_m} \right), \quad i = 1, 2, \ldots, 99$$

and that for jth quartile is given by

$$Q_j = L_m + h \left(\frac{jN/4 - C}{f_m} \right), \quad j = 1, 2, 3$$

where

$$N = \text{total frequency} = \sum_{i=1}^{n} f_i$$

n = number of classes

f_m = frequency of the class where the percentile or quartile lies

L_m = lower class boundary of the class where percentile or quartile lies

C = cumulative frequency of class below the class where the percentile or quartile lies

h = width of the class interval.

The percentiles and quartiles are computed as follows:

1. The f_i-value, i.e., frequency of the ith class in the data table is computed:

$$f_i = \frac{i-1}{n-1}$$

where n the number of values.
2. The first quartile is obtained by interpolation between the f-values immediately below and above 0.25, to arrive at the value corresponding to the f-value 0.25.
3. The third quartile is obtained by interpolation between the f-values immediately below and above 0.75, to arrive at the value corresponding to the f-value 0.75.

Definition 3.4 *(Mode)* The **mode** of a sample $x_1, x_2, \ldots, x_n$ is its most frequent value, namely **modal value** (mode(x)).

$$mode(x) = l + \left(\frac{f_s}{f_p + f_s} \times h \right)$$

where

$$l = \text{lower limit of modal class}$$
$$f_s = \text{frequency in the class succeeding modal class}$$
$$f_p = \text{frequency in the class preceding modal class}$$
$$h = \text{width of class interval}$$

where modal class is the class interval with highest frequency. When the sample is continuous, usually its mode is defined as the most frequent class of the sample instead of a single point. In categorical data, many analysts prefer the mode of a sample, or its median instead of its mean, since they are easier to relate to samples due to its nature.

Example 3.9 (*Example* 3.1 *Continued*) The commanders now want to know what are the most frequent values for variable height of their officers, and see if it has a relationship to its mean and median.

For instance, the mode of the variable E.L is 2 (graduate level). The mode of variables 1, 3, and 4 are summarized in Table 3.7.

Note that the mode of a variable cannot always be a single point. For instance, variables 1 and 4 have the continuous values in some intervals. This illustrates how different values in categorical variables can be modes, and interpretation of modal categories gives an idea of the popular behavior in a sample.

In our example, modal values are still close to the means and medians of all variables, except for the height which is a little far from its mean. This means that the officers have a well-defined central tendency on all the variables measured, so the commanders have more information for planning its supplies.

Note that in a perfectly symmetric data, all the three measures coincide.

Definition 3.5 (*Geometric mean*) The **geometric mean** of a sample $x_1, x_2, \ldots, x_n$ namely $(\overline{x}_g)$ is the nth root of the product of the observations.

$$\overline{x}_g = \sqrt[n]{\prod_{i=1}^{n} x_i} \ . \tag{3.2}$$

Table 3.7 Sample modes for Example 3.1

Variable	Height (m)	E.L	Weight (kg)
Mode	(1.715–1.805)	2	(89.155–91.175)

Table 3.8 Geometric means for Example 3.1

Variable	Height (m)	Weight (kg)
Geometric mean	1.855	89.016

If the sample contains a zero, then its geometric mean $\bar{x}_g$ is zero. In categorical data, it is common to have categories including the value zero. Thus, the geometric mean is not the best central tendency measure for this kind of data.

Example 3.10 (*Example* 3.1 *Continued*) The commanders want to know more about the mean height of the officers, so they request for computing the geometric mean of all variables. For instance, the geometric mean of the variable weight is 89.016 kg.

$$\bar{x}_g = \sqrt[20]{83.1 \times 91.8 \times \cdots \times 89.9 \times 88.3} = 89.016.$$

The geometric mean of variables 1 and 4 is summarized in Table 3.8.

The geometric means of the variables 1 and 4 are still closer to its sample mean, which indicates a stable central tendency of data samples.

3.3.2 Variability Measures

Variability refers to the *spread* of a sample, its range, or distribution. A variability measure gives us an idea of how a data sample is spread around a value. Two important variability measures are the *variance* and the *standard deviation*, which are defined as follows.

Definition 3.6 (*Sample Variance*) The **sample variance** s^2 of a sample composed of the observations $x_1, x_2, \ldots, x_n$ is the arithmetic mean of the squared distances between each observation and its sample mean,

$$s^2 = \frac{\sum_{i=1}^{n}(x_i - \bar{x})^2}{n-1} = \frac{n\sum_{i=1}^{n}x_i^2 - \left(\sum_{i=1}^{n}x_i\right)^2}{n(n-1)} \qquad (3.3)$$

where $n - 1$ is the degrees of freedom of the sample variance, one degree of freedom is lost since $\sum_{i}^{n} x_i$ is known.

The variance is always nonnegative, and it has the squared units of the sample values. If the data observations are more spread, the variance is higher and vice versa.

Another important property of the variance is its sensitivity to extreme values (outliers), since they can change the value of the variance dramatically. Just one extreme value can push the variance to be higher than expected. So the analyst has to be careful when having extreme or unexpected values in the sample.

In practice, variance is one of the most useful measures since it gives us a good idea about how the variable is spread in the sample, and information about the behavior of the sample and its variability as well. The smaller variance a sample has, the more stable it is.

Remark 3.2 Usually, the variance is computed using the mean as central tendency value, but it is possible to compute the variance using other central tendency measures such as the median or mode.

Definition 3.7 (*Sample standard deviation*) The **sample standard deviation** of a sample composed of the observations $x_1, x_2, \ldots, x_n$ is the squared root of its sample variance, denoted by s,

$$s = \sqrt{s^2}. \tag{3.4}$$

The standard deviation is always positive, and it has the same units as the sample. The behavior of the sample standard deviation is very similar to the sample variance.

In practice, the standard deviation is very useful since it gives us an idea about the variability of a sample, in the same sample units. This is very useful when analyzing data because the analyst often wishes to speak in the units same as his sample for all his analysis, and sometimes it is hard to think in squared units when performing reports and/or presentations.

Remark 3.3 As the variance can be computed using the other central tendency measures, the sample standard deviation can also be computed using other central tendency measures.

Example 3.11 (*Example 3.1 Continued*) The commanders now want to know more information about the variability in the heights shown by their officers. To do so, they request for computing the sample variance of the provided data.

For instance, the variance of the variable height is $0.01736 \, m^2$ as shown below:

$$s^2 = \frac{(1.76 - 1.859)^2 + (1.83 - 1.859^2) + \cdots + (1.99 - 1.859)^2}{20 - 1} = \frac{0.32978}{19} = 0.01736 \, m^2.$$

The variance of variables 1 and 4 are summarized in Table 3.9.

In this case, the weight of the officers shows more variability than the other variables. The height seems to be very stable since its variance is small.

The recommendation to the commanders is to keep in mind that the weight of the officers varies considerably, and the supplies should satisfy this condition.

Table 3.9 Sample variances for Example 3.1

Variable	Height (m)	Weight (kg)
Variance	0.01736	6.2447

Table 3.10 Standard
deviations for Example 3.1

Variable	Height (m)	Weight (kg)
Standard deviation	0.1317	2.4989

Example 3.12 (*Example* 3.1 *Continued*) The commanders now want to know about the variability shown by their officers, in the same units as the original variables to see how they are spread. To do so, we compute the standard deviation.

For instance, the standard deviation of the variable height is 0.1317 m, as follows

$$s = \sqrt{s^2} = \sqrt{0.01736\,\mathrm{m}^2} = 0.1317\,\mathrm{m}.$$

The standard deviation of variables 1 and 4 is summarized in Table 3.10.

At first glance, the commanders can see that the weight of their officers is more spread, showing more variability than other variables, as variances as well. Hence, the supplies should be adjusted accordingly.

3.3.3 Coefficient of Variation

Another interesting question arises when analyzing multiple data sets to compare variance/standard deviation of different variables. As each variable has different unit, their variances/standard deviations have different units as well which makes it difficult to compare **volatility** or **stability** of variables. This means that the analyst should be able to say which variables (or samples) are more volatile than others, leaving behind the original units of each variable.

The coefficient of variation addresses this problem through a simple computation which gives us an idea of the behavior of the standard deviation of a sample with regard to its mean.

Definition 3.8 (*Coefficient of variation*) The **coefficient of variation** of a sample composed of the observations $x_1, x_2, \ldots, x_n$ is the ratio of its standard deviation and the absolute value of its mean denoted as c_v

$$c_v = \frac{\sigma}{\mu}. \tag{3.5}$$

The coefficient of variation is always positive without units (to simplify the analysis). Higher the c_v of a variable, more volatile/spread the variable is. The c_v has the property of having no units effect. Hence, it is a standardized measure which gives us an idea of the real spread of a variable, no matter what units (kg, Tons, m³, etc.) the sample have, which is very convenient to make analysis without confusion.

Example 3.13 (*Example* 3.1 *Continued*) The commanders are interested in having a measure of the real variability of each variable, since they are a little bit confused

Table 3.11 Coefficients of variation for Example 3.1

Variable	Height (m)	Weight (kg)
Coefficient of variation	0.071	0.028

with different squared units, and so they requested for the coefficient of variation c_v of all samples. As an example, the c_v of the variable 4 (weight) is:

$$c_v = \frac{0.13175}{1.859} = 0.071.$$

The coefficient of variation of variables 1 and 4 is summarized in Table 3.11.

Note that the more spread variable is the height which has the smaller unit and the less spread variable is the weight which has the largest unit. This shows to the commanders that the supplies have to be more spread in terms of weight than height since the weight is clearly more variable than height. Finally, the suppliers must cover the height needs in term of more sizes available for their officers, more variety in shoes, jackets, uniforms, etc.

Skewness

Skewness is a measure of asymmetry or in a more mathematical sense is a measure of how asymmetric the sample under consideration is. The skewness of a sample composed of the observations $x_1, x_2, x_3, \ldots, x_n$ is the third central moment:

$$\mu_3 = \frac{\sum_{i=1}^{n} (x_i - \overline{x})^3}{n}. \tag{3.6}$$

The sign of μ_3 gives us the direction of skewness. If $\mu_3 > 0$, we say that the sample is positively skewed, whereas $\mu_3 = 0$ corresponds to symmetry and $\mu_3 < 0$ means that the sample is negatively skewed as shown in Fig. 3.16. In order to compare different samples, statisticians use coefficient of skewness, denoted by c_s, which can be calculated by dividing skewness by cube of sample standard deviation.

$$c_s = \frac{\mu_3}{s^3}. \tag{3.7}$$

Unlike skewness whose units are cube of the units of sample under consideration, coefficient of skewness is a dimensionless measure. If a population is symmetric about its mean, skewness is always zero. If skewness is positive, it might be because of some very large values called outliers or maybe the sample has a right tail.

Kurtosis

Kurtosis refers to how peaked the sample is. For the same reason, some statisticians also refer to it as peakedness. The coefficient of Kurtosis of a sample composed

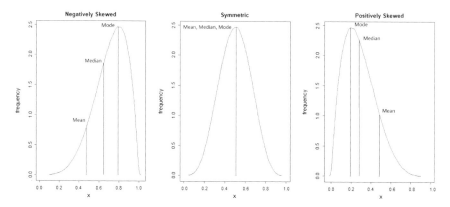

Fig. 3.16 Skewness

of the observations $x_1, x_2, x_3, \ldots, x_n$ is the arithmetic mean of the fourth powers of the distances between each observation and its mean, denoted by β.

$$\beta = \frac{\frac{1}{n} \sum_{i=1}^{n} (x_i - \overline{x})^4}{s^4} = \frac{\mu_4}{s^4}. \tag{3.8}$$

Statisticians use another measure called excess kurtosis, denoted by γ, to describe whether a sample is flatter or peaked in nature as compared to a normal distributed sample.

$$\gamma = \beta - 3. \tag{3.9}$$

For a normal distributed sample, $\beta = 3$ implies $\gamma = 0$.

A high value of kurtosis means that the sample has sharper peaks and fat tails. Also if the value of $\gamma > 0$, the sample is said to be leptokurtic. If $\gamma = 0$, it is called mesokurtic and if $\gamma < 0$, it is called platykurtic and the sample has wider peak about the mean and thinner tails as shown in Fig. 3.17.

3.3.4 Displaying the Measures and Preparing Reports

Usually, a well-prepared frequency table combined with an explanatory graph where the c_v, s, and s^2 are located with their sizes can be very useful when describing data samples. The idea is to show the analyst the sense and information that all measures provide, to make a good decision based on structured information.

As a guide for practitioners, we provide a reference graph which can be useful when preparing reports or further analysis about samples. Figure 3.18 shows where the sample mean is located mostly. It also shows the size of c_v, s, and s^2 and their relationship to $\overline{x}$.

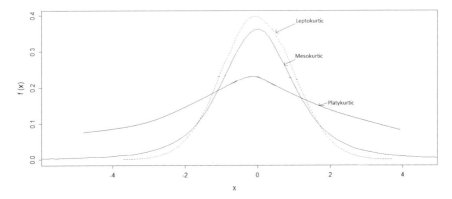

Fig. 3.17 Kurtosis

Fig. 3.18 Measures of a sample

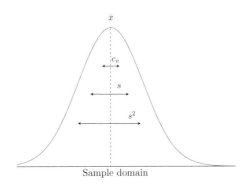

Another important aspect of applied statistics is the way in which an analyst makes a report to summarize the measures. The analyst has to present their findings in an easy and concise format, in order to provide the best understanding about the results.

Example 3.14 (*Example 3.1 Continued*) After finishing the computation of all main results, the commanders want to summarize the obtained statistics into a comprehensive report. Thus, the analyst prepares a summary report of the obtained results, as shown in Table 3.12.

The information provided in Table 3.12 shows the average of variables 1 and 4, which is fundamental in planning supplies such as uniforms, shoes, equipment, and food. The range, variance, and standard deviation give an idea of the variety in sizes, materials, and qualities for the supplies, and the coefficient of variation provides valuable information about the quantities of each kind of supply.

3.3.4.1   Useful Hints When Performing Reports

Writing a report is an important part of a statistical study since it is what the customer reads. A statistical report is the business card of the analyst and the obtained results.

Table 3.12 Summary report for Example 3.1

Variable	Height (m)	Weight (kg)
Max	2.13	93.2
Min	1.71	83.1
Range R_j	0.42	10.1
Mean $\bar{x}$	1.859	89.05
Variance s^2	0.01736	6.2447
Std. deviation s	0.1317	2.4989
Coeff. of variation c_v	0.071	0.028
Median	1.82	89.15
Mode	(1.715–1.805)	(89.155–91.175)
Geom. mean $\bar{x}_g$	1.855	89.016
Skewenss μ_3	0.7612	−0.4624
Kurtosis β	2.4220	3.0126

Thus, the better a report is presented, the better the information provided. Thus, the following hints are recommended when preparing statistical reports:

1. **Be orderly**: Any statistical report should include several ordered main sections, namely *Description of the information*, *Main report* presented in tables, *Descriptive graphs*, and *Description of the results*. Additionally, a statistical report can include some recommendations for implementing the obtained results.
2. **Be clear**: Take care of the language usage since any report should be clear to the reader. Always write in an easy way, using both technical and nontechnical language, so as to be understood by the reader. Always include key information about the study and avoid useless information.
3. **Be brief**: Do not extend a report. Be concise and just write what is necessary (or what it was requested by the customer). Sometimes, extensive text can be confusing to the reader which is undesirable. Do not exclude interesting findings or important information and just include all needed information supplemented by a clear and brief explanation of the report.
4. **Good presentation**: It is highly desirable to show the obtained results in good presentation, language, and standard format. A pleasant report is easier to understand and read. Do not allow changes in the text format of graphs but use a standard format according to the customer's requirements.

The above recommendations can help the analyst to present a clear and good report that the customer may consider as valuable. The main idea is to provide key information to be implemented by the customer. The interested readers may refer to Ross (2014).

Problems

3.1 Marks (out of 100) obtained by a class of students in course "Probability and Statistics" is given in Table 3.13.

1. Write the frequency table for interval of ten marks, i.e., 1–10, 11–20, and so on.
2. Draw the histogram of the distribution.
3. Comment on the symmetry and peakedness of the distribution after calculating appropriate measures.

3.2 Let X equal the duration (in min) of a telephone call that is received between midnight and noon and reported. The following times were reported

3.2, 0.4, 1.8, 0.2, 2.8, 1.9, 2.7, 0.8, 1.1, 0.5, 1.9, 2, 0.5, 2.8, 1.2, 1.5, 0.7, 1.5, 2.8, 1.2

Draw a probability histogram for the exponential distribution and a relative frequency histogram of the data on the same graph.

3.3 Let X equal the number of chips in a chocolate chip cookies. Hundred observations of X yielded the following frequencies for the possible outcome of X.

$Outcome(x)$	0	1	2	3	4	5	6	7	8	9	10
$Frequency$	0	4	8	15	16	19	15	13	7	2	1

1. Use these data to graph the relative frequency histogram and the Poisson probability histogram.
2. Do these data seem to be observations of a Poisson random variable with mean λ. Find λ.

3.4 Show that the total sum of the distances between each data and its mean, namely $d, d_i = x_i - \bar{x}$, is zero.

3.5 Calculate the value of commonly used statistics to find measure of spread for the runs scored by the Indian Cricket Team based on their scores in last 15 One Day Internationals while batting first. The data are shown in Table 3.14.

Table 3.13 Data for Problem 3.1

68	34	56	23	45	78	67	96	76	45	75	34	89	92	50	47	56	72	59	55	92	88	53	71	49
53	19	45	34	23	76	45	67	43	56	78	94	74	38	58	52	64	53	89	48	58	54	39	66	62

Table 3.14 Data for Problem 3.5

281	307	251	429	241	189	256	194	267	385	228	299	247	331	389

Table 3.15 Frequency table for Problem 3.6

Age(years)	0–14	15–19	20–29	30–39	40–49	50–79
Frequency	14	40	28	27	24	17

3.6 A mobile phone company examines the ages of 150 customers to start special plans for them. Consider frequency table shown in Table 3.15.

1. Draw the histogram for the data.
2. Estimate the mean age for these policyholders.
3. Estimate the median age for these policyholders.

3.7 A sample of 10 claims in an insurance company had mean and variance of 5,478 and 1,723, respectively. On reconciliation, it was found that one claim of 3,250 was wrongly written as 4,250. Calculate the mean and standard deviation of the sample with correct values.

3.8 Suppose a state government wants to analyze the number of children in families for improving their immunization program. They analyze a group of 200 families and report their findings in the form of a frequency distribution shown in Table 3.16

1. Draw the bar chart for the following data and calculate the total number of children.
2. Calculate mean, mode, and median of the data.
3. Calculate coefficient of kurtosis and coefficient of skewness in the above data.

3.9 An insurance company wants to analyze the claims for damage due to fire on its household content's policies. The values for a sample of 50 claims in Rupees are shown in Table 3.17.
   Table 3.18 displays the grouped frequency distribution for the considered data.

Table 3.16 Frequency table for Problem 3.8

No. of children	0	1	2	3	4	5	6	7
No. of families	18	30	72	43	25	8	3	1

Table 3.17 Data for Problem 3.9

57000	115000	119000	131000	152000	167000	188000	190000	197000	201000
206000	209000	213000	217000	221000	229000	247000	250000	252000	253000
257000	257000	258000	259000	260000	261000	262000	263000	267000	271000
277000	285000	287000	305000	307000	309000	311000	313000	317000	321000
322000	327000	333000	351000	357000	371000	399000	417000	433000	499000

Table 3.18 Grouped frequency distribution of the data given in Table 3.17

Claim size (In 1000's of Rupees)	50–99	100–149	150–199	200–249	250–299	300–349	350–399	400–449	450–500
Frequency	1	3	5	8	16	10	4	2	1

1. What is the range of the above data?
2. Draw a bar graph for Table 3.18.
3. For the data given in Table 3.18 if instead of equal-sized groups, we had a single group for all value below 250, how would this bar be represented?
4. Calculate the mean, median, mode, and sample geometric mean.
5. Calculate the sample standard deviation and sample variance.
6. Calculate the coefficient of variation.

3.10 Suppose the Dean of a college wants to know some basic information about the height of the students of the last six year groups. To do so, twenty students were selected from each group and their heights were measured, so a total of 120 observations divided into six groups are taken. The obtained results are presented in Table 3.19.

The Dean needs to know if there is any evidence that some students have the required height to play sports such as basketball, volleyball, and swimming. On the other hand, he is looking for some useful information to plan some requirements such as uniforms, shoes and caps. Provide a descriptive report of the information found in the provided data. If possible, also provide graphics and recommendations.

3.11 Show that the variance of any variable s^2 can be expressed as

$$s^2 = \frac{\sum_{i=1}^{n} x_i^2 - \frac{\left(\sum_{i=1}^{n} x_i\right)^2}{n}}{n-1}.$$

Test out this result against all data provided in Problem 3.10.

3.12 The department of analysis of a taxicab company has the records of 15 drivers, which are shown in Table 3.20. Those records include the following information: distance (Km), amount of crashes, amount of fines received per driver, amount of visits to garage for repairs (V.T.G), and days of operation (D.O.O).

Provide a statistical report of the information provided by data. The department of analysis of the company is looking for information useful to identify good drivers and the efficiency of the fleet.

3.13 Consider Problem 3.12. The department of analysis of the company suggests to compare all variables to each others using graphical analysis. The idea is to find

Table 3.19 Data for Problem 3.10

Group 1	Group 2	Group 3	Group 4	Group 5	Group 6
175	172	162	174	177	176
174	179	182	170	185	162
190	170	166	154	177	183
187	177	175	168	179	158
181	186	192	186	191	169
173	183	199	200	183	184
200	188	168	164	171	166
189	169	188	184	178	170
173	178	187	182	182	171
186	179	178	164	170	182
176	163	196	169	183	177
170	171	158	184	154	144
177	178	190	175	152	164
186	198	165	177	173	180
178	184	159	167	189	179
154	203	174	165	190	174
192	193	160	194	174	185
188	174	177	160	182	148
185	175	181	186	183	188
194	172	170	162	155	187

Table 3.20 Recorded data for Problem 3.12

km	Crash	Fines	V.T.G	D.O.O
13,381	0	0	1	240
12,200	0	0	0	240
30,551	1	1	2	240
26,806	0	3	0	240
11,984	0	0	0	240
32,625	0	1	0	240
60,308	1	1	2	240
24,683	0	0	0	240
8,167	0	0	1	360
11,198	0	0	0	360
26,120	0	0	0	360
186,632	2	4	3	360
7,147	1	2	2	360
18,731	0	1	1	360
2,129	2	2	2	360

some patterns whose information should be useful to identify possible relationships among variables. Provide only interesting graphics alongside with a description of their findings.

Reference

Ross SM (2014) Introduction to probability and statistics for engineers and scientists. Academic Press, London

Chapter 4
Sampling Distributions and Estimation

Now, we are ready to discuss the relationship between probability and statistical inference. The two key facts to statistical inference are (a) the population parameters are fixed numbers that are usually unknown and (b) sample statistics are known for any sample. For different samples, we get different values of the statistics and hence this variability is accounted for identifying distributions called sampling distributions. In this chapter, we discuss certain distributions that arise in sampling from normal distribution. The other topics covered in this chapter are as follows: unbiasedness, mean square error, consistency, relative efficiency, sufficiency, minimum variance, Fisher information for a function of a parameter, Cramer–Rao lower bound, efficiency, method of moments, maximum likelihood estimation to find estimators analytically and numerically, and asymptotic distributions of maximum likelihood estimators.

4.1 Introduction

If X is a random variable with certain probability distribution and $X_1, X_2, \ldots, X_n$ are independent random variables each having the same distribution as X, then $(X_1, X_2, \ldots, X_n)$ is said to constitute a random sample from the random variable X. Thus, a sample can be imagined as a subset of a population. A population is considered to be known when we know the distribution function (probability mass function $p(x)$ or probability density function $f(x)$) of the associated random variable X. If, for example, X is normally distributed, we say that the population is normally distributed or that we have a normal population. Each member of a sample has the same distribution as that of the population.

© Springer Nature Singapore Pte Ltd. 2018
D. Selvamuthu and D. Das, *Introduction to Statistical Methods,*
Design of Experiments and Statistical Quality Control,
https://doi.org/10.1007/978-981-13-1736-1_4

Sampling from Infinite Population

Consider an infinite population. Suppose the experiment is performed and x_1 is obtained. Then, the experiment is performed second time and x_2 is observed. Since the population is infinite, removing x_1 does not affect the population, so x_2 is still a random observation from the same population. In terms of random variable, $X_1, X_2, \ldots, X_n$ are mutually independent and identically distributed random variables and their realizations are $x_1, x_2, \ldots, x_n$.

Sampling from Finite Population

Consider the sampling from a finite population. A finite population is a finite collection $\{x_1, \ldots, x_N\}$. Suppose we have to draw a sample of size n from this population. This can be performed in the following ways:

1. **With Replacement**: Suppose a value is chosen from the population such that each value is equally likely to be chosen. After each selection, the chosen item is replaced and a fresh selection is made. This kind of sampling is known as sampling with replacement because item drawn at any stage is replaced in the population. In terms of random variables, the sample $X_1, \ldots, X_n$ obtained is mutually independent and each X_i can take the values $x_1, \ldots, x_N$ with equal probability. Theoretically, sampling from a finite population with replacement is akin to sampling from an infinite population.

2. **Without Replacement**: Sampling without replacement from a finite population is performed as follows. A value is chosen from $x_1, \ldots, x_N$ in such a way that each value is equally likely to be chosen. The value obtained is first sample point. Now, second sample point is chosen from rest $N - 1$ in such a way that each point is equally likely with probability $\frac{1}{N-1}$. This procedure continues giving sample $x_1, \ldots, x_n$. We observe that a value chosen once is not available for choice at any later stage. In terms of random variables, the sample obtained $X_1, \ldots, X_n$ is not mutually independent in this case, but they are identically distributed.

The following examples shall make the above assertions amply clear:

- A pair of coin is tossed. Define the random variable $X_1 =$ number of heads obtained. In a sense, X_1 can be thought of as a sample of size one from the population of all possible tossing of that coin. If we tossed the coin a second time and defined the random variable X_2 as the number of heads obtained on the second toss, X_1, X_2 could presumably be considered as a sample of size two from the same population.
- The total yearly rainfall in a certain locality for the year 2013 could be described as a random variable X_1. During successive years, random variables $X_2, \ldots, X_n$ could be defined analogously. Again, we may consider $(X_1, X_2, \ldots, X_n)$ as a sample of size n obtained from the population of all possible yearly rainfalls at the specified locality. And it might be realistically supposed that the $X_i's$ are independent, identically distributed (i.i.d) random variables.
- The life length of a light bulb manufactured by a certain process in a factory is studied by choosing n bulbs and measuring their life lengths say $T_1, T_2, \ldots, T_n$.

We may consider $(T_1, T_2, \ldots, T_n)$ as a random sample from the population of all possible life lengths of bulbs manufactured in a specified way.

- Consider testing of a new device for measuring blood pressure. Try it out on n people, and record the difference between the value returned by the device and the true value as recorded by standard techniques. Let X_i, $i = 1, 2, \ldots, n$ be the difference for the ith person. The data will consist of a random vector $(X_1, X_2, \ldots, X_n)$, and $(x_1, x_2, \ldots, x_n)$ is a particular vector of real numbers. A possible model would be to assume that $X_1, X_2, \ldots, X_n$ are independent random variables each having $N(0, \sigma^2)$ where σ^2 is some unknown positive real number.

One possible way to do statistical inference for finite populations is by ensuring that every member of the population is equally likely to be selected in the sample, which is often known as a random sample. For populations of relatively small size, random sampling can be accomplished by drawing lots or equivalently by using a table of random numbers specially constructed for such purposes.

Once we have obtained the values of a random sample, we usually want to use them in order to make some inference about the population represented by the sample which in the present context means the probability distribution of the random variable being sampled. Since the various parameters which characterize a probability distribution are numbers, it is natural that we would want to compute certain pertinent numerical characteristics obtainable from the sample values which might help us to make appropriate statements about the parameter values which are often not known.

Definition 4.1 (*Statistics*) Let $X_1, X_2, \ldots, X_n$ be a random sample from a population described by the random variable X, and let $x_1, x_2, \ldots, x_n$ be the values assumed by the sample. Let H be a function defined for the n-tuple $(x_1, x_2, \ldots, x_n)$. Then, $Y = H(X_1, X_2, \ldots, X_n)$ is said to be a statistic provided that it is not a function of any unknown parameter(s).

Thus, a statistic is a real-valued function of the sample. Sometimes, the term statistic is also used to refer to the value of the function. Another important conclusion drawn from the above definition is that a statistic is a random variable. Hence, it is meaningful to consider the probability distribution of a statistic, its expectation, and variance. More often than not for a random variable which is a statistic, we speak about its sampling distribution rather than its probability distribution. The probability distribution of a sample statistics is often called the sampling distribution of the statistics.

Some Important Statistics

There are certain statistics which we encounter frequently. A few of them are as follows:

1. The mean of the sample or sample mean is a random variable defined as

$$\bar{X} = \frac{\sum\limits_{i=1}^{n} X_i}{n}.$$

$$\sum_{i=1}^{n}(X_i - \bar{X})^2$$

2. The sample variance defined as $s^2 = \dfrac{\sum_{i=1}^{n}(X_i - \bar{X})^2}{n-1}$.
3. The minimum of the sample defined as $K = \min(X_1, X_2, \ldots, X_n)$. ($K$ represents the smallest observed value.)
4. The maximum of the sample defined as $M = \max(X_1, X_2, \ldots, X_n)$. ($M$ represents the largest observed value.)

Let us now discuss some of them in detail:

Sample Mean

Consider a population where each element of the population has some numerical quantity associated with it. For instance, the population might consist of the undergraduate students of a country, and the value associated with each student might be his/her height, marks, or age and so on. We often assume the value attached to any element of the population to be the value of a random variable with expectation(μ) and variance(σ^2). The quantities μ and σ^2 are called the population mean and the population variance, respectively. Let $X_1, X_2, \ldots, X_n$ be a sample of values from this population. Also, let the population considered follow a normal distribution. Thus, $X_i \sim N(\mu, \sigma^2)$, $i = 1, 2, \ldots, n$. The sample mean is defined by

$$\bar{X} = \frac{X_1 + X_2 + \cdots + X_n}{n}.$$

Since the value of the sample mean $\bar{X}$ is determined by the values of the random variables in the sample, it follows that $\bar{X}$ is also a random variable. Now, as each member of the population is an independent random variable, using result on sum of independent normal random variables (see Sect. 2.4.2), we have

$$\sum_{i=1}^{n} X_i \sim N(n\mu, n\sigma^2).$$

Therefore, expectation of the sample mean becomes

$$E(\bar{X}) = \frac{n\mu}{n} = \mu. \tag{4.1}$$

That is, the expected value of the sample mean, $\bar{X}$, is equal to the population mean, μ. If the population is infinite and the sampling is random or if the population is finite and sampling is performed with replacement, then the variance of the sample mean turns out to be

$$Var(\bar{X}) = \frac{n\sigma^2}{n^2} = \frac{\sigma^2}{n} \tag{4.2}$$

where σ^2 is the population variance. Thus, $\bar{X} \sim N(\mu, \frac{\sigma^2}{n})$.

If the population is of size N and if sampling is performed without replacement, then (4.2) is replaced by

$$Var(\bar{X}) = \frac{1}{n^2} Var\left(\sum_{i=1}^{n} X_i\right) = \frac{1}{n^2}\left[\sum_{i=1}^{n} Var(X_i) + \sum_{i \neq j}^{n} Cov(X_i, X_j)\right]$$

$$= \frac{1}{n^2}\left[n\sigma^2 + n(n-1)\left(\frac{-\sigma^2}{N-1}\right)\right] = \frac{n\sigma^2}{n^2}\left(\frac{N-n}{N-1}\right) = \frac{\sigma^2}{n}\left(\frac{N-n}{N-1}\right)$$

(4.3)

whereas $E(\bar{X})$ is still given by Eq. (4.1). The term $\frac{N-n}{N-1}$ is also known as the correction factor. Note that (4.3) reduces to (4.2) as $N \to \infty$.

Therefore, we observe that the expectation of sample mean $\bar{X}$ is same as that of an individual random variable X_i, but its variance is smaller than that of the individual random variable by a factor of $\frac{1}{n}$, where n is the sample size. Therefore, we can conclude that sample mean is also symmetric about the population mean μ, but its spread, i.e., variance, decreases with the increase in the sample size. Thus,

$$\bar{X} \sim N\left(\mu, \frac{\sigma^2}{n}\right).$$

By the central limit theorem, for large sample size,

$$Z = \frac{(\bar{X} - \mu)}{(\sigma/\sqrt{n})}$$

(4.4)

has approximately the standard normal distribution, $N(0, 1)$, i.e.,

$$\lim_{n\to\infty} P\left(\frac{(\bar{X} - \mu)}{(\sigma/\sqrt{n})} \leq z\right) = \frac{1}{\sqrt{2\pi}} \int_{-\infty}^{z} e^{-u^2/2} du.$$

Sample Variance

Suppose that $X_1, X_2, \ldots, X_n$ is a random sample from a random variable X with expectation μ and variance σ^2. Then, the sample variance s^2 is given by

$$s^2 = \frac{1}{n-1} \sum_{i=1}^{n} (X_i - \bar{X})^2$$

where $\bar{X}$ is the sample mean. Now, let us derive the expectation of the sample variance.

Consider

$$\sum_{i=1}^{n}(X_i - \bar{X})^2 = \sum_{i=1}^{n}(X_i - \mu + \mu - \bar{X})^2 = \sum_{i=1}^{n}[(X_i - \mu)^2 + 2(\mu - \bar{X})(X_i - \mu) + (\mu - \bar{X})^2]$$

$$= \sum_{i=1}^{n}(X_i - \mu)^2 + 2(\mu - \bar{X})\sum_{i=1}^{n}(X_i - \mu) + n(\mu - \bar{X})^2$$

$$= \sum_{i=1}^{n}(X_i - \mu)^2 - 2n(\mu - \bar{X})^2 + n(\mu - \bar{X})^2 = \sum_{i=1}^{n}(X_i - \mu)^2 - n(\bar{X} - \mu)^2.$$

Therefore,

$$E\left(\frac{1}{n-1}\sum_{i=1}^{n}(X_i - \bar{X})^2\right) = \frac{1}{n-1}\left[n\sigma^2 - n\frac{\sigma^2}{n}\right] = \sigma^2. \qquad (4.5)$$

Note that although s^2 is defined as the sum of squares of n terms, these n terms are not independent. In fact,

$$(X_1 - \bar{X}) + (X_2 - \bar{X}) + \cdots + (X_n - \bar{X}) = \sum_{i=1}^{n}X_i - n\bar{X} = 0.$$

Hence, there is a linear relationship among these n terms which means that as soon as any of the $(n-1)$ of these are known, the nth one is determined.

4.2 Statistical Distributions

Now, we discuss chi-square distribution, student's t-distribution, and F-distribution that arise in sampling from normal distribution. We will study the importance of these distributions later in this section.

4.2.1 The Chi-Square Distribution

Consider a random variable X which has the following probability density function

$$f(x) = \begin{cases} \frac{2^{\frac{-\nu}{2}}}{\Gamma(\nu/2)}x^{(\nu/2)-1} e^{-x/2}, & 0 < x < \infty \\ 0 & \text{otherwise} \end{cases} \qquad (4.6)$$

where $\Gamma(x)$ is the gamma function. Then, X is said to have chi-square distribution with ν degrees of freedom. The degrees of freedom of a distribution are a positive

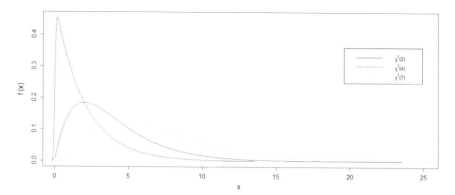

Fig. 4.1 PDF of χ^2 distribution with different degrees of freedom

integer which is same as the number of independent values or quantities which can be assigned to the concern distribution. In notation, we write $X \sim \chi_\nu^2$. It is same as gamma distribution with $r = \nu/2$, where ν is a positive integer and $\lambda = 1/2$.

The χ^2 distribution is asymmetric and changes shape with degrees of freedom as shown in Fig. 4.1.

Mean and Variance of χ^2 Distribution:

$$E(X) = \int_0^\infty x \frac{(1/2)^{\nu/2}}{\Gamma(\nu/2)} x^{\nu/2-1} e^{-x/2} dx = \nu.$$

Similarly, we can see that $E(X^2) = \nu^2 + 2\nu$. Therefore,

$$Var(X) = E[X^2] - [E(X)]^2 = \nu^2 + 2\nu - (\nu)^2 = 2\nu.$$

The first four moments of χ_ν^2 are

$$\mu_1 = \nu, \mu_2 = 2\nu, \mu_3 = 8\nu, \mu_4 = 48\nu + 12\nu^2.$$

The χ_ν^2 distribution is tabulated for values of $\nu = 1, 2, \ldots, 30$. For $\nu > 30$, we can approximate it by normal distribution, i.e.,

$$\frac{X - \nu}{2\nu} \sim N(0, 1).$$

We will write $\chi_{\nu,\alpha}^2$ for the point corresponding to right tail probability α of the χ_ν^2 distribution, i.e.,

$$P(\chi_\nu^2 > \chi_{\nu,\alpha}^2) = \alpha. \tag{4.7}$$

Table A.8 gives the values of $\chi_{\nu,\alpha}^2$ for some selected values of ν and α.

Example 4.1 Let X be the temperature at which certain chemical reaction takes place and be χ^2 random variable with 32 degrees of freedom. Find $P(X \leq 34.382)$. Also, approximate the same using central limit theorem.

Solution: From Table A.8, we can obtain $P(X \leq 34.382) = 0.6457$. On the other hand, using CLT we have

$$P(X \leq 34.382) = P\left(\frac{X - 32}{\sqrt{64}} \leq \frac{34.382 - 32}{\sqrt{64}} \right) = P(Z \leq 0.2978) = 0.6171.$$

Remark 4.1   1. Let $X_1, X_2, \ldots, X_n$ be n independent random variables following chi-square distribution with $v_1, v_2, \ldots, v_n$ degrees of freedom, respectively. Then, the random variable $W = X_1 + X_2 + \cdots + X_n$ also follows chi-square distribution with $v_1 + v_2 + \cdots + v_n$ degrees of freedom. In other words, we can say that chi-square distribution satisfies the reproductive property.
   2. Let X_1 and X_2 be two independent random variables such that X_1 follows chi-square distribution with v_1 degrees of freedom, while $Y = X_1 + X_2$ follows chi-square distribution with v degrees of freedom. Assume that $v > v_1$. Then, X_2 also follows chi-square distribution with $v - v_1$ degrees of freedom.
   3. Given a standard normal random variable Z, Z^2 follows χ^2 distribution with one degree of freedom (see Example 2.17). Similarly, let $Z_1, Z_2, \ldots, Z_v$ be independent standard normal random variables, and $\sum_{i=1}^{v} Z_i^2$ follows χ^2 distribution with v degrees of freedom.
   4. It is worthwhile to note that a χ^2 variable with $v = 2$ is same as the exponential variable with parameter $\lambda = 1/2$.
   5. If $X \sim U(0, 1)$, then $-2 \log(X)$ follows chi-square distribution with two degrees of freedom.
   6. Another interesting point to note is that if X is gamma-distributed random variable with parameters v and λ, then the random variable $2\lambda X$ is chi-square distributed with $2v$ degrees of freedom. This comes handy for calculating the probabilities associated with gamma distribution as χ^2 tables are easily available.
   7. For large v ($v \geq 30$), we can show that $\sqrt{2\chi^2} - \sqrt{2v - 1}$ is approximately normal distributed with mean zero and variance one.

4.2.2 Student's t-Distribution

Consider a random variable X which has the probability density function as follows:

$$f(x) = \frac{\Gamma(\frac{v+1}{2})}{\sqrt{\pi v}\,\Gamma(\frac{v}{2})} \left(1 + \frac{x^2}{v} \right)^{-\frac{v+1}{2}}, \quad -\infty < x < \infty. \tag{4.8}$$

Then, X is said to have student's t-distribution[1] with v degrees of freedom, i.e., $X \sim t_v$.

Given a random variable X which is χ^2-distributed with degrees of freedom v and a standard normal distributed random variable Z which is independent of χ^2, the random variable Y defined as:

$$Y = \frac{Z}{\sqrt{X/v}}$$

follows t-distribution with degrees of freedom v (Verify!).

If v is large ($v \geq 30$), the graph of $f(x)$ closely approximates the standard normal curve as indicated in Fig. 4.2. Since the t-distribution is symmetrical, $t_{y-1} = -t_y$. We will write $t_{v,\alpha}$ for the value of t for which we have

$$P(T > t_{v,\alpha}) = \alpha. \tag{4.9}$$

By symmetry, it follows that

$$P(|T| > t_{v,\frac{\alpha}{2}}) = \alpha. \tag{4.10}$$

Table A.10 gives the values of $t_{v,\alpha}$ for some selected values of v and α. For example, $P(T > t_{5,0.05}) = 2.015$.

Mean and Variance of t-Distribution:

Like the standard normal distribution, t-distribution is also symmetric about 0. Hence, its mean is 0. Also, it is flatter than the normal distribution as shown in Fig. 4.2.

$$E(Y) = 0, \qquad Var(Y) = \frac{v}{v-2}, \quad v > 2.$$

The t-distribution is used in hypothesis testing of mean in sampling distribution and finding the confidence intervals which will be discussed in later chapters. Further, in order to calculate the probabilities, we use tables as integrating the PDF is a very tedious job.

4.2.3   F-Distribution

Consider a random variable X which has the probability density function as follows:

$$f(x) = \begin{cases} \frac{\Gamma(\frac{v_1+v_2}{2})}{\Gamma(\frac{v_1}{2})\Gamma(\frac{v_2}{2})} \left(\frac{v_1}{v_2}\right)^{\frac{v_1}{2}} . x^{\frac{v_1}{2}-1} \left(1 + \frac{v_1}{v_2}x\right)^{-(v_1+v_2)/2} & 0 < x < \infty \\ 0 & \text{otherwise} \end{cases} \tag{4.11}$$

---

[1] William Sealy Gosset (June 13, 1876–October 16, 1937) was an English statistician. He published under the pen name Student and developed the student's t-distribution.

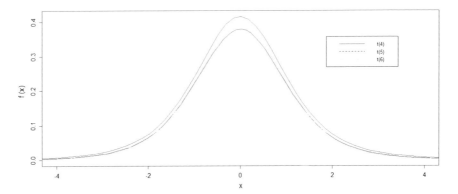

Fig. 4.2 PDF of t-distribution with different values of ν

Then, X is said to have the F-distribution[2] with ν_1 and ν_2 degrees of freedom, i.e., $X \sim F(\nu_1, \nu_2)$. The PDF of X is shown in Fig. 4.3. Consider two independent random variables Y_1 and Y_2 which are χ^2-distributed with degrees of freedoms ν_1 and ν_2, respectively. The random variable X defined as follows:

$$X = \frac{Y_1/\nu_1}{Y_2/\nu_2}$$

will have an F-distribution with ν_1 and ν_2 degrees of freedom, i.e., $X \sim F(\nu_1, \nu_2)$ (Verify!).

Mean and Variance of F-Distribution:

For $k > 0$,

$$E(X^k) = \left(\frac{\nu_1}{\nu_2}\right)^k \frac{\Gamma\left(k + \frac{\nu_2}{2}\right)\Gamma\left(\frac{\nu_1}{2} - k\right)}{\Gamma\left(\frac{\nu_2}{2}\right)\Gamma\left(\frac{\nu_1}{2}\right)}, \nu_1 > 2k.$$

In particular,

$$E(X) = \frac{\nu_1}{\nu_1 - 2}, \nu_1 > 2, \quad Var(X) = \frac{\nu_1^2(2\nu_2 + 2\nu_1 - 4)}{\nu_2(\nu_1 - 2)^2(\nu_1 - 4)}, \nu_1 > 4.$$

Typical F-distribution is given in Fig. 4.3:

F-distribution is asymmetric and is right-skewed. Again, we will calculate the probabilities using the tables rather than integrating the PDF. F-distribution is important in carrying out hypothesis testing and calculating the confidence intervals while comparing variances of two different populations.

---

[2]George Waddel Snedecor (October 20, 1881–February 15, 1974) was an American mathematician and statistician. He contributed to the foundations of analysis of variance, data analysis, experimental design, and statistical methodology. Snedecor's F-distribution and the George W. Snedecor Award of the American Statistical Association are named after him.

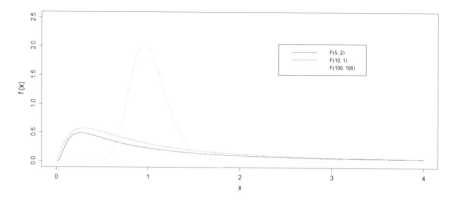

Fig. 4.3 The PDF of F-distribution with different parameters ν_1 and ν_2

We write $F_{\nu_1,\nu_2,\alpha}$ for the upper α percent point of the $F(\nu_1, \nu_2)$ distribution, i.e.,

$$P(F_{\nu_1,\nu_2} > F_{\nu_1,\nu_2,\alpha}) = \alpha. \tag{4.12}$$

For a given confidence level $(1 - \alpha)100\%$, we have

$$F_{\nu_1,\nu_2,\alpha} = \frac{1}{F_{\nu_2,\nu_1,1-\alpha}}. \tag{4.13}$$

Table A.11 gives the values of $F_{\nu_1,\nu_2,\alpha}$ for some selected values of ν_1, ν_2 and α. Table 4.1 gives the mean, variance, and MGF of the standard sampling distributions.

4.2.4 Some Important Results on Sampling Distributions

1. If $X_1, X_2, \ldots, X_n$ constitute a random sample drawn from a population with mean μ and variance σ^2, then the sample mean($\bar{X}$) has $E[\bar{X}] = \mu$ and $Var(\bar{X}) = \frac{\sigma^2}{n}$.
2. If $\bar{X}$ is the sample mean of random sample of size n drawn without replacement from a population of size N with mean μ and variance σ^2, then

$$E(\bar{X}) = \mu, \quad Var(\bar{X}) = \frac{\sigma^2(N - n)}{n(N - 1)}.$$

Further $E(s^2) = \frac{N}{N-1}\sigma^2$. As $N \to \infty$, $E(s^2)$ tends to the population variance σ^2.
3. If $X_1, X_2, \ldots, X_n$ constitute a random sample drawn from a normal population with mean μ and variance σ^2 and the sample variance s^2 is given by

Table 4.1 Standard sampling distributions

S.No.	Distribution and notation	PDF $f(x)$	Mean $E(X)$	Variance $V(X)$	MGF $M(t)$
1	Normal $X \sim N(\mu, \sigma^2)$	$\frac{1}{\sigma\sqrt{2\pi}}e^{-\frac{1}{2}\left(\frac{x-\mu}{\sigma}\right)^2}, \ -\infty < x < \infty.$	μ	σ^2	$e^{\mu t + \frac{1}{2}\sigma^2 t^2}$
2	Chi-square $X \sim \chi_\nu^2$	$\begin{cases} \frac{2^{\nu/2}}{\Gamma(\nu/2)}x^{(\nu/2)-1}e^{-x/2}, & 0 < x < \infty \\ 0 & \text{otherwise} \end{cases}$	ν	2ν	$(1-2t)^\nu$
3	Student's t $X \sim t_\nu$	$\begin{cases} \frac{\Gamma(\frac{\nu+1}{2})}{\sqrt{\pi\nu}\,\Gamma(\frac{\nu}{2})}\left(1+\frac{x^2}{\nu}\right)^{-\frac{\nu+1}{2}}, & x > 0 \\ 0, & \text{otherwise} \end{cases}$	0	$\frac{\nu}{\nu-2}$	Does not exist
4	F distribution $X \sim F_{\nu_1, \nu_2}$	$\begin{cases} \frac{\Gamma(\frac{\nu_1+\nu_2}{2})}{\Gamma(\frac{\nu_1}{2})\Gamma(\frac{\nu_2}{2})}\left(\frac{\nu_1}{\nu_2}\right)^{\frac{\nu_1}{2}}\cdot x^{\frac{\nu_1}{2}-1}\left(1+\frac{\nu_1}{\nu_2}x\right)^{-(\nu_1+\nu_2)/2}, & 0 < x < \infty \\ 0 & \text{otherwise} \end{cases}$	$\frac{\nu_2}{\nu_2-2}, \ \nu_2 > 2$	$\frac{\nu_2^2(2\nu_1+2\nu_2-4)}{\nu_1(\nu_2-2)^2(\nu_2-4)}, \ \nu_2 > 4$	Does not exist

$$s^2 = \frac{\sum_{i=1}^{n}(X_i - \bar{X})^2}{n - 1}.$$

then $\frac{(n-1)s^2}{\sigma^2}$ has χ^2 distribution with $n - 1$ degrees of freedom (see Problem 4.2).

4. **Joint Distribution of Sample Mean and Sample Variance**

Let $X_i \sim N(\mu, \sigma^2)$, $i = 1, 2, \ldots, n$ be a random sample from $N(\mu, \sigma^2)$. We want to find the joint distribution of $\bar{X} = \frac{\sum_{i=1}^{n} X_i}{n}$ and $S^2 = \frac{\sum_{i=1}^{n}(X_i - \bar{X})^2}{n-1}$.

We will show that $\bar{X}$ and S^2 are independent and hence their joint PDF can be obtained as a product of their individual PDFs. In order to show that $\bar{X}$ and S^2 are independent, it is sufficient to show that $\bar{X}$ and $(X_1 - \bar{X}, X_2 - \bar{X}, \ldots, X_n - \bar{X})$ are independent.

The MGF of $\bar{X}$ and $(X_1 - \bar{X}, X_2 - \bar{X}, \ldots, X_n - \bar{X})$ is given by

$$
\begin{aligned}
M(t, t_1, t_2, \ldots, t_n) &= E\left(\exp\{t\bar{X} + t_1(X_1 - \bar{X}) + \cdots + t_n(X_n - \bar{X})\}\right) \\
&= E\left(\exp\left\{\sum_{i=1}^{n} t_i X_i - \left(\sum_{i=1}^{n} t_i - t\right)\bar{X}\right\}\right) \\
&= E\left(\exp\left\{\sum_{i=1}^{n} X_i\left(t_i - \frac{t_1 + t_2 + \cdots + t_n - t}{n}\right)\right\}\right) \\
&= E\left(\Pi_{i=1}^{n} \exp\left\{\frac{X_i(nt_i - n\bar{t} + t)}{n}\right\}\right)\left(\text{where } \bar{t} = \sum_{i=1}^{n} \frac{t_i}{n}\right) \\
&= \Pi_{i=1}^{n} E\left(\exp\left\{\frac{X_i(nt_i - n\bar{t} + t)}{n}\right\}\right) \\
&= \Pi_{i=1}^{n} \exp\left\{\frac{\mu(t + n(t_i - \bar{t}))}{n} + \frac{\sigma^2}{2n^2}(t + n(t_i - \bar{t}))^2\right\} \\
&= \exp\{\mu t\}\exp\left\{\frac{\sigma^2}{2n^2}\left(nt^2 + n^2\sum_{i=1}^{n}(t_i - \bar{t})^2\right)\right\} \\
&= \exp\left\{\mu t + \frac{\sigma^2}{2n}t^2\right\}\exp\left\{\frac{\sigma^2}{2}\sum_{i=1}^{n}(t_i - \bar{t})^2\right\} \\
&= M_{\bar{X}}(t) M_{X_1 - \bar{X}, \ldots, X_n - \bar{X}}(t_1, t_2, \ldots, t_n)
\end{aligned}
$$

5. If the population variance σ^2 is unknown, then replace σ in Eq. (4.4) by the random variable s. Take $T = \frac{\bar{X} - \mu}{s/\sqrt{n}}$. One can prove using Remark 3 and definition of student's t-distribution that T has student's t-distribution with $n - 1$ degrees of freedom whenever the population random variable is normal distributed.

6. Let two independent random samples of size v_1 and v_2, respectively, be drawn from the two normal populations with variances σ_1^2 and σ_2^2, respectively. Then if the variances of the random samples are given by s_1^2, s_2^2, respectively, the statistic

$$F = \frac{(\nu_1 - 1)s_1^2/(\nu_1 - 1)\sigma_1^2}{(\nu_2 - 1)s_2^2/(\nu_2 - 1)\sigma_2^2} = \frac{s_1^2/\sigma_1^2}{s_2^2/\sigma_2^2}$$

has the F-distribution with $\nu_1 - 1$, $\nu_2 - 1$ degrees of freedom. It follows using Remark 3 and definition of F-distribution.

Example 4.2 Bombay Stock Exchange (BSE) is the world's tenth largest stock market by market capitalization. Given that there are 5,000 companies that are listed on BSE. Assume that the pairs of stocks are uncorrelated and the average price of a stock is 500 rupees, with a standard deviation of 100 rupees. Suppose you draw a random sample of 50 stocks. What is the probability that the average price of a sampled stock will be less than 497 rupees?

Solution:

Let average price of a sampled stock be represented by $\bar{X}$.

$$E(\bar{X}) = \mu, \quad Var(\bar{X}) = \frac{\sigma^2}{n}. \tag{4.14}$$

Hence, from the values given in the problem, we have $\mu = 500$, $\sigma = 100$, and $n = 5000$.

$$P(\bar{X} \leq 497) = P\left(\frac{\bar{X} - \mu}{\frac{\sigma}{\sqrt{n}}} \leq \frac{497 - \mu}{\frac{\sigma}{\sqrt{n}}}\right) = \Phi\left(\frac{497 - \mu}{\frac{\sigma}{\sqrt{n}}}\right)$$

$$= 1 - \Phi\left(\frac{\mu - 497}{\frac{\sigma}{\sqrt{n}}}\right) = 1 - \Phi\left(\frac{3}{\frac{100}{\sqrt{5000}}}\right) = 1 - \Phi(2.12) = 0.017$$

where the value of $\Phi(2.12)$ comes from Table A.7 in Appendix.

4.2.5 Order Statistics

Let $(X_1, \ldots, X_n)$ be an n-dimensional random variable and $(x_1, \ldots, x_n)$ be its realization. Arrange $(x_1, \ldots, x_n)$ in increasing order of magnitude so that

$$x_{(1)} \leq x_{(2)} \leq \cdots \leq x_{(n)}$$

where $x_{(1)} = \min\{x_1, x_2, \ldots, x_n\}$, $x_{(2)}$ is the second smallest value and so on, $x_{(n)} = \max\{x_1, x_2, \ldots, x_n\}$. If any two x_i, x_j are equal, their order does not matter.

Definition 4.2 The function $X_{(k)}$ of $(X_1, \ldots, X_n)$ that takes the value $x_{(k)}$ in each possible sequence $(x_1, \ldots, x_n)$ of realizations of $(X_1, \ldots, X_n)$ is known as kth order statistic or statistic of order k. $\{X_{(1)} \leq X_{(2)} \leq \cdots \leq X_{(n)}\}$ is called the set of order statistics for $(X_1, \ldots, X_n)$.

Example 4.3 Let $X_1, \ldots, X_n$ be n independent random variables with PDF

$$f_{X_i}(x_i) = \begin{cases} \lambda_i e^{-\lambda_i x_i} & \text{if } x_i > 0, \\ 0 & \text{if } x_i \leq 0. \end{cases} \quad i = 1, 2, \ldots, n.$$

with $\lambda_i > 0$ $\forall$ i. Find the distribution of $Y = \min\{X_1, X_2, \ldots, X_n\}$ and $Z = \max\{X_1, X_2, \ldots, X_n\}$.

Solution:

For $y > 0$, we have

$$P(Y > y) = P(\min\{X_1, X_2, \ldots, X_n\} > y) = P(X_1 > y, X_2 > y, \ldots, X_n > y)$$

$$= P(X_1 > y)P(X_2 > y)\cdots P(X_n > y) = \Pi_{i=1}^{n} e^{-\lambda_i y} = e^{-y \sum\limits_{i=1}^{n} \lambda_i}.$$

Therefore, CDF of Y is given by, for $y > 0$

$$F_Y(y) = 1 - P(Y > y) = 1 - e^{-y \sum\limits_{i=1}^{n} \lambda_i}.$$

Hence, Y follows exponential distribution with parameter $\sum\limits_{i=1}^{n} \lambda_i$. Similarly, CDF of Z is given by, for $z > 0$

$$P(Z \leq z) = P(\max\{X_1, X_2, \ldots, X_n\} \leq z) = P(X_1 \leq z, X_2 \leq z, \ldots, X_n \leq z)$$
$$= P(X_1 \leq z)P(X_2 \leq z)\cdots P(X_n \leq z) = \Pi_{i=1}^{n}(1 - e^{-\lambda_i z}).$$

4.3 Point Estimation

In the previous section on sampling distributions, we mentioned that very often a sample from a random variable X may be used for the purpose of estimating one or more of several unknown parameters associated with the PMF/PDF of X. Suppose, for example, that one is interested to arrive at a conclusion regarding the proportion of females in a particular region who prefer to read a particular fashion magazine. It would be tedious or sometimes impossible to ask every woman about her choice in order to obtain the value of the population proportion denoted by parameter p.

Instead, one can select a large random sample and calculate the proportion $\hat{p}$ of women in this sample who favors that particular fashion magazine. The obtained value of $\hat{p}$ can now be used to make inferences regarding the true population proportion p.

The main purpose of selecting random samples is to elicit information about the unknown population parameters. That is, an analyst uses the information in a sample $X_1, X_2, \ldots, X_n$ to make inferences about an unknown parameter θ of the population. A sample of size n is a long list of numbers that may be hard to infer anything. Thus, the analyst wishes to summarize the information contained in the sample by determining some characteristics of sample values which is usually performed by computing statistics.

Let us consider the problem where a manufacturer supplies 10000 objects. If X is the number of objects that are acceptable, then

$$X \sim B(10000, p) \tag{4.15}$$

where p is the probability that a single object is acceptable.

The probability distribution of X depends on the parameter p in a very simple way. The question is: Can we use the sample $X_1, X_2, \ldots, X_n$ in some way in order to estimate p? Is there some statistic that may be used to estimate p?

Note: It should be noted that for any sample of size $n < 10000$, the estimate for p is most likely not equal to its actual value.

Thus, when we propose $\hat{p}$ as an estimator for p, we do not really expect value of $\hat{p}$ to be equal to p (recall that $\hat{p}$ is a *random variable* and can thus take many values).

We shall now discuss the characteristics of a "good estimate" and various methods to estimate the parameters, etc.

4.3.1 Definition of Point Estimators

Let us begin by defining some concepts which will help us to resolve the problems suggested above.

Definition 4.3 (*Estimator*) Let X be a random variable corresponding to the population. Assume that the distribution of X is $F(x, \theta)$, where θ is the parameter. Further, assume that θ is unknown. Let $(X_1, X_2, \ldots, X_n)$ be a random sample of size n drawn from the population. Any function $T(X_1, X_2, \ldots, X_n)$, independent of θ, is an estimator of $\psi(\theta)$.

In other words, an estimator is a statistic involving sample values (such as sum of the sample $\sum_{i=1}^{n} X_i$, variance of the sample, etc.) which is used to estimate parameters of a distribution.

For example, if x_i are the observed grade point averages of a sample of 88 students, then

$$\bar{x} = \frac{1}{88} \sum_{i=1}^{88} x_i = 3.12$$

is a point estimate of μ, the mean grade point average of all the students in the population.

Estimators are generally depicted with a "hat" on top, such as $\hat{p}$, $\hat{\theta}$. The values assumed by the estimator are known as the estimates of the unknown parameter.

4.3.2 Properties of Estimators

Generally, in questions related to estimation, we are given the distribution as well as a sample. It is our job to choose an estimator that can accurately determine the parameters of the given random variable.

Let us take an example of a Poisson-distributed random variable with parameter λ with sample $X_1, X_2, \ldots, X_n$, and we have to come up with an estimator for λ. Note that for a Poisson distribution, both variance and mean are equal to λ. Hence, we have two possible estimators:

1. $\frac{\sum_{i=1}^{n} X_i}{n}$.
2. $\frac{\sum_{i=1}^{n}(X_i - \bar{X})^2}{n-1}$.

We can think of several others, but let us stick with these for now. Now, the question is: *Which of these is a better estimator of* λ? This question shall be answered in the following way.

Definition 4.4 (*Unbiased Statistic*) A statistic $T(X)$ is said to be unbiased for the function $\psi(\theta)$ of unknown parameter θ if $E(T(X)) = \psi(\theta)$.

For instance, sample mean $\bar{X}$ is an unbiased estimator of the population mean μ, because

$$E(\bar{X}) = E\left(\sum_{i=1}^{n} X_i\right) = \sum_{i=1}^{n} E(X_i) = \mu.$$

An estimator that is not unbiased is called biased. The bias, denoted by $b(T, \psi)$, is given by

$$b(T, \psi) = E_\theta(T(X)) - \psi(\theta).$$

It can be observed that if $T(X)$ is an unbiased estimator of $\psi(\theta)$ and g is a linear or affine function then $g(T(X))$ is also an unbiased estimator of $g(\psi(\theta))$. Consider the quantity $(T(X) - \psi(\theta))^2$ which might be regarded as a measure of error or loss involved in using $T(X)$ to estimate $\psi(\theta)$. The quantity $E((T(X) - \psi(\theta))^2)$ is called the mean square error (MSE) of the estimator $\psi(\theta)$. Thus, the MSE decomposes as $MSE = Var(T(X)) + b(T, \psi)^2$. Note that the MSE of an unbiased estimator is equal to the variance of the estimator.

Definition 4.5 (*Locally Minimum Variance Unbiased Estimator (LMVUE)*) Let $\theta_0 \in \Theta$ and $U(\theta_0)$ be the class of all unbiased estimators $T(X)$ of θ_0 such that $E_{\theta_0}(T)^2 < \infty$. Then, $T_0 \in U(\theta_0)$ is called LMVUE at θ_0 if

$$E_{\theta_0}(T_0 - \theta_0)^2 \le E_{\theta_0}(T - \theta_0)^2 \text{ for all } T \in U(\theta_0).$$

Definition 4.6 (*Uniformly Minimum Variance Unbiased Estimator (UMVUE)*) Let $U(\theta)$ be the class of all unbiased estimators $T(X)$ of $\theta \in \Theta$ such that $E_\theta(T)^2 < \infty$, for all $\theta \in \Theta$. Then, $T_0 \in U(\theta)$ is called UMVUE of θ if

$$E_\theta(T_0 - \theta)^2 \le E_\theta(T - \theta)^2 \text{ for all } \theta \in \Theta \text{ and every } T \in U(\theta).$$

Now, we consider an inequality which provides a lower bound on the variance of an estimator and can be used to show that an unbiased estimator is UMVUE.

Cramér[3] and Rao inequality Suppose the family $\{f(x; \theta) : \theta \in \Theta\}$ where Θ is an open set satisfies the following regularity conditions:

1. The set S where $f(x; \theta)$ is nonzero, called support set, is independent of θ. Thus, $S = \{x : f(x; \theta) > 0\}$ does not depend on θ.
2. For $x \in S$ and $\theta \in \Theta$, the derivative $\frac{\partial f(x;\theta)}{\partial \theta}$ exists and is finite.
3. For any statistic $S(X)$ with $E_\theta|S(X)| < \infty$ for all θ, the following holds,

$$\frac{\partial}{\partial \theta} \int S(x) f(x; \theta) dx = \int S(x) \frac{\partial}{\partial \theta} f(x; \theta) dx$$

whenever right-hand side exists.

Let $T(X)$ be a statistic with finite variance such that $E_\theta(T(X)) = \psi(\theta)$. If

$$I(\theta) = E_\theta \left(\frac{\partial}{\partial \theta} \log f(x; \theta) \right)^2$$

satisfies $0 < I(\theta) < \infty$, then

$$Var_\theta T(X) \ge \frac{(\psi'(\theta))^2}{I(\theta)}.$$

Remark 4.2 1. The quantity $I(\theta)$ is known as Fisher information in the random sample $X_1, X_2, \ldots, X_n$.
2. $I(\theta) = nI_1(\theta)$ where $I_1(\theta)$ is Fisher information in X_1 and is given by

$$I_1(\theta) = E_\theta \left(\frac{\partial}{\partial \theta} \log f(X_1; \theta) \right)^2.$$

3. The Fisher information in X_1, $I_1(\theta)$ can also be given by

---

[3]Harald Cramér (1893–1985) was a Swedish mathematician, actuary, and statistician, specializing in mathematical statistics and probabilistic number theory. John Kingman described him as "one of the giants of statistical theory."

$$I_1(\theta) = -E\left(\frac{\partial^2}{\partial\theta^2}\log f(x_1;\theta)\right). \tag{4.16}$$

4. If $Var(T(X))$ attains the lower bound, then $T(X)$ is UMVUE for $\psi(\theta)$.

Example 4.4 Let X follow Poisson distribution with parameter λ. Find the UMVUE of λ.

Solution:

$I(\lambda) = E_\theta\left(\frac{\partial}{\partial\theta}\log f_\theta(X)\right)^2 = \frac{n}{\lambda}$. Let $\psi(\lambda) = \lambda$. Then, the lower bound is $\frac{\lambda}{n}$. Let $T(X) = \bar{X}$, and we have

$$Var(\bar{X}) = Var\left(\sum_{i=1}^{n}X_i\right) = \frac{\lambda}{n}.$$

Since $Var(\bar{X})$ attains the lower bound, hence $\bar{X}$ is UMVUE of λ.

Let us get back to our question: *Which of these is an efficient estimator?*

Definition 4.7 (*Relative Efficient Estimator*) Let $T_1(X)$ and $T_2(X)$ be unbiased estimators of θ, i.e., $E(T_1(X)) = \theta = E(T_2(X))$. Then, the relative efficiency of estimators $T_1(X)$ and $T_2(X)$ is the ratio of their variances, i.e.,

$$\frac{Var(T_1(X))}{Var(T_2(X))}.$$

We say that $T_1(X)$ is relatively more efficient as compared to $T_2(X)$ if the ratio is less than 1, i.e., if $Var(T_1(X)) \le Var(T_2(X))$.

For instance, when $X \sim P(\lambda)$ and λ is unknown, the sample mean and the sample variance are unbiased estimators. Since, $Var(\bar{X}) \le Var(s^2)$, we conclude that $\bar{X}$ is a relatively more efficient estimator.

Example 4.5 Suppose $X \sim N(\mu, \sigma^2)$ where σ is known. Let $X_1, X_2, \ldots, X_n$ be a random sample of size n. Prove that $\mu_1 = \bar{X}$ is the relatively efficient estimator of μ with $\mu_2 = X_2$ and $\mu_3 = \frac{X_2+\bar{X}}{2}$.

Solution:

We can easily check
$$E(\mu_1) = E(\mu_2) = E(\mu_3) = \mu.$$

Hence, all three estimators are unbiased. Note that unbiasedness is not a property that helps us choose between estimators. To do this, we must examine mean square error. Now, we have

$$Var(\mu_1) = \sigma^2/n, \quad Var(\mu_2) = \frac{\sigma^2/n + \sigma^2}{4}, \quad Var(\mu_3) = \sigma^2.$$

Thus, $\bar{X}$ is the relatively efficient estimator of μ.

Definition 4.8 (*Efficient Estimator*) Let $T(X)$ be an unbiased estimator of $\psi(\theta)$, i.e., $E(T(X)) = \psi(\theta)$. Then, $T(X)$ is said to be an efficient estimator of $\psi(\theta)$, $\forall\, \theta \in \Theta$ if its variance archives equality in the Cramer–Rao lower bound, i.e.,

$$Var(T(X)) = \frac{(\psi'(\theta))^2}{I(\theta)} \tag{4.17}$$

where

$$I(\theta) = E_\theta\left(\frac{\partial}{\partial\theta}\log f(x;\theta)\right)^2. \tag{4.18}$$

Note that an efficient estimator is always an UMVUE, but the converse need not be true. The reason for this is fairly intuitive. The variance of a random variable measures the variability of the random variable about its expected value. For if the variance is small, then the value of the random variable tends to be close to its mean, which in the case of an unbiased estimate means close to the value of the parameter.

Example 4.6 Let X be a normally distributed random variable with finite expectation μ and variance σ^2. Let $\bar{X}$ be the sample mean. Show that $\bar{X}$ is an efficient estimator of μ.

Solution: First, we can easily say that $\bar{X}$ is unbiased as

$$E(\bar{X}) = E\left(\frac{\sum\limits_{i=1}^{n} X_i}{n}\right) = \mu. \tag{4.19}$$

Similarly, variance of $\bar{X}$ is given by

$$Var(\bar{X}) = \frac{\sigma^2}{n}.$$

The CRLB for $\bar{X}$ is $\frac{\sigma^2}{n}$. Hence, $\bar{X}$ is an efficient estimator of θ.

How to distinguish a good estimator from a bad one? Ideally, as we obtain more observations, we have more information and our estimator should become more accurate. This is not a statement about a single estimator, but one about a sequence of estimators $T_1(X)$, $T_2(X)$, ...

Definition 4.9 (*Consistent Estimator*) Let $\{T_n(X), n = 1, 2, \ldots\}$ be a sequence of estimators based on a sample $(X_1, X_2, \ldots, X_n)$ of the parameter θ. We say that $T_n(X)$ is a consistent estimator of θ if for every $\varepsilon > 0$

$$\lim_{n \to +\infty} P(|\hat{T}_n - \theta| > \varepsilon) = 0. \tag{4.20}$$

This probability is really hard to compute from sample data. Using Chebyshev's inequality, we can simplify the expression and prove that, *for a consistent estimator*, $\lim_{n \to \infty} E(T_n(X)) = \theta$ and $\lim_{n \to \infty} Var(T_n(X)) = 0$. Thus, we can be sure of consistency provided $\lim_{n \to \infty} E(T_n(X)) = \theta$ and $\lim_{n \to \infty} Var(T_n(X)) = 0$.

Example 4.7 Let X be a random variable with finite expectation μ and variance σ^2. Let $\bar{X}$ be the sample mean, based on a random sample of size n. Prove that $\bar{X}$ is an unbiased and consistent estimator of μ.

Solution:

First, we can easily say that:

$$E(\bar{X}) = E\left(\frac{\sum_{i=1}^{n} X_i}{n}\right) = \frac{E\left(\sum_{i=1}^{n} X_i\right)}{n} = \frac{n\mu}{n} = \mu. \tag{4.21}$$

Thus, the estimator is unbiased. Similarly, variance of $\bar{X}$ is given by

$$Var(\bar{X}) = \frac{\sigma^2}{n}. \tag{4.22}$$

Note that $\lim_{n \to \infty} Var(\bar{X}) = 0$, and thus sample mean is a consistent and unbiased estimator.

Definition 4.10 (*Sufficient Estimator*) Let $\mathbf{X} = (\mathbf{X_1}, \mathbf{X_2}, \dots, \mathbf{X_n})$ be a sample from $\{F(x; \theta) : \theta \in \Theta\}$. A statistic $T = T(X)$ is sufficient for θ or for the family of distributions $\{F(x; \theta) : \theta \in \Theta\}$ if and only if the conditional distribution of X, given $T = t$, does not depend on θ.

Also, it can be shown that a statistic $T(X)$ is sufficient for family of distributions $\{F(x; \theta) : \theta \in \Theta\}$ if and only if one of the following conditions holds:

1. $P(X_1, X_2, \dots, X_n \mid T(X))$ is independent of θ.
2. Condition expectation $E(Z/T)$ is independent of θ for every random variable Z such that $E(Z)$ exists.
3. The conditional distribution of every random variable Z given $T = t$, which always exists, is independent of θ.

Example 4.8 Consider a random sample $X_1, X_2, \dots, X_n$ from Bernoulli's distribution $B(1, p)$. Prove that $\sum_{i=1}^{n} X_i$ is a sufficient estimator for p.

Solution:

Let $T(X_1, X_2, \ldots, X_n) = \sum_{i=1}^{n} X_i$; then if $\sum_{i=1}^{n} x_i = x$, we have

$$P\left(X_1 = x_1, X_2 = x_2, \ldots, X_n = x_n / \sum_{i=1}^{n} X_i = x\right) = \frac{P\left(X_1 = x_1, X_2 = x_2, \ldots, X_n = x_n, \sum_{i=1}^{n} X_i = x\right)}{P\left(\sum_{i=1}^{n} X_i = x\right)}$$

and 0 otherwise. Thus, we have for $\sum_{i=1}^{n} x_i = x$,

$$P\left(X_1 = x_1, X_2 = x_2, \ldots, X_n = x_n / \sum_{i=1}^{n} X_i = x\right) = \frac{p^{\sum_{i=1}^{n} x_i}(1-p)^{n-\sum_{i=1}^{n} x_i}}{\binom{n}{x} p^x (1-p)^{n-x}} = \frac{1}{\binom{n}{x}}$$

which is independent of unknown parameter p. Hence, $\sum_{i=1}^{n} X_i$ is sufficient estimator for p.

Now, we present the result for sufficient statistic in case of discrete random variables as a theorem. Its proof is beyond the scope of this text. The interested readers may refer to Rohatgi and Saleh (2015).

Theorem:Factorization Theorem Let $(X_1, X_2, \ldots, X_n)$ be discrete random variables with joint PMF $p(x_1, x_2, \ldots, x_n; \theta), \theta \in \Theta$. Then, $T(X_1, X_2 \ldots, X_n)$ is sufficient for θ if and only if we have

$$p(x_1, x_2, \ldots, x_n; \theta) = h(x_1, x_2, \ldots, x_n)g(T(x_1, x_2, \ldots, x_n, \theta))$$

where h is a nonnegative function of $x_1, x_2, \ldots, x_n$ and is independent of θ, and g is a nonnegative non-constant function of θ and $T(X_1, X_2, \ldots, X_n)$.

Now, using factorization theorem we conclude that $\sum_{i=1}^{n} X_i$ is sufficient statistics for p without using conditional probability argument which is used in Example 4.8.

$$P(X_1 = x_1, X_2 = x_2, \ldots, X_n = x_n) = p^{\sum_{i=1}^{n} x_i}(1-p)^{n-\sum_{i=1}^{n} x_i}.$$

Take $h(x_1, x_2, \ldots, x_n) = 1$ and $g(x_1, x_2, \ldots, x_n, p) = (1 - p)^n (\frac{p}{1-p})^{\sum_{i=1}^{n} x_i}$. Hence, we observe that $\sum_{i=1}^{n} X_i$ is sufficient.

Example 4.9 Suppose that $X_1, \ldots, X_n$ form a random sample from a Poisson distribution with unknown mean μ, $(\mu > 0)$. Prove that $T = \sum_{i=1}^{n} X_i$ is a sufficient statistic for μ.

Solution:

For every set of nonnegative integers $x_1, \ldots, x_n$, the joint PMF $f_n(x; \mu)$ of $X_1, \ldots, X_n$ is as follows:

$$f_n(x; \mu) = \prod_{i=1}^{n} \frac{e^{-\mu} \mu^{x_i}}{x_i!} = \left(\prod_{i=1}^{n} \frac{1}{x_i!} \right) e^{-n\mu} \mu^{\sum_{i=1}^{n} x_i}.$$

It can be seen that $f_n(x|\mu)$ has been expressed as the product of a function that does not depend on μ and a function that depends on μ but depends on the observed vector x only through the value of $\sum_{i=1}^{n} x_i$. By factorization theorem, it follows that $T = \sum_{i=1}^{n} X_i$ is a sufficient statistic for μ.

Exponential Family of Probability Distribution

A study about the properties of some probability distributions gives that the dimension of sufficient statistics is same as that of the parameter space no matter what is the size of the sample. Such an observation led to the development of an important family of distributions known as the exponential family of probability distributions. Some of the common distributions from exponential family include the binomial, the normal, the gamma, and the Poisson distribution.

One-parameter members of the exponential family have PDF or PMF of the form

$$f(x; \theta) = \exp [c(\theta) T(x) + d(\theta) + S(x)].$$

Suppose that $X_1, \ldots, X_n$ are i.i.d. samples from a member of the exponential family, then the joint PDF is

$$f(x; \theta) = \prod_{i=1}^{n} \exp\left[c(\theta)T(x_i) + d(\theta) + S(x_i)\right] \tag{4.23}$$

$$= \exp\left[c(\theta) \sum_{i=1}^{n} T(x_i) + nd(\theta)\right] \exp\left[\sum_{i=1}^{n} S(x_i)\right]. \tag{4.24}$$

From this result, it is apparent by the factorization theorem that $\sum_{i=1}^{n} T(X_i)$ is a sufficient statistic.

For example, the joint PMF of Bernoulli distribution is

$$P(X = x) = \prod_{i=1}^{n} \theta_i^x (1 - \theta)^{1-x_i} \quad \text{for } x_i = 0 \text{ or } x = 1$$

$$= \exp\left[\log\left(\frac{\theta}{1-\theta}\right) \sum_{i=1}^{n} x_i + \log(1 - \theta)\right]. \tag{4.25}$$

It can be seen that this is a member of the exponential family with $T(x) = x$, and we can also see that $\sum_{i=1}^{n} X_i$ is a sufficient statistic, which is the same as that obtained in Example 4.8.

Remark 4.3 1. Any one-to-one function of a sufficient statistics is also a sufficient statistics. Hence, there are numerous sufficient statistics in a population distribution.

Definition 4.11 (*Minimal Sufficient Statistic*) A sufficient statistics $T(X)$ is called a minimal sufficient statistic if for any other sufficient statistic $T'(X)$, $T(X)$ is a function of $T'(X)$.

Definition 4.12 (*Complete Statistic*) Let $f(x; \theta) : \theta \in \Theta\}$ be a family of PDFs (or PMFs). We say that this family is complete if $E_\theta g(T) = 0$ for all $\theta \in \Theta$ implies that $P_\theta\{g(T) = 0\} = 1$ for all $\theta \in \Theta$ where g is a Borel measurable function. A statistic $T(X)$ is said to be complete if the family of distributions of T is complete.

A statistic $T(X_1, X_2, \ldots, X_n)$ is said to be complete if family of distributions of T is complete.

Example 4.10 Let $X_1, X_2, \ldots, X_n$ be a random sample from Bernoulli's $B(1, p)$. Prove that $\sum_{i=1}^{n} X_i$ is complete statistic.

Solution: Consider a statistic $T(X) = \sum_{i=1}^{n} X_i$. It is a sufficient statistic as shown in Example 4.9. The family of distribution of $T(X)$ is binomial family $\{B(n, p),$

$0 < p < 1\}$ since sum of independent Bernoulli random variables is a binomial random variable. Consider

$$E_p(g(T)) = \sum_{k=0}^{n} g(k) \binom{n}{k} p^k(1-p)^{n-k} = 0 \text{ for all } p \in (0, 1)$$

which is same as

$$(1-p)^n \sum_{k=0}^{n} g(k) \binom{n}{k} s \left(\frac{p}{1-p}\right)^k = 0 \; \forall \; 0 < p < 1$$

which is a polynomial in $(\frac{p}{1-p})$ with uncountable many roots; hence, the coefficients must vanish. It follows that

$$g(k) = 0 \text{ for all } k = 0, 1, \ldots, n.$$

Therefore, $\sum_{i=1}^{n} X_i$ is a complete statistic.

Definition 4.13 (*Ancillary Statistic*) A statistic $A(X)$ is said to be ancillary if its distribution does not depend on the underlying model parameter θ.

The definition implies that an ancillary statistic alone contains no information about θ. But these statistics are important because when used in conjunction with another statistic, they may contain valuable information for inferences about θ. Some examples of ancillary statistic are range which is given by $X_{(n)} - X_{(1)}$, $X_1 - \bar{X}$. Now, we present an important result which connects complete sufficient statistics and ancillary statistics as following two theorems without proof. Its proof is beyond the scope of this text. The interested readers may refer to Rohatgi and Saleh (2015).

Theorem 4.1 (Lehmann–Scheffe Theorem) *If $T(X)$ is a complete sufficient statistic and $W(X)$ is an unbiased estimator of $\psi(\theta)$, then $\phi(T) = E(W/T)$ is an UMVUE of $\psi(\theta)$. Furthermore, $\phi(T)$is the unique UMVUE in the sense that if T^* is any other UMVUE, then $P_\theta(\phi(T) = T^*) = 1$ for all θ.*

Corollary: If U is a complete sufficient statistics for θ and $g(U)$, a function of U, is such that $E(g(U)) = \psi(\theta)$, then $g(U)$ is a UMVUE for $\psi(\theta)$.

Remark 4.4 1. The above theorem confirms that the estimators based on sufficient statistics are more accurate in the sense that they have smaller variances.
2. Let T and S be two sufficient statistics for θ such that S is a function of T, i.e., $S = g(T)$, and let $\hat{\phi}(T)$ be an unbiased estimator of $\phi(\theta)$ based on T. Then,

$$\hat{\phi}(S) = E(\hat{\phi}(\hat{T})/S)$$

is also an unbiased estimator of $\phi(\theta)$ with

$$Var(\hat{\phi}(S)) \le Var(\hat{\phi}(T))$$

with equality if and only if $\hat{\phi}(T)$ is already a function of S.

Example 4.11 Consider Example 4.10, and find a UMVUE for the parameter p and for p^2.

Solution:

We already know that $T = \sum_{i=1}^{n} X_i$ is a complete sufficient statistics and $T \sim B(n, p)$. Also, it is known that X_1 is an unbiased estimator of p. By Lehmann–Scheffe Theorem, we have $E(X_1/T) = \bar{X}$ which is the UMVUE for θ. As we know that UMVUE, if exists, will be a function of the complete sufficient statistics, T, only. Therefore, to find UMVUE of p^2, we start with $\bar{X}^2 = \frac{T^2}{n^2}$. Now,

$$E(\bar{X}^2) = Var(\bar{X}) + (E(\bar{X}))^2 = \frac{p(1-p)}{n} + p^2 = \frac{p}{n} + p^2\left(1 - \frac{1}{n}\right).$$

Consider

$$\phi(t) := \frac{1}{n(n-1)}(t^2 - t).$$

Then, $E(\phi(T)) = p^2$. Hence, by corollary to Lehmann–Scheffe Theorem, we have $\phi(T)$ which is the UMVUE for p^2.

Theorem 4.2 (Basu's Theorem) *If $T(X)$ is a complete sufficient statistic for θ, then any ancillary statistic $S(X)$ is independent of $T(X)$.*

Note that converse of Basu's theorem is not true; that is, a statistic which is independent of every ancillary statistic need not be complete.

Example 4.12 Let $(X_1, X_2, \ldots, X_n)$ denote a random sample of size n from a distribution that is $N(\mu, \sigma^2)$. Prove that the sample mean $\bar{X}$ and sample variance σ^2 are independent.

Solution:

We know that for every known σ^2 the mean $\bar{X}$ of the sample is a complete sufficient statistics for the parameter $\mu, -\infty < \mu < \infty$. Consider the statistics $s^2 = \frac{1}{n-1}\sum_{i=1}^{n}(X_i - \bar{X})^2$ which is location invariant. Thus, s^2 must have a distribution that does not depend upon μ; hence by Basu's Theorem, s^2 and $\bar{X}$, the complete sufficient statistics for μ, are independent.

Example 4.13 The data obtained from a Geiger counter experiment are given in Table 4.2. In the given table, k is the number of particles observed in $\frac{1}{8}$th of a minute, while n_k is the number of emissions in which k particles were observed. Assuming

Table 4.2 Data for Example 4.13

k	0	1	2	3	4	5	6	7	8	9	10	11	Total
n_k	57	203	383	525	532	408	273	139	49	27	10	6	2612

the distribution Poisson, estimate the parameter λ. It is also given to you that λ is defined in such a way that

$$P(X = k) = \frac{e^{-(1/8)\lambda}\left(\frac{1}{8}\lambda\right)^k}{k!}, k = 0, 1, 2, \ldots \tag{4.26}$$

Let us find an estimator for λ. As the samples were noted for an eighth of a minute, we can simply take the sample mean of the data and multiply it by 8. Thus, $\hat{\lambda} = 8\bar{X}$. Now, the sample mean would be simply the total number of particles(0*number of emissions with 0 particles+1*number of emissions with 1 particle and so on) divided by total number of emissions.

$$\bar{X} = \frac{\sum\limits_{k=0}^{11} k n_k}{\sum\limits_{k=0}^{11} n_k} = 3.87 \text{ particles.}$$

Multiply by 8, and we get $\hat{\lambda} = 30.96$.

Example 4.14 Consider the following two-parameter normal family

$$f(x; \mu, \sigma) = \frac{1}{\sqrt{2\pi}\sigma}e^{-\frac{1}{2}\frac{(x-\mu)^2}{\sigma^2}}, \quad -\infty < x < \infty, -\infty < \mu < \infty, \sigma > 0$$

Find the consistent estimators for μ and σ.

Solution:

We know that $E(\bar{X}) = \mu$ and $E(\bar{X}^2) = \mu^2 + \sigma$. Clearly

$$\bar{x}_n = \frac{x_1 + x_2 + \cdots + x_n}{n} \quad \text{and} \quad t_n = \frac{x_1^2 + x_2^2 + \cdots + x_n^2}{n}$$

are consistent for μ and $\mu^2 + \sigma$. Then, $\bar{x}_n$ and $u_n = t_n - \bar{x}_n^2$ are consistent for μ and σ.

Definition 4.14 (*Asymptotically Most Efficient Estimator*) Often, we deal with estimators which are not most efficient but their efficiency satisfies the condition $\lim_{n \to \infty} \hat{\theta} = 1$ and they are at least asymptotically most efficient estimates.

Table 4.3 Data

Claims	0	1	2	3	4	5
Frequency	7	12	19	18	11	7

From the practical point of view, they are most efficient estimates from large samples. For example, let $X \sim N(\mu, \sigma^2)$. Consider $U = \frac{ns^2}{n-1}$; then, U is an unbiased estimator, but it is neither sufficient nor most efficient. But for large sample, we have

$$\lim_{n \to \infty} e = \lim_{n \to \infty} \frac{n}{n-1} = 1.$$

Hence, U is asymptotically most efficient estimator.

4.4 Methods of Point Estimation

Sometimes, we need to find an unknown parameter of the population. As an example, consider monthly claims arriving from an insurance portfolio and the data are shown in Table 4.3.

We know that Poisson distribution with parameter λ is one of the most obvious choices for modeling number of claims. Then again, most importantly we need to answer "What should be the value of the parameter λ for modeling this problem?" This chapter gives two methods to estimate the value of the unknown parameter based on the information provided by a given sample.

4.4.1 Method of Moments

The first method is the method of moments in which we simply equate sample moments with their population counterparts and solve for the parameter(s) we need to estimate. The other method known as the method of maximum likelihood involves differentiation to obtain the parameter value maximizing the probability of obtaining the given sample. There are other methods for estimating the population parameter from a given sample, but in this chapter we will only discuss above-mentioned methods. An important thing to note is that these two methods might not always give the same value for the parameter. The expression "point estimation" refers to the problem of finding a single number to estimate the parameter value. This contrasts with "confidence interval estimation" where we wish to find a range of possible values, which we will discuss in the next section.

One-Parameter Case

This is the simplest case in which we equate population mean, $E(X)$ to sample mean, and solve for the parameter to be estimated, i.e.,

$$E[X] = \frac{1}{n} \sum_{i=1}^{n} x_i.$$

Note that for some populations the mean does not involve the parameter, such as the uniform distribution defined over $(-\theta, \theta)$ or the normal $N(0, \sigma^2)$, in which case a higher-order moment must be used. However, such cases are rarely of practical importance.

Consider a uniform distributed random variable X on the interval $(-\theta, \theta)$. As we can see $E(X) = (1/2)(-\theta + \theta) = 0$. Equating this to the sample mean is futile as it does not consider any term involving our parameter θ. Instead, we go for second moment or simply variance and obtain the value for our parameter, i.e., $Var(X) = [\theta - (-\theta)]^2/12 = \theta^2/3$, as this term involves parameter θ, equating this to sample variance will give us a value for our parameter.

It is very important to highlight that estimate and estimators are two different things. "Estimate" is a numerical value calculated from a formula. On the other hand, "estimator" is a function that maps the sample space to a set of sample estimates. As a convention, we write the estimator in upper case, as it is essentially a random variable and thus will have a sampling distribution. Estimate being just a numerical value is written in lower case.

Two-Parameter Case

Now, we need to estimate two parameters and hence two equations are required. The most natural choice is to equate first- and second-order moments of the population with their sample counterparts to obtain two equations and solve this pair of equations to find the solution for our parameters. It is somewhat easier to work with variance instead of second-order moment. Actually working with both will lead us to exactly same answers which is also true for other higher-order moments about the mean (skewness, kurtosis, etc) being used instead of higher-order moments about zero ($E[X^3]$, $E[X^4]$, etc).

The first-order equation is the same as the one-parameter case:

$$E(X) = \frac{1}{n} \sum_{i=1}^{n} x_i.$$

The second equation is:

$$E(X^2) = \frac{1}{n} \sum_{i=1}^{n} x_i^2$$

or equivalently

$$Var(X) = \frac{1}{n} \sum_{i=1}^{n} (x_i - \bar{x})^2 = \frac{1}{n} \sum_{i=1}^{n} x_i^2 - \bar{x}^2.$$

It is important to note that while calculating the sample variance s^2 we use $(n-1)$ as divisor instead of n and thus our equation changes to

$$Var(X) = \frac{1}{n-1} \sum_{i=1}^{n} (x_i - \bar{x})^2 = \frac{1}{n-1} \sum_{i=1}^{n} x_i^2 - \bar{x}^2.$$

Obviously using this definition will give us a different result, but as the sample size increases the differences fade out. This definition is better as (s^2) is an unbiased estimator of population variance.

Remark 4.5 1. The method of moments is the oldest method of deriving point estimators. It almost always produces some asymptotically unbiased estimators, although they may not be the best estimators.
2. In some cases, however, a moment estimator does not exist.

4.4.2 Method of Maximum Likelihood

The most important aspect of parametric estimation is the method of maximum likelihood which is simple to state, universal, and asymptotically optimal. The method of maximum likelihood is widely regarded as the best general method of finding estimators. "Asymptotic" here just means when the samples are very large.

One-Parameter Case

The most important stage in applying the method is that of writing down the likelihood:

$$L(x; \theta) = \prod_{i=1}^{n} f(x_i; \theta); \quad \theta \in \Theta$$

for a random sample $x_1, x_2, ..., x_n$ from a population with density or probability function $f(x; \theta)$.

$\prod$ means product, so $\prod_{i=1}^{n} f(x_i; \theta)$ means $f(x_1) \times f(x_2) \times \cdots f(x_n)$.

The above expression simply means that the likelihood function is simply the product of the PDFs of the sample values. The notation $f(x_i; \theta)$ is given just to emphasize that the PDF depends on the value of θ, which we are trying to estimate.

The likelihood is the probability of observing the sample in the discrete case and is proportional to the probability of observing values in the neighborhood of the sample in the continuous case.

Notice that the likelihood function is a function of the unknown parameter θ. Thus, different values of θ would give different values for the likelihood. The maximum likelihood approach is to find the value of θ that would have been most likely to give us the particular sample we got. In other words, we need to find the value of θ that maximizes the likelihood function.

For a continuous distribution, the probability of getting any exact value is zero. However, we have the following for infinitesimally small $\varepsilon > 0$

$$P(x - \varepsilon < X \leq x + \varepsilon) \approx \int_{x-\varepsilon}^{x+\varepsilon} f(t).dt \simeq 2\varepsilon f(x).$$

In most cases, taking logarithms greatly simplifies the determination of the maximum likelihood estimator (MLE) $\hat{\theta}$. Differentiating the likelihood or log-likelihood with respect to the parameter and setting the derivative to zero give the maximum likelihood estimator for the parameter.

$$\frac{\partial \ln L(x; \theta)}{\partial \theta} = \sum_{i=1}^{n} \frac{\partial \ln L(x_i; \theta)}{\partial \theta} = 0.$$

It is necessary to check, either formally or through simple logic, that the turning point is a maximum. Generally, the likelihood starts at zero, finishes at or tends to zero, and is nonnegative. Therefore, if there is one turning point, it must be a maximum. The formal approach would be to check that the second derivative is negative.

One can observe that while writing the log-likelihood function, any term that does not contain the parameter (l in this case) can be treated as a constant since while differentiating the log-likelihood function the derivative will be zero for that term. Furthermore, MLEs satisfy the invariance property in the sense that if $\hat{\theta}$ is the MLE of θ then $g(\theta)$ is MLE of the function $g(\hat{\theta})$.

Example 4.15 Let $(X_1, X_2, \ldots, X_n)$ be a random sample from the exponential distribution with PDF

$$f(x; \lambda) = \lambda e^{-\lambda x}, \quad 0 < x < \infty, \quad \lambda \in (0, \infty).$$

Find the maximum likelihood estimator for λ.

Solution:

The likelihood function is given as follows.

$$L(\lambda) = L(\lambda; x_1, x_2, \ldots, x_n) = \lambda^n e^{-\lambda \sum_{i=1}^{n} x_i}, \quad 0 < \lambda < \infty$$

$$\ln L(\lambda) = n \ln \lambda - \lambda \sum_{i=1}^{n} x_i .$$

Thus,

$$\frac{\partial \ln L(\lambda)}{\partial \lambda} = \frac{n}{\lambda} - \sum_{i=1}^{n} x_i = 0.$$

The solution of this equation for λ is

$$\frac{1}{\lambda} = \frac{1}{n} \sum_{i=1}^{n} x_i = \bar{x} .$$

At $\lambda = \frac{1}{\bar{x}}$, $\ln L(\lambda)$ has maximum. Thus, the maximum likelihood estimator for λ is $\hat{\lambda} = \frac{1}{\bar{X}}$, i.e., $\frac{1}{\hat{\lambda}} = \bar{X}$.

Two-Parameter Case

In this case, to maximize the log-likelihood function with respect to two parameters, one needs to use the concept of maximizing a function of two variables. Although this is straightforward method in principle, in some of the cases the solution of the resulting equations may be complicated requiring an iterative or numerical solution. The idea is to partially differentiate the log-likelihood function with respect to each parameter and equate each of them to zero, and the following resulting system of simultaneous equation is solved.

$$\frac{\partial \ln L(x; \theta_1, \theta_2)}{\partial \theta_j} = \sum_{i=1}^{n} \frac{\partial \ln L(x_i; \theta_1, \theta_2)}{\partial \theta_j}; \quad j = 1, 2.$$

Hence, in summary, the steps for finding the maximum likelihood estimator in straightforward cases are:

1. Obtain the likelihood function, L, and write the simplified expression for log-likelihood function $\ln L$.
2. Partially differentiate $\ln L$ with respect to each parameter that is to be estimated.
3. Equate each of the partial derivatives to zero, and obtain the system of simultaneous equations.
4. Solve these equations to obtain the MLE of the unknown parameters.

Example 4.16 Consider the following two-parameter normal family with PDF

$$f(x; \mu, \sigma) = \frac{1}{\sqrt{2\pi}\sigma} e^{-\frac{1}{2}\frac{(x-\mu)^2}{\sigma^2}}, \quad -\infty < x < \infty, -\infty < \mu < \infty, \sigma > 0.$$

Find the MLE for μ and σ^2

Solution:

We have

$$\ln L = -\frac{n}{2}\ln(2\pi) - \frac{n}{2}\ln(\sigma^2) - \frac{1}{2\sigma^2}\sum_{i=1}^{n}(x_i - \mu)^2.$$

Partially differentiating with respect to μ and σ^2, we have the following likelihood equations:

$$\frac{\partial \ln L}{\partial \mu} = \frac{1}{\sigma^2}\sum_{i=1}^{n}(x_i - \mu)$$

$$\frac{\partial \ln L}{\partial \sigma^2} = -\frac{n}{2\sigma^2} + \frac{1}{2\sigma^4}\sum_{i=1}^{n}(x_i - \mu)^2$$

Solving the likelihood equations by equating them to zero, we get

$$\hat{\mu} = \frac{1}{n}\sum_{i=1}^{n}x_i$$

$$\hat{\sigma}^2 = \frac{1}{n}\sum_{i=1}^{n}(x_i - \bar{x})^2$$

i.e., the sample mean and the sample variance are the MLE for μ and σ, respectively. Note that the MLE of σ^2 differs from the sample variance s^2 in that the denominator is n rather than that $n - 1$. However, for large n, these two estimators of σ^2 will be approximately equal.

Example 4.17 Find the MLE for μ for the family of Cauchy distribution

$$f(x; \mu) = \frac{1}{2\pi(1 + (x - \mu)^2)}, \quad -\infty < x < \infty, -\infty < \mu < \infty$$

Solution:

The likelihood equation is as follows.

$$\sum_{i=1}^{n}\frac{x_i - \mu}{1 + (x_i - \mu)^2} = 0$$

which cannot be solved explicitly. The method still works, but the estimator can only be computed numerically.

4.4.3 Asymptotic Distribution of MLEs

Given a random sample of size n from a distribution with density (or probability function in the discrete case) $f(x; \theta)$, the maximum likelihood estimator $\hat{\theta}$ is such that, for large n, $\hat{\theta}$ is approximately normal, and is unbiased with variance given by the Cramer–Rao lower bound (CRLB), i.e.,

$$\hat{\theta} \sim N(\theta, CRLB)$$

where

$$CRLB = \frac{1}{nE\left\{\left[\frac{\partial}{\partial \theta} \ln f(x; \theta)\right]^2\right\}}.$$

The MLE can therefore be called asymptotically efficient in that, for large n, it is unbiased with a variance equal to the lowest possible value of unbiased estimators. The quantity $I(\theta) = E\left[\frac{\partial}{\partial \theta} \ln f(x; \theta)\right]^2$ in the denominator is known as information function.

This is potentially a very useful result as it provides an approximate distribution for the MLE when the true sampling distribution may be unknown or impossible to determine easily and hence may be used to obtain approximate confidence intervals (which we will discuss in the next section).

The result holds under very general conditions with only one major exclusion: It does not apply in cases where the range of the distribution involves the parameter, such as the uniform distribution. This is due to a discontinuity, so the derivative in the formula does not make sense.

The second formula is normally easier to work (as we would have calculated the second derivative of the log-likelihood when checking that we get a maximum).

Now, we will briefly discuss minimum variance unbiased estimator (MVUE). It is an unbiased estimator having lower variance than any other unbiased estimator for all the possible values of the parameter. An efficient estimator is also the minimum variance unbiased estimator because an efficient estimator attains the equality in the Cramer–Rao inequality for all values of the parameter. In other words, an efficient estimator attains the minimum variance for all possible values of the parameters. However, the MVUE, even if it exists, is not necessarily an efficient estimator since "minimum" does not imply that the equality holds in the Cramer–Rao inequality. In conclusion, an efficient estimator need not exist, but if it exists, it is the MVUE.

Example 4.18 Consider the family of normal distributions with the following PDF.

$$f(x; \mu) = \frac{1}{\sqrt{2\pi}}e^{-\frac{(x-\mu)^2}{2}}, \quad -\infty < x < \infty, -\infty < \mu < \infty.$$

Solution:

The sample mean $\bar{X}$ is clearly an unbiased estimator of μ. Its variance is $\frac{1}{n}$. Now,

$$\frac{\partial \ln L}{\partial \mu} = x - \mu.$$

The information function $I(\mu)$ is calculated as 1. Hence, the Cramer–Rao lower bound is attained.

4.5 Interval Estimation

Often in problems of statistical inference, the experimenter is interested in finding an estimate of a population parameter lying in an interval. Such an estimate is known as interval estimation of a parameter. Interval estimates are preferred over point estimates because they also indicate the precision and accuracy in estimation.

Most common point estimators are the sample mean and the sample variance because they give a point estimate to the unknown parameter instead of giving a range of values. However, when a sample is drawn it is highly unlikely that the value of a point estimate $\hat{\theta}$ is equal to the true value of the unknown parameter θ. In fact, for a continuous random variable X, it is generally the case that $P(\hat{\theta} = \theta) = 0$ for any sample of finite size. Therefore, rather than estimating the exact value of a parameter, it is better to simply estimate the upper and lower bounds of the parameter over an appropriate probability. That is, we can say that the parameter *lies between two values with a certain probability*. This is called interval estimation, and the interval obtained is known as confidence interval. For example, we can say that the mean of any variable lies between 2 and 3 with a probability 0.9. In this chapter, we will derive the confidence interval for the unknown parameter of a normal population when the sample sizes are small and large.

4.5.1 Confidence Interval

Theorem 4.3 *If $\bar{x}$ is the mean of a sample of size n from a normally distributed population with known variance σ^2 and unknown mean μ, then the $(1 - \alpha)100\%$ confidence interval for the population mean is given by*

$$\bar{x} - z_{\frac{\alpha}{2}} \frac{\sigma}{\sqrt{n}} \leq \mu \leq \bar{x} + z_{\frac{\alpha}{2}} \frac{\sigma}{\sqrt{n}}. \tag{4.27}$$

where $z_{\frac{\alpha}{2}}$ is such that

$$P\left(Z > z_{\frac{\alpha}{2}}\right) = \frac{\alpha}{2}$$

Fig. 4.4 Tail probabilities
for standard normal
distribution

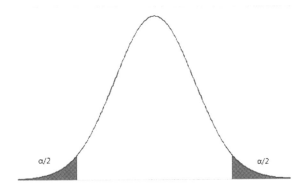

for the standard normal random variable Z.

Proof Consider the statistic

$$\bar{X} = \frac{\sum\limits_{i=1}^{n} X_i}{n}.$$

We know that if $X_i \sim N(\mu, \sigma^2)$, then $\bar{X} \sim N(\mu, \frac{\sigma^2}{n})$. Hence,

$$\frac{\bar{X} - \mu}{\frac{\sigma}{\sqrt{n}}} \sim N(0, 1).$$

For a given α, find a point $z_{\frac{\alpha}{2}}$ (Fig. 4.4) such that

$$P\left(Z > z_{\frac{\alpha}{2}}\right) = \frac{\alpha}{2}.$$

$$\Rightarrow \quad P\left(-z_{\frac{\alpha}{2}} \leq \frac{\bar{X} - \mu}{\frac{\sigma}{\sqrt{n}}} \leq z_{\frac{\alpha}{2}}\right) = 1 - \alpha$$

$$\Rightarrow \quad P\left(\bar{X} - \frac{z_{\frac{\alpha}{2}}\sigma}{\sqrt{n}} \leq \mu \leq \bar{X} + \frac{z_{\frac{\alpha}{2}}\sigma}{\sqrt{n}}\right) = 1 - \alpha.$$

The interval is known as $(1 - \alpha)\%$ confidence interval.

What do we mean when we say that the probability is $(1 - \alpha)$? It simply means that, if we take 100 samples, from the same population, we expect $100(1 - \alpha)$ times that the confidence interval will contain the unknown parameter.

Remark 4.6 In case the population variance is not given, but the sample variance is given, we have to replace normal distribution with student's t-distribution. The confidence interval becomes

$$\bar{x} - t_{n-1,\frac{\alpha}{2}} \frac{s}{\sqrt{n}} < \mu < \bar{x} + t_{n-1,\frac{\alpha}{2}} \frac{s}{\sqrt{n}}$$

where

$$P\left(t > t_{n-1,\frac{\alpha}{2}}\right) = \frac{\alpha}{2}.$$

Example 4.19 The average return of a risky portfolio is 10.5% per year with a standard deviation of 18%. If returns are assumed to be approximately normal, what is the 95% confidence interval for portfolio return next year?

Solution:

Here, population mean μ and population variance σ are 10.5 and 18%, respectively. Thus, the 95% confidence interval for the return, R, is given as

$$10.5 \pm 1.96(18) = (-24.78, 45.78\%). \tag{4.28}$$

which can be interpreted as:

$$P(-24.78 < R < 45.78) = 0.95. \tag{4.29}$$

In other words, one can say the annual return is expected to be within this interval 95% of the time or 95 out of 100 days.

Example 4.20 Consider a queuing system which describes the number of ongoing telephone calls in a particular telephone exchange. Determine a 95% confidence interval for the average talk time (in seconds) of the customers in the system if a random selection of 15 customers yielded the following data

$$54, 53, 58, 72, 49, 92, 70, 71, 104, 48, 65, 72, 65, 70, 85.$$

Solution:

Here, since the population variance is not given, we will use the sample variance to estimate the population variance. Thus, we will be using the student's t-distribution. From the given data, we get

$$\bar{x} = 68.53, \quad s = 15.82.$$

From Table A.10 in Appendix, we have $t_{14,0.025} = 2.145$. Hence, the 95% confidence interval is

$$\left(68.53 - \frac{2.145 \times 15.82}{\sqrt{15}}, 68.53 + \frac{2.145 \times 15.82}{\sqrt{15}}\right)$$

which reduces to $(59.77, 77.29)$.

Example 4.21 Let X_1, X_2, ..., X_{11} be a random sample of size 11 from a normal distribution with unknown mean μ and variance $\sigma^2 = 9.9$. If $\sum\limits_{i=1}^{11} x_i = 132$, then for what value of the constant k is $(12 - k\sqrt{0.9}, 12 + k\sqrt{0.9})$ a 90% confidence interval for μ?

Solution:

The 90% confidence interval for μ when the variance is given is

$$\left(\bar{x} - \frac{\sigma}{\sqrt{n}}, \bar{x} + \frac{\sigma}{\sqrt{n}} \right).$$

Thus, we need to find $\bar{x}$, $\sqrt{\dfrac{\sigma^2}{n}}$ and $z_{\frac{\alpha}{2}}$ corresponding $1 - \alpha = 0.9$. Hence,

$$\bar{x} = \frac{\sum\limits_{i=1}^{11} x_i}{11} = \frac{132}{11} = 12 \qquad \sqrt{\frac{\sigma^2}{n}} = \sqrt{\frac{9.9}{11}} = \sqrt{0.9}.$$

We know that $z_{0.05} = 1.64$ (from Table A.7 in Appendix). Hence, the confidence interval for μ at 90% confidence level is

$$(12 - (1.64)\sqrt{0.9}, 12 + (1.64)\sqrt{0.9}).$$

Comparing this interval with the given interval, we get $k = 1.64$ and the corresponding 90% confidence interval is $(10.444, 13.556)$.

Example 4.22 The mean and standard deviation for the quality grade point averages of a random sample of 36 college seniors are given to be 2.6 and 0.3, respectively. Find the 95 and 99% confidence intervals for the mean of the entire senior class.

Solution:

The point estimate of μ is $\bar{x} = 2.6$. Since the sample size is large, the σ can be approximated by $s = 0.3$. The z value leaving an area of 0.025 to the right and therefore an area of 0.975 to the left is $z_{0.025} = 1.96$ (see Table A.7 in Appendix). Hence, the 95% confidence interval is

$$2.6 - 1.96 \left(\frac{0.3}{\sqrt{36}} \right) < \mu < 2.6 + 1.96 \left(\frac{0.3}{\sqrt{36}} \right)$$

which reduces to $2.50 < \mu < 2.70$. To find a 99% confidence interval, we find the z value $z_{0.005} = 2.575$, and the 99% confidence interval is

$$2.6 - 2.575 \left(\frac{0.3}{\sqrt{36}} \right) < \mu < 2.6 + 2.575 \left(\frac{0.3}{\sqrt{36}} \right)$$

which reduces to

$$2.47 < \mu < 2.73.$$

Thus, we see that a longer interval is required to estimate μ with a higher degree of accuracy. The shortest length of this confidence interval is $2z_{\alpha/2}\frac{\sigma}{\sqrt{n}}$.

Difference Between the Means of Two Populations

Now, suppose we wish to determine confidence intervals for the difference of means of two populations, $\mathbb{N}\left(\mu_1, \sigma_1^2\right)$ and $\mathbb{N}\left(\mu_2, \sigma_2^2\right)$. It is given that

$$\bar{X}_1 \sim N\left(\mu_1, \frac{\sigma_1^2}{n_1}\right), \quad \bar{X}_2 \sim N\left(\mu_2, \frac{\sigma_2^2}{n_2}\right).$$

$$\Rightarrow \quad \bar{X}_1 - \bar{X}_2 \sim N\left(\mu_1 - \mu_2, \frac{\sigma_1^2}{n_1} + \frac{\sigma_2^2}{n_2}\right)$$

$$\Rightarrow \quad \bar{X}_1 - \bar{X}_2 \sim N\left(\mu_1 - \mu_2, \frac{\sigma_1^2 n_2 + \sigma_2^2 n_1}{n_1 n_2}\right).$$

Hence, the confidence interval for the difference between μ_1 and μ_2 is given by

$$\bar{x}_2 - \bar{x}_1 - z_{\frac{\alpha}{2}}\sqrt{\frac{\sigma_1^2}{n_1} + \frac{\sigma_2^2}{n_2}} < \mu_1 - \mu_2 < \bar{x}_2 - \bar{x}_1 + z_{\frac{\alpha}{2}}\sqrt{\frac{\sigma_1^2}{n_1} + \frac{\sigma_2^2}{n_2}}.$$

The above inequality is not useful if we do not know the population variances. Then, we use another quantity, s_p, where $s_p^2 = \frac{(n_1-1)s_1^2+(n_2-1)s_2^2}{n_1+n_2-2}$, which is nothing but the weighted mean of the individual sample variance called pooled variance. Hence, the resulting expression changes to

$$\bar{x}_2 - \bar{x}_1 - t_{\frac{\alpha}{2}, n_1+n_2-2}s_p\sqrt{\frac{1}{n_1} + \frac{1}{n_2}} < \mu_1 - \mu_2 < \bar{x}_2 - \bar{x}_1 + t_{\frac{\alpha}{2}, n_1+n_2-2}s_p\sqrt{\frac{1}{n_1} + \frac{1}{n_2}}.$$

Example 4.23 In sampling from a nonnormal distribution with a variance of 25, how large must the sample size be so that the length of a 95% confidence interval for the mean is 1.96?

Solution: The confidence interval when the sample is taken from a normal population with a variance of 25 is

$$\left(\bar{x} - \frac{\sigma}{\sqrt{n}}, \bar{x} + \frac{\sigma}{\sqrt{n}}\right).$$

Thus, the length of the confidence interval is

$$l = 2z_{\frac{\alpha}{2}}\sqrt{\frac{\sigma^2}{n}} = 2z_{0.025}\sqrt{\frac{25}{n}} = 2(1.96)\sqrt{\frac{25}{n}}.$$

But, we are given that the length of the confidence interval is $l = 1.96$. Thus,

$$1.96 = 2(1.96)\sqrt{\frac{25}{n}}$$

$$\Rightarrow n = 100.$$

Hence, the sample size should be 100 so that the length of the 95% confidence interval will be 1.96.

Theorem 4.4 *Let s^2 be the value of sample variance based on a random sample of size n from a normally distributed population. A $(1 - \alpha)100\%$ confidence interval for the population variance is given by*

$$\frac{(n-1)s^2}{\chi^2_{\frac{\alpha}{2},n-1}} < \sigma^2 < \frac{(n-1)s^2}{\chi^2_{1-\frac{\alpha}{2},n-1}}. \tag{4.30}$$

Proof Consider the following statistic

$$V = (n-1)\frac{s^2}{\sigma^2}.$$

Note that $(n-1)\frac{s^2}{\sigma^2}$ follows χ^2_{n-1} distribution.

$$\Rightarrow \quad P\left(\chi^2_{1-\frac{\alpha}{2},n-1} < V < \chi^2_{\frac{\alpha}{2},n-1}\right) = 1 - \alpha$$

$$\Rightarrow \quad P\left(\chi^2_{1-\frac{\alpha}{2},n-1} < (n-1)\frac{s^2}{\sigma^2} < \chi^2_{\frac{\alpha}{2},n-1}\right) = 1 - \alpha$$

$$\Rightarrow \quad P\left(\frac{\chi^2_{1-\frac{\alpha}{2},n-1}}{(n-1)s^2} < \frac{1}{\sigma^2} < \frac{\chi^2_{\frac{\alpha}{2},n-1}}{(n-1)s^2}\right) = 1 - \alpha$$

$$\Rightarrow \quad P\left(\frac{(n-1)s^2}{\chi^2_{1-\frac{\alpha}{2},n-1}} < \sigma^2 < \frac{(n-1)s^2}{\chi^2_{\frac{\alpha}{2},n-1}}\right) = 1 - \alpha.$$

Example 4.24 Let $X_1, X_2, \ldots, X_{13}$ be a random sample from a normal distribution. If $\sum_{i=1}^{13} x_i = 246.61$ and $\sum_{i=1}^{13} x_i^2 = 4806.61$. Find the 90% confidence interval for σ^2?

Solution:

We have

$$\bar{x} = 18.97$$

$$s^2 = \frac{1}{n-1}\sum_{i=1}^{13}(x_i - \bar{x})^2 = \frac{1}{n-1}\sum_{i=1}^{13}(x_i^2 - 13\bar{x}^2) = \frac{1}{12}128.41.$$

For $\alpha = 0.10$, from Tables A.8 and A.9 in Appendix, $\chi^2_{\frac{\alpha}{2},n-1}$ and $\chi^2_{1-\frac{\alpha}{2},n-1}$ are 21.03 and 5.23, respectively. Hence, the 90% confidence interval for σ^2 is $\left(\dfrac{128.41}{21.03}, \dfrac{128.41}{5.23} \right)$, i.e., (6.11, 24.55).

Ratio of Variances

Theorem 4.5 *Let s_1^2 and s_2^2 be the values of sample variance of random samples of size n_1 and n_2, respectively, from two normal populations with variance σ_1^2 and σ_2^2. A (1-α)% confidence interval for the ratio of the two population variances is given by*

$$\frac{s_1^2}{F_{\frac{\alpha}{2},n_1-1,n_2-1}s_2^2} < \frac{\sigma_1^2}{\sigma_2^2} < \frac{s_1^2}{F_{1-\frac{\alpha}{2},n_1-1,n_2-1}s_2^2}. \tag{4.31}$$

Example 4.25 The following table gives the number of e-mails sent per week by employees at two different companies. Find a 95% confidence interval for $\dfrac{\sigma_1}{\sigma_2}$.

| Company 1 | 81 | 104 | 115 | 111 | 85 | 121 | 95 | 112 | 100 | 117 | 113 | 109 | 101 | |
| Company 2 | 99 | 100 | 104 | 98 | 103 | 113 | 95 | 107 | 98 | 95 | 101 | 109 | 99 | 93 | 105 |

Solution:

The sample variances are $s_1^2 = 148.577$ and $s_2^2 = 31.067$. For $\alpha = 0.05$, from Table A.11 in Appendix, we get $F_{\frac{\alpha}{2},n_1-1,n_2-1} = F_{0.025,12,14} = 3.0501$ and $F_{1-\frac{\alpha}{2},n_1-1,n_2-1} = F_{0.975,12,14} = 3.2062$. Thus, using Eq. (4.31), we get a 95% confidence interval for $\dfrac{\sigma_1^2}{\sigma_2^2}$ is (1.568,15,334) and 95% confidence interval for $\dfrac{\sigma_1}{\sigma_2}$ is (1.252, 3.916).

Things to Remember: As a general rule of thumb, follow the steps to do questions of point estimation.

1. Figure out distribution, and hence, PDF of the random variable involved.
2. List all the quantities we can formulate, e.g., sample mean, moments of the random variable.
3. Use the formulas of the above quantities to get an equation that results in the value of the parameter.

The interested readers to know about more on sampling distribution may refer to Cassella and Berger (2002), Rao (1973), Rohatgi and Saleh (2015) and Shao (2003).

Problems

4.1 Suppose $X \sim B(n, p)$ show that the sample proportion X/n is an unbiased estimator of p.

4.2 Let $X_1, X_2, \ldots, X_n$ be a random sample from normal distributed population with mean μ and variance σ^2. Let s^2 be the sample variance. Prove that $\frac{(n-1)s^2}{\sigma^2}$ has χ^2 distribution with $n-1$ degrees of freedom.

4.3 Let $X_1, X_2, \ldots, X_n$ be a normal distribution with mean $E(X_i) = \mu$, $i = 1, 2, \ldots, n$, $Var(X_i) = \sigma^2$, $i = 1, 2, \ldots, n$ and $Cov(X_i, X_j) = \rho\sigma^2$, $i \neq j$. Let $\bar{X}$ and s^2 denote the sample mean and sample variance, respectively. Then, prove that $\frac{(1-\rho)t}{1+(n-1)\rho}$ has student's t-distribution with $n-1$ degrees of freedom.

4.4 Let X and Y be two independent random variables. Show that $X + Y$ is normally distributed if and only if both X and Y are normal.

4.5 Let X_i, $i = 1, 2, \ldots, n$ be a random sample of size n drawn from a population of size N with population distribution given by

$$f(x) = \frac{1}{\sigma\sqrt{2\pi}}e^{\frac{-(x-\mu)^2}{2\sigma^2}}, \quad -\infty < x < \infty.$$

Show that $\sum_{i \neq j} Cov(X_i, X_j) = n(n-1)\left(\frac{-\sigma^2}{N-1}\right)$.

4.6 Let (X, Y) be a random vector of continuous type with joint PDF $f(x, y)$. Define $Z = X + Y$, $U = X - Y$, $V = XY$, $W = \frac{X}{Y}$. Then, prove that the PDFs of Z, U, V, and W are, respectively, given by

$$f_Z(z) = \int_{-\infty}^{\infty} f(x, z - x)dx$$

$$f_U(u) = \int_{-\infty}^{\infty} f(u + y, y)dy$$

$$f_V(v) = \int_{-\infty}^{\infty} f(x, v/x)\frac{1}{|x|}dx$$

$$f_W(w) = \int_{-\infty}^{\infty} f(xw, x)|x|dx.$$

4.7 Let X and Y be two independent $N(0, 1)$ random variables. Show that $X + Y$ and $X - Y$ are independent.

4.8 If X and Y are independent random variables with the same distribution and $X + Y$, $X - Y$ are independent. Show that all random variables $X, Y, X + Y, X - Y$ are normally distributed.

4.9 If X and Y are independent exponential random variable with parameter λ, then show that $\frac{X}{X+Y}$ follows uniform distribution on $(0, 1)$.

4.10 If X_1 and X_2 are independent random variables such that $X_1 \sim Exp(\lambda)$ and $X_2 \sim Exp(\lambda)$, show that $Z = \min\{X_1, X_2\}$ follows $Exp(2\lambda)$. Hence, generalize this result for n independent exponential random variables.

4.11 Let $Z \sim N(0, 1)$ and $X \sim \chi_n^2$. Prove that $Y = \frac{Z}{\sqrt{\frac{X}{n}}} \sim t(n)$.

4.12 Let $X \sim \chi_{v_1}^2$ and $Y \sim \chi_{v_2}^2$. Prove that $Z = \frac{X/v_1}{Y/v_2} \sim F(v_1, v_2)$.

4.13 Let $X_i \sim U(0, 1)$, $i = 1, 2, \ldots, n$ such that X_i are independent. Then, find the distribution of $-\sum_{i=1}^{n} \log(X_i)$.

4.14 Let $X_1, X_2, \ldots, X_n$ be a random sample from a Bernoulli distribution with the parameter p, $0 < p < 1$. Prove that the sampling distribution of the sample variance S^2 is

$$P\left(S^2 = \frac{i(n-i)}{n(n-1)}\right) = {}^nC_i p^i(1-p)^{n-i} + {}^nC_{n-i} p^{n-i}(1-p)^i$$

where $i = 0, 1, 2, \ldots, [\frac{n}{2}]$

4.15 Let $X_1, X_2 \ldots, X_n$ be a random sample form a distribution function F, and let $F_n^*(x)$ be the empirical distribution function of the random sample. Prove that $Cov(F_n^*(x), F_n^*(y)) = \frac{1}{n} F(x)(1 - F(y))$.

4.16 Find $Cov(\bar{X}, S^2)$ in terms of population moments. Under what conditions is $Cov(\bar{X}, S^2) = 0$? Discuss this result when the population is normal distributed.

4.17 Find the probability that the maximum of a random sample of size n from a population exceeds the population median.

4.18 Consider repeated observation on a $m-$dimensional random variable with mean $E(X_i) = \mu, i = 1, 2, \ldots, m$, $Var(X_i) = \sigma^2, i = 1, 2, \ldots, m$ and $Cov(X_i, X_j) = \rho\sigma^2, i \neq j$. Let the ith observation be $(x_{1i}, \ldots, x_{mi})$, $i = 1, 2, \ldots, n$. Define

$$\bar{X}_i = \frac{1}{m} \sum_{j=1}^{m} X_{ji},$$

$$W_i = \sum_{j=1}^{m} (X_{ji} - \bar{X}_i)^2,$$

$$B = m \sum_{i=1}^{n} (\bar{X}_i - \bar{X})^2,$$

$$W = W_1 + \cdots + W_n.$$

where B is sum of squares between and W is sum of squares within samples.

1. Prove (i) $W \sim (1 - \rho)\sigma^2 \chi^2(mn - n)$ and (ii) $B \sim (1 + (m - 1)\rho)\sigma^2 \chi^2(n - 1)$.

2. Suppose $\frac{(1-\rho)B}{(1+(m-1)\rho)W} \sim F_{(n-1),(mn-n)}$. Prove that when $\rho = 0$, $\frac{W}{W+B}$ follows beta distribution with parameters $\frac{mn-n}{2}$ and $\frac{n-1}{2}$.

4.19 Show that the joint PDF of $(X_{(1)}, X_{(2)}, \ldots, X_{(n)})$ is given by

$$f(x_{(1)}, \ldots, x_{(n)}) = \begin{cases} n! \Pi_{i=1}^{n} f_{X_i}(x_{(i)}) & \text{if } x_{(1)} \le x_{(2)} \cdots \le x_{(n)} \\ 0 & \text{otherwise} \end{cases}.$$

Also, prove that the marginal PDF of $X_{(r)}$ $r = 1, 2, \ldots, n$, where $X_{(r)}$ is the rth minimum, is given by

$$f_{X_{(r)}}(x_{(r)}) = \frac{n!}{(r-1)!(n-r)!}[F(x_{(r)})]^{r-1}[1 - F(x_{(r)})]^{n-r} f(x_{(r)}).$$

4.20 Let $X_1, X_2, \ldots, X_n$ be a random sample from Poisson distribution with parameter λ. Show that $\alpha \bar{X} + (1-\alpha)s^2, 0 \le \alpha \le 1$, is a class of unbiased estimators for λ. Find the UMVUE for λ. Also, find an unbiased estimator for $e^{-\lambda}$.

4.21 Let $X_1, X_2, \ldots, X_n$ be a random sample from Bernoulli distribution $B(1, p)$. Find an unbiased estimators for p^2 if it exists.

4.22 Suppose that 200 independent observations $X_1, X_2, \ldots, X_{200}$ are obtained from random variable X. We are told that $\sum_{i=1}^{200} x_i = 300$ and that $\sum_{i=1}^{200} x_i^2 = 3754$. Using these values, obtain unbiased estimates for $E(X)$ and $Var(X)$.

4.23 If X_1, X_2 and X_3 are three independent random variables having the Poisson distribution with the parameter λ, show that

$$\hat{\lambda_1} = \frac{X_1 + 2X_2 + 3X_3}{6}$$

is an unbiased estimator of λ. Also, compare the efficiency of $\hat{\lambda}_1$ with that of the alternate estimator.

$$\hat{\lambda_2} = \frac{X_1 + X_2 + X_3}{3}.$$

4.24 Let X be Cauchy-distributed random variable with PDF

$$f(x; \theta) = \frac{1}{\pi} \frac{1}{(1 + (x - \theta)^2)}, \quad -\infty < x < \infty, -\infty < \theta < \infty.$$

Find the Cramer–Rao lower bound for the estimation of the location parameter θ.

4.25 Consider the normal distribution $N(0, \sigma^2)$. With a random sample $X_1, X_2, \ldots, X_n$, we want to estimate the standard deviation σ. Find the constant c so that $Y = c \sum_{i=1}^{n} |X_i|$ is an unbiased estimator of σ and determine its efficiency.

4.26 If $X_1, X_2, \ldots, X_n$ is a random sample from $N(\mu, 1)$. Find a lower bound for the variance of an estimator of μ^2. Determine an unbiased minimum variance estimator of μ^2, and then compute its efficiency.

4.27 Let $X_1, X_2, \ldots, X_n$ be random sample from a Poisson distribution with mean λ. Find the unbiased minimum variance estimator of λ^2.

4.28 Prove that $\bar{X}$ the mean of a random sample of size n from a distribution that is $N(\mu, \sigma^2)$, $-\infty < \mu < \infty$ is an efficient estimator of μ for every known $\sigma^2 > 0$.

4.29 Assuming population to be $N(\mu, \sigma^2)$, show that sample variance is a consistent estimator for population variance σ^2.

4.30 Let $X_1, X_2, \ldots, X_n$ be a random sample from uniform distribution on an interval $(0, \theta)$. Show that $\left(\prod_{i=1}^{n} X_i \right)^{1/n}$ is consistent estimator of θe^{-1}.

4.31 The number of births in randomly chosen hours of a day is as follows.

$$4, 0, 6, 5, 2, 1, 2, 0, 4, 3.$$

Use this data to estimate the proportion of hours that had two or fewer births.

4.32 Consider the number of students attended probability and statistics lecture classes for 42 lectures. Let $X_1, X_2, \ldots, X_{30}$ denote the number students attended in randomly chosen 30 lecture classes. Suppose the observed data are given in Table 4.4.

1. Using method of moments, find the estimators for the population mean and population variance.
2. Assume that X_i's are i.i.d random variables each having discrete uniform distribution with interval 70 and 90, and find ML estimators for the population mean and population variance.

4.33 Using method of moments, find the estimators of the parameters for the following population distributions
(i) $N(\mu, \sigma^2)$ (ii) $Exp(\lambda)$ (iii) $P(\lambda)$.

4.34 Let $X_1, X_2, \ldots, X_n$ be i.i.d from the uniform distribution $U(a, b)$, $-\infty < a < b < \infty$. Prove that, using the method of moments, the estimators of a and b are, respectively,

Table 4.4 Data for Problem 4.32

100	90	85	95	88	82	92	84	88	87	82	88	82	91	92
82	90	82	87	92	70	84	79	88	81	82	78	81	82	90

$$\hat{a} = \bar{X} - \sqrt{\frac{3(n-1)}{n} S^2}$$

$$\hat{b} = \bar{X} + \sqrt{\frac{3(n-1)}{n} S^2}.$$

4.35 Let $X_1, X_2, \ldots, X_n$ be i.i.d from the binomial distribution $B(n, p)$, with unknown parameters n and p. Prove that, using the method of moments, the estimators of n and p are, respectively,

$$\hat{n} = \frac{\bar{X}}{1 - \frac{n-1}{n} \frac{S^2}{\bar{X}}}$$

$$\hat{p} = 1 - \frac{n-1}{n} \frac{S^2}{\bar{X}}.$$

4.36 Let $X_1, X_2, \ldots, X_n$ be i.i.d from the gamma distribution $G(r, \lambda)$, with unknown parameters r and λ. Prove that, using the method of moments, the estimators of r and λ are, respectively,

$$\hat{r} = \frac{\bar{X}^2}{\frac{1}{n} \sum_{i=1}^{n} X_i^2 - \bar{X}^2}$$

$$\hat{\lambda} = \frac{\frac{1}{n} \sum_{i=1}^{n} X_i^2 - \bar{X}^2}{\bar{X}}.$$

4.37 A random variable X has PDF

$$f(x; \theta) = \frac{1}{2} e^{-|x-\theta|}, \quad -\infty < x < \infty, -\infty < \theta < \infty.$$

Obtain the ML estimates of θ based on a random sample $X_1, X_2, \ldots, X_n$.

4.38 Consider a queueing system in which the arrival of customers follows Poisson process. Let X be the distribution of service time, which has gamma distribution; i.e., the PDF of X is given by

$$f(x; \lambda, r) = \begin{cases} \frac{\lambda (\lambda x)^{r-1} e^{-\lambda x}}{\Gamma(r)}, & 0 < x < \infty \\ 0, & \text{otherwise} \end{cases}$$

Suppose that r is known. Let $X_1, X_2, \ldots, X_n$ be a random sample on X. Obtain the ML estimate of λ based on this sample. Further, assume that r is also unknown. Determine the ML estimators of λ and r.

4.39 Prove that for the family of uniform distribution on the interval $[0, \theta]$, $max(X_1, X_2, \ldots, X_n)$ is the MLE for θ.

4.40 Suppose that the random sample arises from a distribution with PDF

$$f(x; \theta) = \begin{cases} \theta x^{\theta-1}, & 0 < x < 1, \; 0 < \theta < \infty \\ 0 & \text{otherwise} \end{cases}.$$

Show that $\hat{\theta} = -\dfrac{n}{\ln \prod_{i=1}^{n} X_i}$ is the MLE of θ. Further, prove that in a limiting sense, $\hat{\theta}$ is the minimum variance unbiased estimator of θ and thus θ is asymptotically efficient.

4.41 Find the maximum likelihood estimator based on a sample of size n from the two-sided exponential distribution with PDF

$$f(x; \theta) = \frac{1}{2} e^{-|x-\theta|}, \quad -\infty < x < \infty.$$

Is the estimator unbiased?

4.42 Prove that method of moment estimator is consistent for the estimation of $r > 0$ in the gamma family

$$f(x; r) = \begin{cases} \frac{e^{-x} x^{r-1}}{\Gamma(r)} & 0 < x < \infty \\ 0, & \text{otherwise} \end{cases}.$$

4.43 Let $X_1, X_2, \ldots, X_n$ be a random sample from the normal distribution with both mean and variance equal to an unknown parameter θ.

1. Is there a sufficient statistics?
2. What is the MLE for θ?
3. What is the Cramer–Rao lower bound?

4.44 Let $X_1, X_2, \ldots, X_n$ be random sample from the geometric distribution with PMF

$$p(x; q) = \begin{cases} q^x(1-q), & x = 0, 1, 2, \ldots; \; 0 \le q \le 1 \\ 0 & \text{otherwise} \end{cases}.$$

1. Find the MLE $\hat{q}$ of q.
2. Show that $\sum_{i=1}^{n} X_i$ is a complete sufficient statistics for q.
3. Determine the minimum variance unbiased estimator of q.

4.45 Let $X_1, X_2, \ldots, X_n$ be random sample from a Poisson distribution with mean λ. Find the minimum variance unbiased estimator of λ^2.

4.46 Theorem (Lehmann–Scheffe Theorem): An unbiased estimator that is a complete sufficient statistics is the unique UMVUE. Using Lehmann–Scheffe theorem, prove that $\bar{X}$ is the unique UMVUE for θ of a Bernoulli distribution.

4.47 Let $X_1, X_2, \ldots, X_n$ be a random sample from the Poisson distribution with parameter λ. Show that $\displaystyle\sum_{i=1}^{n} X_i$ is a minimal sufficient statistics for λ.

4.48 Let $X_1, X_2, \ldots, X_n$ be a random sample from the normal distribution $N(0, \theta)$. Show that $\displaystyle\sum_{i=1}^{n} X_i^2$ is a minimal sufficient statistics for θ.

4.49 If $X_1, X_2, \ldots, X_n$ is a random sample from $N(\theta, 1)$. Find a lower bound for the variance of an estimator of θ^2. Determine the minimum variance unbiased estimator of θ^2, and then compute its efficiency.

4.50 Let $X_1, X_2, \ldots, X_n$ be a random sample from the modified geometric distribution with PMF

$$p(x; q) = (1 - q)^{x-1} q, \quad x = 1, 2, \ldots.$$

Prove that maximum likelihood estimator of q is

$$\hat{q} = \frac{n}{\displaystyle\sum_{i=1}^{n} x_i} = \frac{1}{\bar{x}}.$$

4.51 Let $\bar{X}$ be the mean of a random sample of size n from $N(\mu, 25)$. Find the smallest sample size n such that $(\bar{X} - 1, \bar{X} + 1)$ is a 0.95 level confidence interval for μ.

4.52 The contents of seven similar containers of sulfuric acid are 9.8, 10.2, 10.4, 9.8, 10.0, 10.2, and 9.6 liters. Find a 95% confidence interval for the mean of all such containers, assuming an appropriate normal distribution.

4.53 If the standard deviation of the lifetimes of light bulbs is estimated to be 100 hours. What should be the sample size in order to be 99% confident that the error in the estimated average lifetime will not exceed 20 hours? Repeat the exercise for 95 and 99.73% confidence level.

4.54 A company has 500 cables. Forty cables were selected at random with a mean breaking strength of 2400 pounds and a standard deviation of 150 pounds.

(a) What are the 95% confidence limits for estimating the mean breaking strength of remaining 460 cables?
(b) With what degree of confidence could we say that the mean breaking strength of remaining 460 cables is 2400 ± 35 pounds.

4.55 The standard deviation of the breaking strength of 100 cables was 180 pounds. Find 99% confidence limits for standard deviation.

References

Cassella G, Berger RL (2002) Statistical inference. Duxbury, Pacific Grove

Rao CR (1973) Linear statistical inference and its applications. Wiley, New York

Rohatgi VK, Ehsanes Saleh AKMd (2015) An introduction to probability and statistics. Wiley, New York

Shao J (2003) Mathematical statistics. Springer texts in statistics. Springer, Berlin

Chapter 5
Testing of Hypothesis

In this section, we shall discuss another way to deal with the problem of making a statement about an unknown parameter associated with a probability distribution, based on a random sample. Instead of finding an estimate for the parameter, we shall often find it convenient to hypothesize a value for it and then use the information from the sample to confirm or refute the hypothesized value.

5.1 Testing of Statistical Hypothesis

Definition 5.1 Let $f(x; \theta)$ be the probability distribution of a population, where $\theta \in \Theta$, x, and θ may be vectors. Parametric statistical hypothesis is the testing (verification) of an assertion (statement) about the unknown parameter θ with the help of observations obtained from a random sample of size n.

Have a look at the following examples.

1. Examination marks follow normal distribution with mean 15 and variance 3 (Parametric testing).
2. Most of the time the student president is from Shivalik Hostel (Parameter involved is proportion).
3. The average talk time on phone of a person is 3 min and follows exponential distribution (Parameter involved is distribution of population itself).
4. Traffic speed and traffic volume are correlated with correlation coefficient equal to 0.7. (Here, we are testing if two populations are independent or not.)

© Springer Nature Singapore Pte Ltd. 2018
D. Selvamuthu and D. Das, *Introduction to Statistical Methods,*
Design of Experiments and Statistical Quality Control,
https://doi.org/10.1007/978-981-13-1736-1_5

5.1.1 Null and Alternate Hypothesis

The statement (assertion) about an unknown parameter $\theta \in \Theta$ is called a null hypothesis denoted by

$$H_0 : \theta \in \Theta_0, \quad \Theta_0 \subset \Theta,$$

and the complement statement is called an alternative/alternate hypothesis denoted by

$$H_1 : \theta \in \Theta_1 = \Theta \setminus \Theta_0.$$

A point needs to be kept in mind here is that the 'equals to' is always included in the null hypothesis, i.e., the null hypothesis is always formed with the '=', '$\geq$', and '$\leq$' signs.

More generally, if we have a family of distributions $\{P_\theta, \; \theta \in \Theta\}$, the null hypothesis may be that $\theta \in \Theta_0$ and the alternative $\theta \in \Theta_1 = \Theta - \Theta_0$. The hypothesis is said to be simple hypothesis if both Θ_0 and Θ_1 consist of only one point. Any hypothesis that is not simple is called composite.

But how do you decide whether to accept the null hypothesis or reject it? This is done by using a test statistic and checking whether it lies in the critical region or not.

Test Statistic: A test statistic is a statistic whose value is determined using the realization of the sample.

Critical Region: The critical region, also called the rejection region, is the set of values of the test statistic for which the null hypothesis is to be rejected.

The simplest problem of hypothesis testing arises when we have only two possible alternatives models, and we have to choose one of them on the basis of the observed sample. For instance, assume we have two possible probability density functions $f_0(x)$ and $f_1(x)$ and a sample of size one, i.e., x is observed. We have to select one of the PDFs based on observation x. The null hypothesis states that the f_0 is the true pdf, whereas the alternative hypothesis states that f_1 is the true pdf. A decision test or test of hypothesis will be of the following form. A region Ω, called the critical region, is determined, and if $x \in \Omega$, we are critical of the null hypothesis and we reject it in favor of the alternate. If $x \notin \Omega$, we write do not reject the null hypothesis.

The scenario of rejecting the null hypothesis, when in fact it is true, is called Type I error. The probability of committing Type I error is called level of significance of a test, which is denoted by α and is given by $\alpha = \int_\Omega f_1(x)dx$. Another kind of statistical error that can occur is called Type II error, which is failing to reject the null hypothesis when it is actually false. The probability, $\int_{\Omega^c} f_1(x)dx$, of making such an error is denoted by β. Another related concept is the power of a test, which is equal to $1 - \beta$, which basically means the probability of rejecting the null hypothesis when it is indeed false. An ideal test will be one which minimizes the probability of committing both Type I and Type II errors. Just like the problem with every ideal scenario, it is impossible to achieve the objective of simultaneously minimizing both types of errors.

Table 5.1 Table of error types

	H_0 is true	H_0 is false
Do not reject H_0	Correct inference	Type II error
Reject H_0	Type I error	Correct inference

$$\alpha = P(\text{Type I error}) = P(\text{reject } H_0 \mid H_0 \text{ is true}).$$

Similarly,

$$\beta = P(\text{Type II error}) = P(\text{accept } H_0 \mid H_0 \text{ is false}).$$

Table 5.1 shows the relation between truth/falseness of null hypothesis and the outcome of the hypothesis test.

Example 5.1 Let X be a continuous type random variable that follows exponential distribution, i.e., $X \sim Exp(\lambda)$. Consider the following null hypothesis $\mu = 20$ against the alternative hypothesis $\mu = 30$, where $\mu = \frac{1}{\lambda}$. Let a sample of size one, i.e., x is observed from the distribution. The test is to reject the null hypothesis if value of x is greater than 28. Find the probabilities of (i) a Type I error (ii) a Type II error.

Solution:

(i) The probability of a Type I error is given by:
P(reject H_0 when H_0 is true) $= $ P($X > 28$ when X follows $Exp(\frac{1}{20})$) $=$ $1 - F_X(28) = e^{\frac{-28}{20}} = 0.2466$.
(ii) The probability of a Type II error is given by:
P(do not reject H_0 when H_0 is false) $=$ P($X < 28$ when X follows $Exp(\frac{1}{30})$) $=$ $1 - F_X(28) = 1 - e^{\frac{-28}{30}} = 0.6068$.

p-**value**: The *p*-value associated with a test, under the assumption of hypothesis H_0, is the probability of obtaining a result equal to or more extreme than what was actually observed. Rather than selecting the critical region ahead of time, the *p*-value of a test can be reported, and a decision can be made by comparing *p*-value with α. The smaller the *p*-value, the larger the significance, i.e., we reject the null hypothesis if *p*-value is less than α.

5.1.2 Neyman–Pearson Theory

After defining the two types of errors, we can discuss the notion of a "good test." To find a good test, we fix the value of the probability of committing Type I error α and minimize the probability of committing a Type II error, β. We try to find the "most powerful test" by minimizing β for every value of the parameter specified by our alternate hypothesis. A test,denoted by Φ, of size α is said to be the most powerful test if it has more power than power of any other test of the same size.

It is time that we state the Neyman[1]–Pearson[2] Lemma clearly. In case of a simple two hypothesis test, we try to maximize the power of our test by minimizing β. For any given level of significance, we set an upper bound on the likelihood ratio L_0/L_1, where L_0 and L_1 are the likelihood functions of the data under our null hypothesis (H_0) and alternate hypothesis (H_1), respectively.

Mathematically, if we consider a critical region of size α, and if there exists a constant k such that the value of the likelihood ratio is less than k in our critical region and greater than k outside that critical region, we say that the critical region is the most powerful critical region of size α for testing our hypothesis. In other words, Neyman–Pearson test will reject the null hypothesis if the likelihood ratio is less than the critical value, i.e., k.

Theorem 5.1 (Neyman–Pearson Lemma (NP-Lemma)) *Consider $H_0 : \theta = \theta_0$ against $H_1 : \theta = \theta_1$. Then, any test ϕ of the form*

1.

$$\phi(\mathbf{x}) = \begin{cases} 1 & if \ \frac{f_1(\mathbf{x})}{f_0(\mathbf{x})} > k \\ \gamma & if \ \frac{f_1(\mathbf{x})}{f_0(\mathbf{x})} = k \\ 0 & if \ \frac{f_1(\mathbf{x})}{f_0(\mathbf{x})} < k \end{cases}$$

is most powerful of its size. Here, $k > 0$ is constant and $0 < \gamma < 1$, $f_1(\mathbf{x}) = \mathbf{f}_{\theta_1}(\mathbf{x})$ and $f_0(\mathbf{x}) = \mathbf{f}_{\theta_0}(\mathbf{x})$.

2. If $k = \infty$, the test

$$\phi(\mathbf{x}) = \begin{cases} 1 & if \ f_0(\mathbf{x}) = \mathbf{0} \\ 0 & if \ f_0(\mathbf{x}) > \mathbf{0} \end{cases}$$

is most powerful of size 0 for testing H_0 against H_1.

In case, we are unable to find such a region by the NP-Lemma, we use likelihood ratio tests which are essentially the generalizations of the NP-Lemma. Interested readers may refer to Rohatgi and Saleh (2015) for the proof of NP-Lemma.

Example 5.2 Consider the following problem of hypothesis testing based on a sample of size one, i.e., x.

$$H_0 : f \equiv f_0(x) \quad \text{against} \quad f \equiv f_1(x)$$

---

[1] Jerzy Neyman (1894–1981) was a Polish mathematician and statistician who spent the first part of his professional career at various institutions in Warsaw, Poland, and then at University College London, and the second part at the University of California, Berkeley. Neyman first introduced the modern concept of a confidence interval into statistical hypothesis testing.

[2] Egon Sharpe Pearson, (1895–1980) was one of three children and the son of Karl Pearson and, like his father, a leading British statistician. Pearson is best known for the development of the Neyman–Pearson Lemma of statistical hypothesis testing. He was the President of the Royal Statistical Society in 1955–1956 and was awarded its Guy Medal in gold in 1955. He was awarded a CBE in 1946.

where

$$f_0(x) = \begin{cases} 2x, & 0 < x < 1 \\ 0 & \text{otherwise} \end{cases}$$

$$f_1(x) = \begin{cases} 2(1-x), & 0 < x < 1 \\ 0 & \text{otherwise} \end{cases}$$

Using NP-Lemma, construct the uniformly most powerful (UMP) test of size α. Also, determine the power function.

Solution:

The most powerful test is of the form

$$\phi(x) = \begin{cases} 1 \text{ if } \frac{f_1(x)}{f_0(x)} > k \\ 0 \text{ if } \frac{f_1(x)}{f_0(x)} \leq k \end{cases}$$

which is same as

$$\phi(x) = \begin{cases} 1 \text{ if } \frac{2(1-x)}{2x} > k \\ 0 \text{ if } \frac{2(1-x)}{2x} \leq k \end{cases}$$

where k is determined as follows:

$$\alpha = P(\text{reject } H_0 \mid H_0 \text{ is true}) = P_0 \left(\frac{1-X}{X} > k \right)$$

$$= P_0 \left(X < \frac{1}{1+k} \right) = \int_0^{\frac{1}{1+k}} 2t \, dt.$$

This gives $k = \frac{1-\sqrt{\alpha}}{\sqrt{\alpha}}$. Hence, the most powerful test of size α is given by

$$\phi(x) = \begin{cases} 1 \text{ if } X < \sqrt{\alpha} \\ 0 \text{ if } X \geq \sqrt{\alpha} \end{cases}.$$

The power of the test is given by

$$P(\text{accept } H_0 \mid H_1 \text{ is true}) = P_1(X < \sqrt{\alpha}) = \int_0^{\sqrt{\alpha}} 2(1-t) \, dt = 1 - (1-\sqrt{\alpha})^2.$$

Example 5.3 Let $(X_1, X_2, \ldots, X_n)$ be a random sample from normal distribution i.e., $N(0, \theta)$, where the variance θ is unknown. Consider the hypothesis test of the null hypothesis $H_0 : \theta = \theta_0$ where θ_0 is the fixed positive number, against the alternative composite hypothesis $H_1 : \theta > \theta_0$. Find the uniformly most powerful test of size α.

Solution: We have $\Omega = \{\theta; \ \theta \geq \theta_0\}$. The joint pdf of $(X_1, X_2, \ldots, X_n)$ is

$$L(\theta; \ x_1, x_2, \ldots, x_n) = \left(\frac{1}{2\pi\theta}\right)^{\frac{n}{2}} exp\left[-\frac{1}{2\theta}\sum_{i=1}^{n}x_i^2\right].$$

Let θ' be a number greater than θ_0, and let k be a positive constant. Let C be the set of points such that the following holds

$$\left(\frac{\theta'}{\theta_0}\right)^{\frac{n}{2}} exp\left[-\left(\frac{\theta'-\theta}{2\theta_0\theta'}\right)\sum_{i=1}^{n}x_i^2\right] \leq k$$

or, equivalently

$$\sum_{i=1}^{n}x_i^2 \geq \frac{2\theta_0\theta'}{\theta'-\theta_0}\left[\frac{n}{2}\ln\left(\frac{\theta'}{\theta_0}\right) - \ln k\right] = c.$$

It follows from NP-Lemma that the region $C = \{(x_1, x_2, \ldots, x_n): \ \sum_{i=1}^{n}x_i^2 \geq c\}$ is the best rejection region for testing the simple hypothesis $H_0: \ \theta = \theta_0$ against the simple alternative hypothesis $H_1: \ \theta = \theta'$.

Now, we need to determine c, such that the rejection region C is of the desired size α. Under the null hypothesis, the random variable $\frac{1}{\theta_0}\sum_{i=1}^{n}x_i^2$ follows chi-square distribution with n degrees of freedom. Since $\alpha = P\left(\frac{1}{\theta_0}\sum_{i=1}^{n}x_i^2 \geq \frac{c}{\theta_0}\right)$, the constant $\frac{c}{\theta_0}$ may be obtained from Table A.8 or Table A.9 in Appendix, and thus, c can be determined. Then $C = \{(x_1, x_2, \ldots, x_n): \ \sum_{i=1}^{n}x_i^2 \geq c\}$ is the best critical region of size α for testing $H_0: \ \theta = \theta_0$ against the alternative hypothesis $H_1: \ \theta = \theta'$.

Moreover, for each number θ' greater than θ_0, the above argument holds. That is if θ' is the another number greater than θ_0, then $C = \{(x_1, x_2, \ldots, x_n): \ \sum_{i=1}^{n}x_i^2 \geq c\}$ is the best critical region of size α for testing $H_0: \ \theta = \theta_0$ against the alternative hypothesis $H_1: \ \theta = \theta'$. Accordingly, $C = \{(x_1, x_2, \ldots, x_n): \ \sum_{i=1}^{n}x_i^2 \geq c\}$ is a uniformly most powerful critical region of size α for testing $H_0: \ \theta = \theta_0$ against $H_1: \ \theta > \theta_0$. If $(x_1, x_2, \ldots, x_n)$ denotes the experimental value of random sample $(X_1, X_2, \ldots, X_n)$, then $H_0 : \theta = \theta_0$ is rejected at the significance level α, and $H_1 : \theta > \theta_0$ is accepted, if $\sum_{i=1}^{n} x_i^2 \geq c$, otherwise, $H_0 : \theta = \theta_0$ is not to be rejected.

5.1.3 Likelihood Ratio Test

The likelihood ratio test involves finding the critical region by setting an upper bound on the ratio of the maximum value of likelihood under the null hypothesis and the maximum value of the likelihood under all the admissible values. The test rejects the null hypothesis if:

$$\frac{\max(\text{Likelihood under } H_0)}{(\text{Likelihood under } H_0 \text{ and } H_1)} < \text{critical value.}$$

Example 5.4 Consider a random sample $(X_1, X_2, \ldots, X_n)$ from an exponential distribution with parameter λ. Consider

$$H_0 : \lambda = \lambda_0 \quad \text{against} \quad H_1 : \lambda > \lambda_0$$

Determine the likelihood ratio test associated with the test of H_0 against H_1.

Solution:

The likelihood function is

$$L(\lambda; \mathbf{x}) = \prod_{i=1}^{n} f(x_i, \lambda) = \lambda^n e^{-\lambda \sum_i x_i}. \tag{5.1}$$

The maximum of Eq. (5.1) over the null hypothesis is given by $\lambda_0^n e^{-\lambda_0 \sum_i x_i}$. Similarly, the maximum of Eq. (5.1) over H_0 and H_1 is given as follows:

(i) Taking logarithm on both sides of Eq. 5.1, we have $\ln L = n \ln \lambda - \lambda \sum_i x_i$.

(ii) Differentiating with respect to λ and putting equal to zero, we get $\hat{\lambda} = \frac{1}{\bar{x}}$.

(iii) Since $L(\lambda; \mathbf{x})$ is an increasing function for $\lambda < \frac{1}{\bar{x}}$ and decreasing for $\lambda > \frac{1}{\bar{x}}$, we have

$$\max\{L(\lambda; \mathbf{x})\} = \begin{cases} \bar{x}^{-n} e^{-n} & \text{if } \lambda_0 \leq \frac{1}{\bar{x}} \\ \lambda_0^n e^{-\lambda_0 \sum_i x_i} & \text{if } \lambda_0 > \frac{1}{\bar{x}} \end{cases}.$$

(iv) Therefore, the likelihood ratio test is given by

$$
\phi(\mathbf{x}) = \begin{cases} \dfrac{\lambda_0^n e^{-\lambda_0 \sum\limits_{i} x_i}}{\bar{x}^{-n} e^{-n}} & \text{if } \lambda_0 \leq \frac{1}{\bar{x}} \\ 1 & \text{if } \lambda_0 > \frac{1}{\bar{x}} \end{cases} .
$$

Now, let us see some cases wherein all of the above concepts are applied.

5.1.4   Test for the Population Mean

Suppose that $X_1, X_2, \ldots, X_n$ is a random sample from a normal distribution $N(\mu, \sigma^2)$ where μ is unknown and σ^2 is a known quantity. Suppose we want to test the null hypothesis that the mean (μ) is equal to some specified value against the alternative that it is not. That is, we want to test

$$
H_0 : \mu = \mu_0 \quad \text{against} \quad H_1 : \mu \neq \mu_0
$$

for a specified value μ_0. We know from our previous knowledge that the natural point estimator for the population mean is the sample mean

$$
\bar{X} = \frac{\sum\limits_{i=1}^{n} X_i}{n} .
$$

The test statistic in this case is defined as

$$
Z = \frac{\bar{X} - \mu}{\sigma/\sqrt{n}} \sim N(0, 1).
$$

It is reasonable to accept H_0 if $\bar{X}$ is not too far from μ_0. Hence, the critical region is of the form

$$
C = \{(x_1, \ldots, x_n) \mid |\bar{X} - \mu_0| > k\}
$$

where, for given level of significance α, we can determine the value of k as follows:

$$
P\left(|\bar{X} - \mu_0| > k\right) = \alpha.
$$

Under the null hypothesis, $Z = \frac{\bar{X} - \mu_0}{\sigma/\sqrt{n}} \sim N(0, 1)$, so we have

$$
P\left(|Z| > \frac{k\sqrt{n}}{\sigma}\right) = \alpha
$$

which is equivalent to

$$2P\left(Z > \frac{k\sqrt{n}}{\sigma}\right) = \alpha.$$

Since $P(Z > z_{\frac{\alpha}{2}}) = \frac{\alpha}{2}$, we get

$$\frac{k\sqrt{n}}{\sigma} = z_{\frac{\alpha}{2}} \text{ or } k = \frac{z_{\frac{\alpha}{2}}\sigma}{\sqrt{n}}.$$

Now, for a given significance level α, the null hypothesis $H_0 : \mu = \mu_0$ will be accepted when

$$-z_{\frac{\alpha}{2}} < z_0 < z_{\frac{\alpha}{2}}$$

where $z_0 = \frac{\bar{X}-\mu_0}{\frac{\sigma}{\sqrt{n}}}$. Now, we see the following examples.

Example 5.5 Assume that the examination marks of students follow normal distribution with unknown mean (μ) but known variance $(\sigma^2) = 0.25$. Test the hypothesis

$$H_0 : \mu = 8 \quad \text{against} \quad H_1 : \mu \neq 8$$

with the level of significance $\alpha = 0.01$, the sample mean being equal to 7.8 and the sample size being 50.

Solution:

Step 1
$$H_0 : \mu = 8 \quad \text{against} \quad H_1 : \mu \neq 8$$

Step 2 Test statistics:
$$Z = \frac{\bar{x} - \mu_0}{\sigma_0/\sqrt{n}}.$$

Step 3 From observed information with null hypothesis
$$z_0 = \frac{\bar{x} - \mu_0}{\sigma_0/\sqrt{n}} = -2.83.$$

Step 4 For $\alpha = 0.01$, check if

$$-2.57 = -z_{0.005} < z_0 < z_{0.005} = 2.57.$$

Step 5 The null hypothesis is rejected as $z_0 < -z_{0.005}$.

Example 5.6 Suppose that, we have a random sample of 50 elements drawn from a population in which the characteristic X has a normal distribution with standard deviation $\sigma = 5$ and an unknown expected value μ. From the sample of size $n = 50$, we have obtained $\bar{x} = 2$. The problem is to test the hypothesis $H_0 : \mu = 0$ at 95% confidence level.

Solution:

Step 1
$$H_0 : \mu = 0 \quad \text{against} \quad H_1 : \mu \neq 0$$

Step 2   Test statistics:
$$Z = \frac{\bar{x} - \mu_0}{\sigma_0/\sqrt{n}}.$$

Step 3   From observed information with null hypothesis
$$z_0 = \frac{2 - 0}{5/\sqrt{50}} = 2\sqrt{2} = 2.828.$$

Step 4   For $\alpha = 0.05$, check if
$$-1.96 = -z_{0.025} < z_0 < z_{0.025} = 1.96.$$

Step 5   The null hypothesis is rejected as $z_0 > z_{0.025}$.

Example 5.7 A researcher has gathered data on daily returns on crude oil market (Brent Blend) over a recent period of 250-days. The mean daily return has been 0.1%, and the population standard deviation (available from historical data) of daily returns is 0.25%. The researcher believes that the mean daily return of Brent crude oil is not equal to zero. Construct a hypothesis test of the researcher's belief with level of significance $\alpha = 0.1$?

Solution:

The null hypothesis is the one the researcher wants to reject

Step 1
$$H_0 : \mu = 0 \quad \text{against} \quad H_1 : \mu \neq 0$$

Step 2   And the test statistics for the z-test is
$$Z = \frac{\bar{X} - \mu}{\sigma/\sqrt{n}}.$$

Step 3   From observed information with null hypothesis
$$z_0 = \frac{\bar{x} - \mu_0}{\sigma_0/\sqrt{n}} = 6.32.$$

Step 4   Check if
$$-1.64 = -z_{0.05} < z_0 < z_{0.05} = 1.64.$$

Step 5   The null hypothesis is rejected as $z_0 > z_{0.05}$.

Thus, the researcher's hypothesis is true.

Remark 5.1 • The above case can be modified for the situation wherein the variance (σ^2) of the distribution is unknown. The sampling distribution of the test statistic under consideration now is the t-distribution in place of the standard normal distribution and the population variance (σ^2) is replaced by the sample variance (s^2).

$$t = \frac{\bar{X} - \mu}{s/\sqrt{n}}$$

which after substitution of values gives

$$t_0 = \frac{\bar{x} - \mu_0}{s_0/\sqrt{n}}.$$

We reject the null hypothesis when t_0 falls in the critical region. A summary of this hypothesis test is given in Table 5.2 for one-sided test as well. If $n \geq 30$, the t-distribution can be replaced by the standard normal distribution.

• Another modification of the above case is when we have to consider the difference between the two means. Two independent populations are considered with their variances, σ_1^2 and σ_2^2, known. The null hypothesis is defined as

$$H_0 : \mu_1 - \mu_2 = \delta \quad \text{against} \quad H_1 : \mu_1 - \mu_2 \neq \delta.$$

Now the test statistic is

$$Z = \frac{\bar{X} - \bar{Y} - (\mu_1 - \mu_2)}{\sqrt{\frac{\sigma_1^2}{n_1} + \frac{\sigma_2^2}{n_2}}}$$

which after substitution of values with null hypothesis gives

$$z_0 = \frac{\bar{x} - \bar{y} - \delta}{\sqrt{\frac{\sigma_1^2}{n_1} + \frac{\sigma_2^2}{n_2}}}.$$

The null hypothesis is rejected if z_0 falls in the critical region. This hypothesis test is summarized in Table 5.3.

Table 5.2 Hypothesis test of $H_0 : \mu = \mu_0$ against different forms of H_1 when σ^2 is unknown

Alternative hypothesis (H_1)	Reject H_0 if
$\mu \neq \mu_0$	$t_0 < -t_{\alpha/2,n-1}$ or $t_0 > t_{\alpha/2,n-1}$ (Two-sided test)
$\mu > \mu_0$	$t_0 > t_{\alpha,n-1}$ (One-sided test)
$\mu < \mu_0$	$t_0 < -t_{\alpha,n-1}$ (One-sided test)

Table 5.3 Hypothesis test of $H_0 : \mu_1 - \mu_2 = \delta$ against different forms of H_1 when σ^2 is known

Alternative hypothesis (H_1)	Reject H_0 if
$\mu_1 - \mu_2 \neq \delta$	$z_0 < -Z_{\alpha/2}$ or $z_0 > Z_{\alpha/2}$ (Two-sided test)
$\mu_1 - \mu_2 > \delta$	$z_0 > Z_\alpha$ (One-sided test)
$\mu_1 - \mu_2 < \delta$	$z_0 < -Z_\alpha$ (One-sided test)

- We can also modify the above case, when we have to consider the difference between the two means of two independent populations, but with their variances unknown. Thus, the sampling distribution test statistics under consideration would be the t-distribution instead of the standard normal distribution, and the population variance will be replaced by the sample variance. Hence, our test statistics becomes

$$t = \frac{(\bar{X} - \bar{Y}) - (\mu_1 - \mu_2)}{s_p\sqrt{\dfrac{1}{n_1} + \dfrac{1}{n_2}}}$$

where

$$s_p^2 = \frac{(n_1 - 1)s_1^2 + (n_2 - 1)s_2^2}{n_1 + n_2 - 2}.$$

After substitution, we have

$$t_0 = \frac{(\bar{x} - \bar{y}) - \delta}{s_p\sqrt{\dfrac{1}{n_1} + \dfrac{1}{n_2}}}.$$

The null hypothesis is rejected if t_0 falls in the critical region. A summary of this test is given in Table 5.4.

Example 5.8 The average return of a portfolio A of ten stocks was 77%. The average return on a similar portfolio (same stocks but with special weight balancing method) B of ten stocks was 75%. Carry out a statistical test to assess whether the portfolio with special balancing method has the same return as portfolio at 5% level of significance.

Table 5.4 Hypothesis test of $H_0 : \mu_1 - \mu_2 = \delta$ against different forms of H_1 when σ^2 is unknown

Alternative hypothesis (H_1)	Reject H_0 if
$\mu_1 - \mu_2 \neq \delta$	$t_0 < -t_{\frac{\alpha}{2},n_1+n_2-2}$ or $t_0 > t_{\frac{\alpha}{2},n_1+n_2-2}$ (Two-sided test)
$\mu_1 - \mu_2 > \delta$	$t_0 > t_{\frac{\alpha}{2},n_1+n_2-2}$ (One-sided test)
$\mu_1 - \mu_2 < \delta$	$t_0 < -t_{\frac{\alpha}{2},n_1+n_2-2}$ (One-sided test)

You are given that $\sum_{i=1}^{10} x_i^2 = 59420$ and $\sum_{i=1}^{10} y_i^2 = 56390$. Assume that returns are normally distributed and that the variance of the underlying distribution for each portfolio is the same.

Solution:

We are testing: $H_0 : \mu_1 = \mu_2$ against $H_1 : \mu_1 \neq \mu_2$. $T = \frac{\bar{X} - \bar{Y} - (\delta)}{s_p \sqrt{\frac{1}{n_1} + \frac{1}{n_2}}}$, where $s_p^2 = \frac{(n_1-1)s_1^2 + (n_2-1)s_2^2}{n_1+n_2-2}$. It follows $t_{n_1+n_2-2}$. Using the observed values of $n_1 = 10, n_2 = 10$, $\bar{X} = 75$, $\bar{Y} = 77$, and $s_p^2 = 3.873^2$. Then,

$$T_0 = \frac{77 - 75 - (0)}{3.873\sqrt{\frac{1}{10} + \frac{1}{10}}} = 1.15. \tag{5.2}$$

This is less than 1.734 the upper 5% point of the t_{18} distribution. So we have insufficient evidence to reject H_0 at 5% level. Therefore, it is reasonable to conclude that the portfolio with special balancing scheme has same return as portfolio X.

5.1.5 Test for the Variance

Suppose that $X_1, X_2, \ldots, X_n$ is a sample from a normal distribution $N(\mu, \sigma^2)$ where both the parameters μ and σ^2 are unknown. Suppose we want to test the null hypothesis that the variance (σ^2) is equal to some specified value against the alternative that it is not. That is, we want to test

$$H_0 : \sigma^2 = \sigma_0^2 \quad \text{against} \quad H_1 : \sigma^2 \neq \sigma_0^2$$

with the level of significance (α). The test statistic in this case is defined as

$$\frac{(n-1)s^2}{\sigma^2} \sim \chi_{n-1}^2$$

where s^2 is the sample variance obtained from the random sample of size n.

Once the values given in the question are substituted into the test statistic, we obtain

$$\chi_0^2 = \frac{(n-1)s_0^2}{\sigma_0^2}.$$

The null hypothesis will be rejected, if $\chi_0^2 > \chi_{n-1,\alpha/2}^2$ or $\chi_0^2 < \chi_{n-1,1-\alpha/2}^2$.

Remark 5.2 • The above concept can similarly be extended to one-sided tests, a summary of which is shown in Table 5.5.

Table 5.5 Hypothesis test of $H_0 : \sigma^2 = \sigma_0^2$ against different forms of H_1

Alternative hypothesis (H_1)	Reject H_0 if
$\sigma^2 \neq \sigma_0^2$	$\chi_0^2 > \chi_{n-1,\alpha/2}^2$ or $\chi_0^2 < \chi_{n-1,1-\alpha/2}^2$ (Two-sided test)
$\sigma^2 > \sigma_0^2$	$\chi_0^2 > \chi_{n-1,\alpha}^2$ (One-sided test)
$\sigma^2 < \sigma_0^2$	$\chi_0^2 < \chi_{n-1,1-\alpha}^2$ (One-sided test)

- Another variation of the above is discussed below.
  Consider two independent populations with known means (μ_1 and μ_2) but unknown variances (σ_X^2 and σ_Y^2). The null hypothesis is defined as

$$H_0 : \sigma_X^2 = \sigma_Y^2$$

The test statistic in this case is defined as

$$F = \frac{s_X^2/\sigma_X^2}{s_Y^2/\sigma_Y^2}$$

where s_X^2 and s_Y^2 are the sample variance. Once the values are substituted with null hypothesis, we obtain

$$F_0 = \frac{s_X^2}{s_Y^2}.$$

This hypothesis test can be extended to one-sided test also. Table 5.6 summarizes this hypothesis test.

Example 5.9 Chicago Mercantile Exchange is one of the largest futures and options exchanges. It is a dominant venue for the sale of gold futures contract. Historical weekly returns of futures have a standard deviation of 0.4%. A trader wants to verify whether this claim still adequately describes the standard deviation of the contracts. He collected weekly returns data for 24 weeks and measured standard deviation of 0.38%. Determine if the more recent standard deviation is different from the historic standard deviation?

Solution:

The trader wants to test whether the variance has changed, up or down, a two-tailed test should be used. The hypothesis test structure takes the form:

Step 1
$$H_0 : \sigma^2 = 0.16 \quad \text{against} \quad H_1 : \sigma^2 \neq 0.16$$

Step 2   The test statistics is:
$$\chi^2 = \frac{(n-1)s^2}{\sigma^2}.$$

Table 5.6 Hypothesis test of $H_0 : \sigma_X^2 = \sigma_Y^2$ against different forms of H_1

Alternative hypothesis (H_1)	Reject H_0 if
$\sigma_X^2 \neq \sigma_Y^2$	$F_0 < F_{1-\alpha/2,n_1-1,n_2-1}$ or $F_0 > F_{\alpha/2,n_1-1,n_2-1}$
$\sigma_X^2 > \sigma_Y^2$	$F_0 > F_{\alpha,n_1-1,n_2-1}$
$\sigma_X^2 < \sigma_Y^2$	$F_0 < F_{1-\alpha,n_1-1,n_2-1}$

Step 3   There are 23 degrees of freedom. Thus, from the sample data with null hypothesis, we have

$$\chi^2 = 20.7575.$$

Step 4   Let us use 5% level of significance, meaning there will be 2.5% probability in each tail of the distribution. Using Table A.9 in Appendix, decision rule based on the critical region would be reject H_0 if: $\chi^2_{23,0.975} < 11.689$ or $\chi^2_{23,0.025} > 38.076$.

Step 5   Since the test statistics falls between the two critical regions, we fail to reject the null hypothesis that the variance is 0.16%.

Example 5.10 Benchmark crude is the crude oil which serves as a reference price for buyers/sellers of crude oil. The primary benchmarks are WTI, Brent, and OPEC. Mathew is examining two crude oil markets, Brent Blend and West Texas Intermediate (WTI). He suspects that the WTI is more volatile than the Brent. To confirm this suspicion, Mathew has looked at a sample of returns of last 31 days from WTI oil market and 41 days from Brent oil market. He measured the sample standard deviation to be 0.430% for WTI market and 0.380% for Brent. Construct a hypothesis to test Mathew's suspicion?

Solution:

We are concerned whether the variance of returns of WTI is greater than the variance of returns of the Brent oil market (Table 5.6).

Step 1   The test hypothesis is:

$$H_0 : \sigma_1{}^2 \leq \sigma_2{}^2 \quad \text{against} \quad H_1 : \sigma_1{}^2 > \sigma_2{}^2$$

where $\sigma_1{}^2$ is the variance of returns for WTI and $\sigma_2{}^2$ is the variance of returns for the Brent oil market. Note that, this is a one- tailed or one-sided test.

Step 2   For tests of difference between variances, the appropriate test statistics using F-distribution is:

$$F = \frac{s_1{}^2/\sigma_1^2}{s_2{}^2/\sigma_2^2}.$$

Step 3   From the given sample data under the null hypothesis,

$$F_0 = \frac{s_1{}^2}{s_2{}^2} = \frac{(0.438)^2}{(0.380)^2} = 1.2805.$$

Step 4   Let us conduct the hypothesis test at 5% level of significance. Using the
sample sizes from two markets, the critical F-value for this one-sided test is found
to be 1.74. The value is obtained from Table A.13 in Appendix of F-distribution
at 5% level of significance with $n_1 = 30$ and $n_2 = 40$. Hence, the decision rule is
based on reject H_0 if: $F_0 > 1.74$.

Step 5   Since the calculated F-statistics of 1.2805 is less than the critical value
of 1.74; we fail to reject the null hypothesis at 5% level of significance. Mathew
should conclude that the WTI market is not more volatile than Brent Blend market.

5.1.6   Test for the Distribution

In parametric models, one assumes a priori that the distribution has a specific form
with one or more unknown parameters and tries to find out the best or at least reasonably efficient procedures that answer specific questions regarding the parameters. If
the assumptions are violated our procedures might become faulty. One of the interesting problems in statistical inference is to determine whether or not a particular
probabilistic model is appropriate for a given random phenomenon. This problem
often reduces to a problem of testing whether a given random sample comes from a
particular probability distribution (pre-specified partially or completely) or not. For
instance, we may a priori feel that the number of customers coming to a particular
bank is a random sample coming from Poisson distribution. This hypothesis or claim
can be tested by observing the number of people coming to the bank over a period
of a month or few months and taking a decision based on the observations that the
underlying distribution is Poisson or not.

Definition 5.2 (*Goodness of Fit*) Testing the significance of the discrepancy between
the experimental values and theoretical values obtained is called testing for Goodness
of Fit.

In other words, this is the test of the null hypothesis that a set of data comes from a
population having a given distribution against the alternative that the population has
some other distribution.

Remark 5.3   1.   Goodness of Fit is a **nonparametric test**.
   2.   Goodness of Fit test can be applied to both the cases, i.e., when population
        distribution is discrete or continuous.

The data given in the question are in tabulated form. The recorded samples form
Groups. For instance, a sample of 150 light bulbs, their burning time was tabulated
as given in Table 5.7.

Table 5.7 Example 5.11 Life of light bulbs

Burning time	# of Light bulbs (having burning time in the specified range)
0–100	47
100–200	40
200–300	35
≥ 300	28

Mathematical Analysis

Consider the test statistic

$$D^2 = \sum_{i=1}^{k} \frac{(O_i - E_i)^2}{E_i} \sim \chi^2_{k-1}$$

where

$O_i \rightarrow$ observed frequency

$E_i \rightarrow$ theoretical (expected frequency) asserted by the null hypothesis

$n \rightarrow$ total number of observations or sample size

$k \rightarrow$ total number of groups such that $n_1 + n_2 + \cdots + n_k = n$

$P_i \rightarrow$ theoretical probability of occurrence of the event associated with the ith group, $i = 1, 2, \ldots, k$

$E_i = nP_i, i = 1, 2, \ldots, k$.

Steps to Solve the Question

1. H_0: Data follow the given distribution
   H_1: Data do not follow the given distribution.
2. Construct the test statistic
3. Compute the value of the test statistic under the null hypothesis D_0^2.
4. Find the tabled value $\chi^2_{k-1,\alpha}$.
5. Compare the value of the test statistic D_0^2 and the tabled value $\chi^2_{k-1,\alpha}$. If, $D_0^2 > \chi^2_{k-1,\alpha}$, then reject H_0.

Remark 5.4 Few points to be remembered are as follows:

- Construct k groups such that each $nP_i \geq 5$; where $i = 1, \ldots, k$.
  If $nP_i < 5$, then merge the groups where $nP_i < 5$ with other groups.
- D^2 follows χ^2 distribution with $k - 1$ degrees of freedom. This is so because, we have already utilized one information $\sum_{i=1}^{k} n_i = n$.
- If 'r' parameters '$\theta = (\theta_1, \theta_2, \ldots, \theta_n)$' of the distribution of the underlying population are unknown, then estimate these parameters using MLE or Method of Moments.
  In this case, if $D_0^2 > \chi^2_{k-1-r,\alpha}$, then reject H_0.

Example 5.11 Suppose we believe that the life length 'T' (in hours) of light bulb is exponentially distributed with parameter $\lambda = 0.005$. We obtain a sample of 150 bulbs, test them and record their burning time say $T_1, T_2, T_3, \ldots, T_{150}$ and tabulate it as given in Table 5.7. Test the hypothesis that the data represent a random sample from an exponential distribution with $\lambda = 0.005$ at level of significance, $\alpha = 0.01$.

Solution:

Let us consider following hypothesis testing.
H_0 : Given data follow an exponential distribution,
H_1 : Given data do not follow an exponential distribution.
Furthermore, it is given that $n = 150$, $\lambda = 0.005$. With reference to Table 5.8, we have

$$P_1 = \int_0^{100} \lambda e^{-\lambda x} dx = 0.39, \quad P_2 = \int_{100}^{200} \lambda e^{-\lambda x} dx = 0.24,$$

$$P_3 = \int_{200}^{300} \lambda e^{-\lambda x} dx = 0.14, \quad P_4 = \int_{300}^{\infty} \lambda e^{-\lambda x} dx = 0.22.$$

Degrees of freedom $= 4 - 1 = 3$. From Table A.9 in Appendix, $\chi^2_{3,0.01} = 11.341$. As $D_0^2 > \chi^2_{3,0.01}$, H_0 is to be rejected.

Example 5.12 The number of computer malfunctions per day, recorded for 260 days, is listed in Table 5.9. Test the hypothesis that this data follow a Poisson distribution at 5% level of significance.

Table 5.8 Example 5.11 Calculations

Burning Time	O_i (Number of light Bulbs)	$E_i = n P_i$	$(O_i - E_i)^2$	$\frac{(O_i - E_i)^2}{E_i}$
0–100	47	58.5	132.25	2.26
100–200	40	36	16	0.44
200–300	35	21	196	9.33
≥300	28	33	25	0.76
	150			$D_0^2 = 12.79$

Table 5.9 Example 5.12 Observed frequencies of computer malfunctions

Number of malfunctions (x_i)	0	1	2	3	4	5
Number of days (f_i)	77	90	55	30	5	3

Solution:

A Poisson distribution with parameter λ, which is same as mean. Hence, we shall use the sample data (sample mean) to estimate λ. Here, $n = 260$.

$$\bar{x} = \hat{\lambda} = \frac{\sum\limits_{i=1}^{6} f_i x_i}{\sum\limits_{i=1}^{6} f_i} = \frac{325}{260} = 1.25.$$

$$\therefore \hat{\lambda} = 1.25.$$

If $X \sim P(\lambda)$, then $P(X = x) = \frac{e^{-\lambda}\lambda^x}{x!}$, $x = 0, 1, 2, 3 \ldots$.
Consider Table 5.10 for initial calculations. Since a Poisson distribution is valid for all positive values of x, this additional grouping is necessary. $P(X \geq 6) = 1 - P(X \leq 5)$. Since E_i for groups $x_i = 5$ and $x_i \geq 6$ is less than 5, they must be merged with $x_i = 4$, to create a new group, namely $x_i \geq 4$ as shown in Table 5.11.

Step 1 The test hypothesis is: H_0 : Number of daily malfunctions follows Poisson distribution

H_1 : Number of daily malfunctions does not follow Poisson distribution

Step 2 $D^2 = \sum\limits_{i=1}^{k} \frac{(O_i - E_i)^2}{E_i} \sim \chi^2_{k-1} \sim \chi^2(n-1)$.

Step 3 From the sample data with null hypothesis, $D_0^2 = 2.153$.

Step 4 Significance level $\alpha = 0.05$. There are 5 groups. $k = 5$ (refer the above Table 5.11). Hence, degrees of freedom is: $k - 1 - r = 5 - 1 - 1 = 3$. From Table A.9 in Appendix, we get $\chi^2_{3,0.05} = 7.815$.

Step 5 As $D_0^2 \not> \chi^2_{3,0.05}$, H_0 is accepted.

There is no evidence, at 5% level of significance to suggest that the number of computer malfunctions per day does not have a Poisson Distribution.

Table 5.10 Example 5.12 Initial calculations

x_i	$P(X = x_i)$	$E_i = nP_i$
0	0.2865	74.5
1	0.3581	93.1
2	0.2238	58.2
3	0.0933	24.2
4	0.0291	7.6
5	0.0073	1.9
$\geq 6^*$	0.0019	0.5
	1.0000	260.0

Table 5.11 Example 5.12 Calculations

x_i	$O_i = f_i$	$P(X = x_i)$	$E_i = nP_i$	$(O_i - E_i)^2$	$\frac{(O_i - E_i)^2}{E_i}$
0	77	0.2865	74.5	6.25	0.084
1	90	0.3581	93.1	9.61	0.103
2	55	0.2238	58.2	10.24	0.176
3	30	0.0933	24.2	33.64	1.390
≥ 4	8	0.0383	10.0	4.00	0.400
		1.0000			$D_0^2 = 2.153$

5.1.7 Testing Regarding Contingency Tables

Contingency tables provide an easy method to determine the dependency between two variables. The following example should make it clear.

Example 5.13 Fruit trees are subject to a bacteria-caused disease commonly called Blight. One can imagine different treatments for this disease: *Treatment A*: No Action (A Control Group), *Treatment B*: careful removal of affected branches, and *Treatment C*: frequent spraying of the foliage with an antibiotic in addition to careful removal of affected branches. One can also imagine different outcomes from the disease—*Outcome 1*: tree dies in the same year as the disease was noticed, *Outcome 2*: Tree dies 2–4 years after disease was noticed, *Outcome 3*: Tree survives beyond 4 years.

A group of '*N*' trees are classified into one of the treatments and over the next few years the outcome is recorded. If we count the number of trees in a particular treatment/outcome pair, we can display the results in a **Contingency Table** 5.12.

We can now use a χ^2 test to determine whether or not, the two variables treatments and outcomes, are independent.

Mathematical Analysis

- H_0: The variables are independent.
  H_1: The variables are dependent in some way.
  Let e_{ij} represent the expected frequency for the cell in the ith row and jth column. Then, as the variables are independent, we have $e_{ij} = \frac{r_i c_j}{N}$. In words, expected frequency $= \frac{\text{(Row Total)(Column Total)}}{\text{Total Sample Size}}$.

Table 5.12 Contingency Table

Outcome	Treatment A	Treatment B	Treatment C	Row totals
1	#A1	#B1	#C1	total1
2	#A2	#B2	#C2	total2
3	#A3	#B3	#C3	total3
Column totals	Total A	Total B	Total C	Grand total

- Consider the test statistic $D^2 = \sum_{i=1}^{r} \sum_{j=1}^{c} \frac{(n_{ij} - e_{ij})^2}{e_{ij}} \sim \chi^2_{(r-1)(c-1)}$.

  To compute the test statistic D^2 for each cell, take the difference between the observed and the expected frequency and square it, and divide it by the expected frequency.
- The data given in Table 5.12 are the observed frequency.
- We now need to compute the expected frequency. As expected, we mean the expected frequency if the variables are independent. For this purpose, we define the following:

  - r: total number of rows,
  - c: total number of columns,
  - N: total number of observations,
  - n_{ij}: observed count for the cell in row i and column j,
  - r_i: total for row i,
  - c_j: total for column j.

- Reject H_0 if, $D_0^2 > \chi^2_{(r-1)(c-1),\alpha}$, where α is the level of significance.

Example 5.14 A student is interested in determining whether there is a relationship between gender and the major course at her college or not. She randomly sampled some men and women on campus and asked them if their major course was Natural Science (NS), Social Science (SS), or Humanity (H). Her results appear in the Table 5.13. Test at the 5% level of significance, if there is a relationship between the gender and the major course at college.

Solution:

Consider following hypothesis:

H_0: The variables gender and major at college are independent.
H_1: The variables are dependent in some way.

The computation of row and column totals, is shown in Table 5.14. Computation of expected frequency using the formula $e_{ij} = \frac{r_i c_j}{N}$. (e.g., $e_{1,1} = \frac{(34)(21)}{57}$) is shown in Table 5.15. The expected frequencies are mentioned in brackets. Now, computing the test statistic $D_0^2 = 2.229$. Since, $r = 2, c = 3$, df= $(r - 1)(c - 1) = 2$. Given $\alpha = 0.05$, from Table A.9 in Appendix, $\chi^2_{2,0.05} = 5.99$. As $D_0^2 \ngtr \chi_{2,0.05}$, H_0 is rejected.

Table 5.13 Example 5.14 Major course at college

	NS	SS	H
Women	10	14	10
Men	11	8	4

Table 5.14 Example 5.14 Computation of row and column totals

		j = 1 NS	j = 2 SS	j = 3 H	Total
i = 1	Women	10	14	10	34
i = 2	Men	11	8	4	23
	Total	21	22	14	57

Table 5.15 Example 5.14 Computation of expected frequencies

		j = 1 NS	j = 2 SS	j = 3 H	Total
i = 1	Women	10 (12.526)	14 (13.123)	10 (8.35)	34
i = 2	Men	11 (8.474)	8 (8.877)	4 (5.649)	23
	Total	21	22	14	57

Table 5.16 Example 5.15 Contingency table

Gender	Bar in shopping complexes			
	Favor	Oppose	No opinion	Total
Male	250	229	15	494
Female	301	196	9	506
Total	551	425	24	1000

Example 5.15 In a survey of 1000 individuals, they are asked whether they are in favor of, oppose, or have no opinion on a complete ban on smoking at public places. The data are further classified on the basis of gender and is given in Table 5.16. Test the null hypothesis that gender and opinion on bar in shopping complexes are independent at 5% confidence level.

Solution:

Consider the following hypothesis testing

H_0 : The variables gender and opinion on bar in shopping complexes are independent.

H_1 : The variables are dependent in some way

We compute the expected frequency using the formula $e_{ij} = \dfrac{r_i c_j}{N}$.

	Gender	j = 1 Favor	j = 2 Oppose	j = 3 No opinion	Total
i = 1	Male	250	229	15	494
i = 2	Female	301	196	9	506
	Total	551	425	24	1000

We compute the test statistic D^2 as

$$D^2 = \sum_{i=1}^{2} \sum_{j=1}^{3} \frac{(n_{ij} - e_{ij})^2}{e_{ij}}.$$

This follows $\chi^2_{(2-1)(3-1)}$, i.e., χ^2_2,

$$D_0^2 = 1.81 + 1.729 + 0.834 + 1.767 + 1.688 + 0.814 = 8.642.$$

From Table A.9 in Appendix, we have $\chi^2_{2,0.05} = 5.99$. As $D_0^2 > \chi^2_{2,0.05}$, H_0 is to be accepted.

5.1.8 Test Regarding Proportions

In many situations, one is interested to test what fraction of total satisfies some condition or not. For these kind of situations, tests of hypothesis concerning proportions are required. For instance, a candidate competing in next elections is interested in knowing that what proportion of the voters will vote in favor of him/her in the upcoming elections. Another example is that all manufacturing industries are concerned regarding the fraction of defective items in a particular shipment. We shall consider the problem of hypothesis testing that the proportion of successes in a random experiment whose distribution is binomially distributed with some specified parameter value. We are testing the null hypothesis $H_0 : p = p_0$, where p is the parameter of the binomial distribution. The alternative hypothesis may be one of the usual one-sided or two-sided alternatives: $H_1 : p < p_0, p > p_0$, or $p \neq p_0$. The appropriate random variable, denoted by X, on which we base the decision criterion follows $B(n, p)$ and the statistic is $p = X/n$. Note that, it is highly unlikely that one can obtain a critical region of size exactly equal to a pre-specified value of α. Therefore, it is preferable to base the decision on P-values when dealing with samples of small size. To test the hypothesis $H_0 : p = p_0$ against $H_1 : p \neq p_0$ at the α level of significance, we compute the P-value

$$P = 2P(X \leq x \text{ when } p < p_0) \text{ if } x < np_0 \text{ or } P = 2P(X \geq x \text{ when } p < p_0) \text{ if } x > np_0.$$

Reject H_0 if the computed P-value is less than or equal to α.
For one-sided test $H_1 : p < p_0$, we compute

$$P = P(X \leq x \text{ when } p = p_0).$$

Reject H_0 if the computed P-value is less than or equal to α. Similarly, for one-sided test $H_1 : p > p_0$, we compute

$$P = P(X > x \text{ when } p = p_0).$$

Reject H_0 if the computed P-value is less than or equal to α.

Remark 5.5 • When p_0 is not given, p_0 can be estimated from the sample data.
• The binomial probabilities were obtainable from the exact binomial formula or from Table A.2 in Appendix when n is small.
• For large n, approximation procedure is required. When p_0 is very close to 0 or 1, the Poisson distribution may be used. However, approximation into the normal distribution is usually preferred for large n and is very accurate as long as p_0 is not extremely close to 0 or 1.

If we use the normal approximation for large n to test the hypothesis of proportion, consider the test statistic

$$\frac{X - np}{\sqrt{np(1 - p)}} \sim N(0, 1).$$

Substituting the values with null hypothesis, we have

$$z_0 = \frac{x - np_0}{\sqrt{np(1 - p)}}$$

where 'x' is the number of cases favoring out of n the null hypothesis. α is the level of significance. A summary of this test is given in Table 5.17.

Example 5.16 A person complains that out of the total calls he receives at least 45% of the calls are regarding the promotions of some products. To check this claim, a random sample of his 200 calls was selected from his call history. If 70 of these calls are found to be regarding promotional offers, is the person's claim believable at 5% level of significance?

Solution:

Step 1 $H_0 : p \geq 0.45$ against $H_1 : p < 0.45$.
Step 2   Test statistics $Z = \frac{x - np}{\sqrt{np(1-p)}}$.
Step 3   Here, $p_0 = 0.45$, $n = 200$ and $x = 70$. Hence,

$$z_0 = \frac{x - np_0}{\sqrt{np_0(1 - p_0)}} = \frac{70 - (200)(0.45)}{\sqrt{200(0.45)(0.55)}} = -2.842.$$

Table 5.17 Testing of $H_0 : p = p_0$ against H_1

If H_1	Reject H_0 if	
$p \neq p_0$	$z_0 < -z_{\frac{\alpha}{2}}$ or $z_0 > z_{\frac{\alpha}{2}}$	Two-sided test
$p > p_0$	$z_0 > z_\alpha$	One-sided test
$p < p_0$	$z_0 < -z_\alpha$	One-sided test

Step 4   Given $\alpha = 0.05$. From Table A.7 in Appendix, we have $z_{0.025} = 1.96$.
Step 5   As $z_0 < -z_{0.025}$, H_0 is to be rejected.

Situations often arise where we wish to test the hypothesis that two proportions are equal. For example, a vote is to be taken among the students of two lecture classes of the same subject to determine whether a proposed surprise quiz should be conducted. In general, we wish to test the null hypothesis that two proportions are equal. The testing of hypothesis is given by

$$H_0 : p_1 = p_2; \quad H_1 : p_1 < p_2, \; p_1 > p_2, \text{ or } p_1 \neq p_2.$$

In this case, independent samples of size n_1 and n_2 are selected at random from two binomial distributed populations. For large n_1 and n_2, by using the normal approximation, the test statistic becomes

$$Z = \frac{\frac{X_1}{n_1} - \frac{X_2}{n_2}}{\sqrt{p(1-p)\left(\frac{1}{n_1} + \frac{1}{n_2}\right)}}.$$

When H_0 is true, we can substitute $p_1 = p_2 = p$ (say), then

$$z_0 = \frac{x_2/n_1 - x_2/n_2}{\sqrt{p(1-p)(1/n_1 + 1/n_2)}}.$$

When p is not given, upon pooling the data from both samples, the pooled estimate of the proportion p is $\hat{p} = \frac{x_1 + x_2}{n_1 + n_2}$ where x_1 and x_2 are the number of successes in each of the two samples. The critical regions for the appropriate alternative hypothesis are set up as before using critical points of the standard normal curve.

When, we have k populations $(k > 2)$ H_0: $p_1 = p_2 = p_3 \cdots = p_k = p_0$(say)
H_1: at least two of these proportions are not equal
For the ith sample, we have the statistic

$$Z_i = \frac{X_i - n_i p_i}{\sqrt{n_i p_i q_i}} \sim N(0, 1), \; i = 1, 2, 3, \ldots, k; \; q_i = 1 - p_i.$$

Now, consider the test statistic

$$D^2 = \sum_{i=1}^{k} \left(\frac{X_i - n_i p_i}{\sqrt{n_i p_i q_i}}\right)^2 \sim \chi_k^2, \; i = 1, \ldots, k.$$

Note that, $Z_1^2 + Z_2^2 + \cdots + Z_k^2 \sim \chi_k^2$.
n_i: sample size of the ith population
x_i: number of cases of the ith population favoring the null hypothesis.
In this case, reject H_0 if $D_0^2 < \chi_{k,1-\frac{\alpha}{2}}^2$ or $D_0^2 > \chi_{k,\frac{\alpha}{2}}^2$.

Table 5.18 Example 5.17 Test of quality of tire brand

	x_i	$p_i = \widehat{p}$	$n_i p_i$	$n_i p_i q_i$	$(x_i - n_i p_i)^2$	$\frac{(x_i - n_i p_i)^2}{n_i p_i q_i}$
$i = 1$; Brand A	26	$\frac{3}{25}$	24	21.12	4	0.189
$i = 2$; Brand B	23	$\frac{3}{25}$	24	21.12	1	0.047
$i = 3$; Brand C	15	$\frac{3}{25}$	24	21.12	81	3.835
$i = 4$; Brand D	32	$\frac{3}{25}$	24	21.12	64	3.030
						$D_0^2 = 7.101$

Remark 5.6 If p_0 is not given, then it is estimated using the sample values of 'k' populations as follows:

$$\widehat{p_0} = \frac{x_1 + x_2 + \cdots + x_k}{n_1 + n_2 + \cdots + n_k}.$$

This is an unbiased point estimator. In this case, reject H_0 if $D_0^2 < \chi^2_{k-1, 1-\frac{\alpha}{2}}$ or $D_0^2 > \chi^2_{k-1, \frac{\alpha}{2}}$. One degree of freedom is reduced because we have estimated one parameter, i.e., p_0, using the samples.

Example 5.17 If 26 of 200 tires of Brand A failed to last 20,000 miles, while the corresponding figures for 200 tires each of Brand B, C, and D were 23, 15, and 32. Use the level of significance $\alpha = 0.05$ to test the null hypothesis that there is no difference in the quality of the four kinds of tires.

Solution:

Consider the following hypothesis testing.

$H_0: p_1 = p_2 = p_3 = p_4 = p$ (say) H_1: at least two of these proportions are not equal. As the value of p is not given in the question, we shall estimate it.
$\hat{p} = \frac{26+23+15+32}{800} = \frac{3}{25}$
$n_i = 200$; $i = 1, \ldots, 4$

Consider Table 5.18 for detailed calculations. Here $k = 4$ and $\alpha = 0.05$. This is a two-sided test. From Table A.9 in Appendix, we get $\chi^2_{3,0.025} = 9.348$. As $D_0^2 < \chi^2_{3,0.025}$ so H_0 is not to be rejected.

5.2 Nonparametric Statistical Tests

In all the statistical methods discussed so far, it was pre-assumed that a set of observations and some partial information about the probability distribution of the observations is known. Under the assumption that the given information is true, methodologies have been discussed to analyze the data and draw some meaningful inferences about the unknown aspects of the probability distribution. So, these methods are based on the distribution assumption of the observations which are assumed to be known exactly. However, in practice, the functional form of the distribution is seldom known. It is therefore desirable to devise methods that are free from distribution assumption. In this section, we consider a few such methods known as nonparametric methods.

5.2.1 Sign Test

Recall that for a continuous random variable X; the median is the value m such that 50% of the times X lies below m and 50% of the times X lies above m.

Let $X_1, X_2, \ldots, X_n$ be i.i.d. random variables with common pdf f. Consider the hypothesis testing problem

$$H_0 : m = m_0$$

against any of the following alternate hypothesis

$$H_1 : m \neq m_0 \text{ or } m > m_0 \text{ or } m < m_0.$$

Now, consider $X_i - m_0$ for $i = 1, 2, \ldots, n$. Under the null hypothesis, i.e., $m = m_0$, we expect about 50% of the values $x_i - m_0$ to be greater than 0 and rest 50% to be less than 0.

On the other hand, if $m > m_0$ is true, then we expect more than 50% of the values $x_i - m_0$ to be greater than 0 and fewer than 50% to be less than 0. Similarly, for the other case $m < m_0$ (Table 5.19).

This analysis of $X_i - m_0$ under the three situations $m = m_0, m > m_0$, and $m < m_0$ forms the basis of the sign test for a median. Thus, perform the following steps.

1. For $i = 1, 2, \ldots, n$, calculate $X_i - m_0$.
2. Obtain the number of terms with negative signs and call it N^-.
3. Obtain the number of terms with positive sign and call it N^+.

Then, if the null hypothesis is true, i.e., $m = m_0$, then both the variables N^- and N^+ follow a binomial distribution with parameters n and $p = \frac{1}{2}$, i.e.,

$$N^- \sim B\left(n, \frac{1}{2}\right) \quad \text{and} \quad N^+ \sim B\left(n, \frac{1}{2}\right).$$

where α is the level of significance. The rejection region for two-sided and one-sided tests are given in Table 5.19.

Table 5.19 Testing of $H_0 : m = m_0$ against H_1

If H_1	Reject H_0 if	
$m \neq m_0$	$2P(N_{min} \leq \min\{n^-, n^+\}) = p - \text{value} < \alpha$	two-sided test
$m > m_0$	$P(N^- \leq n^-) = p - \text{value} < \alpha$	one-sided test
$m < m_0$	$P(N^+ \leq n^+) = p - \text{value} < \alpha$	one-sided test

Example 5.18 Historical evidences suggested that the median length of time between the arrival of two buses at bus stand was 22 min. It is believed that the median length of time is shorter than 22 min. A random sample of 20 observations is obtained as follows: Based on these data, is there sufficient evidence to conclude that the

9.4	13.4	15.6	16.2	16.4	16.8	18.1	18.7	18.9	19.1
19.3	20.1	20.4	21.6	21.9	23.4	23.5	24.8	24.9	26.8

median is shorter than 22 min at 95% level of significance?

Solution:

We are interested in testing the hypothesis

$$H_0 : m = 22 \ \text{ against } \ H_1 : m < 22.$$

In order to perform the test, we calculate $x_i - 22$, for $i = 1, 2, \ldots, 20$. We observe that number of positive signs are 5. Therefore, the p-value of the test is given by $P(N^+ \leq 5) = 0.0207$ where $N^+ \sim B(20, 0.5)$. Since $0.0207 < 0.05$, we reject the null hypothesis.

5.2.2 Median Test

To test whether the median of two populations are same or not, median test is used. Consider the hypothesis testing problem $H_0 : m_1 = m_2$ against any of the following alternate hypothesis $H_1 : m_1 \neq m_2$. Based on the equality, this test can also be used to conclude whether the two samples come from the same population or not.

Let $X_1, X_2, \ldots, X_m$ and $Y_1, Y_2, \ldots, Y_n$ be two samples of size m and n. This test is based on the 2×2 contingency table as follows:

1. Median M of the combined sample of size $m + n$ is obtained.
2. Number of observations below M and above M for each sample is determined.
3. Then, the following contingency table is analyzed.

	No. of observations in Sample 1	No. of observations in Sample 2	Total
Above Median	a	b	$a + b$
Below Median	c	d	$c + d$
	$a + c = m$	$b + d = n$	$a + b + c + d = N$

Test statistic $= \frac{(ad-bc)^2 N}{(a+c)(b+d)(a+b)(c+d)}$ follows χ^2 distribution with 1 degree of freedom. Thus, decision is to reject the null hypothesis if observed value of test statistics is greater than or equal to the critical value, i.e., $\chi^2 > \chi^2_{1,\alpha}$. It should be noted here that in case of ties, N is adjusted accordingly.

Table 5.20 Example 5.19 Table for two polpulations

x	31.8	32.8	39.2	36.0	30.0	34.5	37.4
y	35.5	27.6	21.3	24.8	36.7	30.0	

Example 5.19 Consider the two populations given in Table 5.20. Examine if the two data sets come from the same population or not using median test at 95% level of significance.

Solution:

Consider the combined sample, we obtain 32.8 as the median, i.e., $M = 32.8$. Since, there is tie, we will consider 12 observations for contingency table. The value of test statistics is given by

	No. of observations in x	No. of observations in y	Total
Above M	4	2	6
Below M	2	4	6
	6	6	12

$$\frac{12(16-4)^2}{6.6.6.6} = 1.33.$$

The critical value $\chi^2_{1,0.05} = 3.84$. Since $1.33 < 3.84$, therefore, do not reject the null hypothesis. Thus, we can conclude that both samples come from the same population.

5.2.3 Kolmogorov Smirnov Test

Consider a Goodness of Fit problem where we want to test the hypothesis that the sample comes from a specified distribution F_0 against the alternative that it is from some other distribution F, where $F(x) \neq F_0(x)$ for some $x \in \mathbb{R}$.

Let $X_1, X_2, \ldots, X_n$ be a sample from distribution F, and let F_n^* be a corresponding empirical distribution function. The statistic

$$D_n = \sup_x \| F_n^*(x) - F(x) \|$$

is called the two-sided Kolmogorov[3]–Smirnov statistic. Similarly, one-sided Kolmogorov–Smirnov statistics are given as

---

[3] Andrey Nikolaevich Kolmogorov (1903–1987) was a twentieth century Russian mathematician who made a significant contribution to the mathematics of probability theory. Kolmogorov was the receipient of numerous awards and honors including Stalin Prize (1941), Lenin Prize (1965), Wolf Prize (1980), and Lobachevsky Prize (1987).

$$D_n^+ = \sup_x [F_n^*(x) - F(x)],$$

and

$$D_n^- = \sup_x [F(x) - F_n^*(x)].$$

The exact distribution of D_n for selected values of n and α has been tabulated by Miller (1956) and Owen (1962). Let $D_{n,\alpha}$ be the upper α-percent point of the distribution of D_n, that is $P(D_n > D_{n,\alpha}) \leq \alpha$. This can be seen from the Table A.14 in Appendix for different α and n. Similarly, the critical values of $D_{n,\alpha}^+$ are also available for selected value of n and α.

To test $H_0 : F(x) = F_0(x)$ for all x at level α, the KS test rejects H_0 if $D_n > D_{n,\alpha}$. Similarly, it rejects $F(x) \geq F_0(x)$ for all x if $D_n^- > D_{n,\alpha}^+$ and rejects $F(x) \leq F_0(x)$ for all x at level α if $D_n^+ > D_{n,\alpha}^+$.

Example 5.20 Consider the data arranged in ascending order given below

$$-0.9772, -0.8027, -0.3275, -0.2356, -0.2016, -0.1601, 0.1514,$$
$$0.2906, 0.3705, 0.3952, 0.4634, 0.6314, 1.1002, 1.4677, 1.9352.$$

Test whether the data come from standard normal distribution or not at the significance level 0.01.

Solution:

x	$F_0(x)$	$F_{15}^*(x)$	$i/15 - F_0(x_i)$	$F_0(x_i) - (i - 1)/15$
-0.9772	0.1642	$\frac{1}{15}$	-0.0976	0.1642
-0.8027	0.2111	$\frac{2}{15}$	-0.0777	0.1444
-0.3275	0.3716	$\frac{3}{15}$	-0.1716	0.2383
-0.2356	0.4068	$\frac{4}{15}$	-0.1401	0.2068
-0.2016	0.4201	$\frac{5}{15}$	-0.0868	0.1534
-0.1601	0.4364	$\frac{6}{15}$	-0.0364	0.1031
0.1514	0.5601	$\frac{7}{15}$	-0.0935	0.1602
0.2906	0.6143	$\frac{8}{15}$	-0.0810	0.146
0.3705	0.6444	$\frac{9}{15}$	-0.0444	0.1112
0.3952	0.6536	$\frac{10}{15}$	0.0130	0.0536
0.4634	0.6785	$\frac{11}{15}$	0.0548	0.0118
0.6314	0.7361	$\frac{12}{15}$	0.0639	0.0028
1.1002	0.8644	$\frac{13}{15}$	0.0023	0.0644
1.4677	0.9289	$\frac{14}{15}$	0.0044	0.0622
1.9352	0.9735	$\frac{15}{15}$	0.0265	0.0402

$D_{15}^- = 0.0639$, $D_{15}^+ = 0.2383$, $D_{15} = 0.2383$. Let the significance level be $\alpha = 0.01$, then $D_{15,0.01} = 0.404$. Since $0.2383 < 0.404$, we cannot reject H_0 at 0.01 level.

5.2.4 Mann Whitney Wilcoxon U Test

Suppose, we have two samples $\{x_1, \ldots, x_m\}$ and $\{y_1, \ldots, y_n\}$ of size m and n, respectively, from two groups from two populations. The Mann Whitney test compares each observation in one sample to each observation in the other sample. The total number of pairwise comparisons is, therefore, mn.

If both the samples have same median, there is equal probability that X_i is greater or smaller than Y_j for each pair (i, j). Thus, we have

$$H_0 : P(X_i > Y_j) = 0.5 \quad \text{against} \quad H_1 : P(X_i > Y_j) \neq 0.5$$

The test statistic is the number of values of X_i, $X_2, \ldots, X_m$ that are smaller than each of $Y_1, Y_2, \ldots, Y_n$, i.e, we count the number of times x_i is less than a y_j. The statistic U is called the Mann-Whitney statistic where

$$U = \sum_{i=1}^{m} \sum_{j=1}^{n} T(X_i, Y_j)$$

where $T(X_i, Y_j)$ is defined as

$$T(X_i, Y_j) = \begin{cases} 1, & \text{if } X_i < Y_j \\ 0, & \text{if } X_i \geq Y_j \end{cases}.$$

For larger values of m, n, i.e., $(m, n > 8)$, the distribution of U is approximated by a normal random variable, i.e., under H_0, we have

$$\frac{\frac{U}{mn} - \frac{1}{2}}{\sqrt{\frac{(m+n+1)}{12mn}}} \to N(0, 1)$$

such that $\frac{m}{(m+n)} \to$ constant.

Example 5.21 Two samples are as follows:

$$\text{Values of } X_i : 1, 2, 3, 5, 7, 9, 11, 18$$

$$\text{Values of } Y_j : 4, 6, 8, 10, 12, 13, 14, 15, 19$$

Test the hypothesis at 95% level of significance that the two samples come from the same population.

Solution:

Thus $m = 8, n = 9$, and $U = 3 + 4 + 5 + 6 + 7 + 7 + 7 + 7 + 8 = 54$. Let us apply the normal approximation. We have the value of test statistics as

$$z_0 = \frac{\frac{54}{8.9} - \frac{1}{2}}{\sqrt{\frac{(8+9+l)}{12.8.9}}} = 1.732.$$

From Table A.7 in Appendix, $P(Z > 1.732) = 0.042$. Since $0.042 < 0.05$, we reject the null hypothesis.

5.3   Analysis of Variance

Consider the following example to understand why analysis of variance is an important hypothesis test which is different from the other tests. An institute is planning to purchase, in quantity, one of the four different tutorial packages designed to teach a particular programming language. Some faculties from the institute claim that these tutorial packages are basically indifferent in the sense that the one chosen will have little effect on the final competence of the students. To test this hypothesis, the institute decided to choose 100 of its students and divided them randomly into four groups of size 25 each. Each member in ith group will then be given ith tutorial package, $i = 1, 2, 3, 4$, to learn the new language. Once the semester is over, a comprehensive exam will be taken. The institute wants to analyze the results of this exam to determine whether the tutorial packages are really indifferent or not. How can they do this?

As the students are distributed randomly into four groups, it is probably reasonable to assume that the score of a student in the exam should follow approximately a normal distribution with parameters that depend on the tutorial package from which he was taught. Also, it is reasonable to assume that the average score of a student will depend on the tutorial package he/she was exposed to, whereas the variability in the test score will result from the inherent variation of 100 students and not from the particular package used.

Analysis of variance or ANOVA is a technique that can be used to check such hypothesis.

Definition 5.3 ANOVA is a technique to test a hypothesis regarding the equality of two or more populations (or treatments) means by analyzing the variances of the observed samples. It allows us to determine whether the difference between the samples are simply due to random errors or due to systematic treatment effects that cause the mean in one group to differ from the mean in another case.

Theory and Mathematical Analysis

Suppose that we have been provided samples of size 'n' from 'k' distinct populations or groups and that we want to test the hypothesis that the 'k' population means are equal. $H_0 : \mu_1 = \mu_2 = \mu_3 = \cdots \cdots = \mu_k$

H_1: At least one pair of sample means is significantly different.

In the ANOVA test, we want to compare between-group variance (BGV— Variation between respective group means) and within-group variance (WGV— Variance of a group/sample).

Let group mean $\bar{x}_i = \frac{x_{i1} + x_{i2} + x_{i3} + \cdots + x_{in}}{n}$ $i = 1, 2, \ldots, k$ (These are group means for each of the 'k' groups). Let grand mean be $\bar{x}_G = \frac{1}{N} \sum_{j=1}^{k} \sum_{i=1}^{n} x_{ij}$ (Mean of values from all 'k' groups), where $N = k.n$.

As all samples are of the same size 'n', $\bar{x}_G = \dfrac{\sum_{i=1}^{k} \bar{x}_i}{k}$. Since there are a total of kn independent observations, it follows that

1. Sum of Squares$_{between}$: It is the sum of the squared deviation of each group mean from the grand mean, multiplied by 'n'—the size of each group. It is also called the treatment term.

$$SS_b = n \sum_{i=1}^{k} (\bar{x}_i - \bar{x}_G)^2.$$

2. Sum of Squares$_{within}$: It is the sum of the squared deviation of each individual value 'x_{ij}' from its respective group mean. It is also called the error term.

$$SS_w = \sum_{i=1}^{k} \sum_{j=1}^{n} (x_{ij} - \bar{x}_i)^2.$$

3. Sum of Squares$_{total}$ or $SS_{total} = \sum_{i=1}^{k} \sum_{j=1}^{n} (x_{ij} - \bar{x}_G)^2.$

We have the following results:

$$SS_b + SS_w = SS_{total}$$

where, degrees of freedom of SS_b: $\mathrm{df}_b = k - 1$, degrees of freedom of SS_w: $\mathrm{df}_w = k(n - 1) = N - k$. The statistics

mean square$_{between}$: $MS_{between} = \dfrac{SS_b}{df_b}$ and mean square$_{within}$: $MS_{within} = \dfrac{SS_w}{df_w}$.

Table 5.21 Example 5.22 Prices of clothes of different brands

Brand A	Brand B	Brand C
15	39	65
12	45	45
14	48	32
11	60	38

The test statistic of ANOVA is the **F-statistic**

$$F_0 = \frac{MS_{between}}{MS_{within}}.$$

For given level of significance α, if the test statistic $F_0 \geq F_{k-1,N-k,\alpha}$, then H_0 **is rejected**. $F_{k-1,N-k,\alpha}$ can be looked up from the F statistic tables in Appendix.

Remark 5.7 • In the above construction of equality of the different means of various distributions, the ANOVA comes under the analysis of one-factor experiment.
• This test works quite well even if the underlying distributions are nonnormal unless they are highly skewed or the variances are quite different. In these cases, we need to transform the observation to make the distribution more symmetric about the variances.

Example 5.22 Prices (in Rs.) of towel clothes from various brands are shown in Table 5.21: Using the method of ANOVA, test the appropriate hypothesis at 5% level of significance to decide if the mean prices of Brand A, B, and C differ.

Solution:

Consider following hypothesis testing
$H_0 : \mu_1 = \mu_2 = \mu_3$
H_1: At least one pair of sample means is significantly different.
Computing group means,

$$\bar{x}_A = \frac{15 + 12 + 14 + 11}{4} = 13$$

$$\bar{x}_B = \frac{39 + 45 + 48 + 60}{4} = 48$$

$$\bar{x}_C = \frac{65 + 45 + 32 + 38}{4} = 45.$$

Computing grand mean

$$\bar{x}_G = \frac{\bar{x}_A + \bar{x}_B + \bar{x}_C}{3} = 35.33.$$

Now, computing SS_b

$$SS_b = n \sum_{i=1}^{k} (\bar{x}_i - \bar{x}_G)^2 = 4((13 - 35.33)^2 + (48 - 35.33)^2 + (45 - 35.33)^2) = 3010.67.$$

and $SS_w = \sum_{i=1}^{k} \sum_{j=1}^{n} (x_{ij} - \bar{x}_i)^2.$

The intermediate calculations are shown in Tables 5.22 and 5.23. In this problem, $SS_w = 10 + 234 + 618 = 862$. There are $k = 3$ groups (A, B, C) and $N = 12$. From Table A.11 in Appendix, for $\alpha = 0.05$, we have $F_{2,9,0.05} = 4.2565$. As $F_0 > F_{2,9,0.05}$, H_0 is to be rejected.

Example 5.23 Consider three hostels Tapti, Narmada, and Krishna at IIT Madras. We want to compare the health of the students in these three hostels on the basis of following weights (in Kg): Let us assume that $\alpha = 0.01$ and

Tapti	77	81	71	76	80
Narmada	72	58	74	66	70
Krishna	76	85	82	80	77

$X_{ij} \sim N(\mu_i, \sigma^2), i = 1, 2, \ldots, k.$

Solution:

Here, $\bar{x}_1 = 77, \bar{x}_2 = 68, \bar{x}_3 = 80$ and $\bar{x}_{..} = 75$. $\alpha = 0.01$; $F_{2,12,0.01} = 6.93$ (from Table A.12)

Table 5.22 Example 5.22 Initial calculations

A	$(x_i - \bar{x}_A)^2$	B	$(x_i - \bar{x}_B)^2$	C	$(x_i - \bar{x}_C)^2$
15	4	39	81	65	400
12	1	45	9	45	0
14	1	48	0	32	169
11	4	60	144	38	49
$\bar{x}_A = 13$	10	$\bar{x}_B = 48$	234	$\bar{x}_C = 45$	618

Table 5.23 Example 5.22 ANOVA calculations

$SS_b = 3010.67$	$df_b = k - 1 = 2$	$MS_b = \frac{SS_b}{df_b} = 1505.34$	$F_0 = \frac{MS_b}{MS_w} = 15.72$
$SS_w = 862$	$df_w = N - k = 9$	$MS_w = \frac{SS_w}{df_w} = 95.78$	

$SS_b = 390$	$df_b = k - 1$	$MS_b = \frac{SS_b}{df_b}$	$F_0 = \frac{MS_b}{MS_w}$
	$= 2$	$= 195$	$= 8.48$
$SS_w = 276$	$df_w = N - k$	$MS_w = \frac{SS_w}{df_w}$	
	$= 12$	$= 23$	

As $F_0 > F_{2,12,0.01}$, H_0 is to be rejected. Hence, we can conclude that the mean weights of the students were not the same for these three hostels.

Example 5.24 Firstborn children tend to develop skills faster than their younger siblings. One possible explanation for this phenomenon is that firstborns have the undivided attention from their parents. If this explanation is correct, then it is reasonable that twins should show slower language development than single children and that triplets should be even slower. The data given in Table 5.24 were obtained from several families. The dependent variable (data) is a measure of language skill at the age of 3 for each child. The higher the score, better the skill.

Using the method of ANOVA, test the appropriate hypothesis at 5% level of significance to decide if the average language skills of single children, twins, and triplets differ.

Solution:

Consider following hypothesis testing
H_0: $\mu_1 = \mu_2 = \mu_3$
H_1: At least one pair of sample means is significantly different
    In this problem, $n = 5$; $N = 15$; $k = 3$. Label 1 for Single Child, 2 for Twins and 3 for Triplet. Computing group means, we get

$$\bar{x}_1 = \frac{8 + 7 + 10 + 6 + 9}{5} = 8$$

$$\bar{x}_2 = \frac{4 + 6 + 7 + 4 + 9}{5} = 6$$

$$\bar{x}_3 = \frac{4 + 4 + 7 + 2 + 3}{5} = 4.$$

Table 5.24 Example 5.24 Language skills data

Single child	Twins	Triplets
8	4	4
7	6	4
10	7	7
6	4	2
9	9	3

Table 5.25 Example 5.24 Initial calculations

Single	$(x_i - \bar{x}_1)^2$	Twin	$(x_i - \bar{x}_2)^2$	Triplet	$(x_i - \bar{x}_3)^2$
8	0	4	4	4	0
7	1	6	0	4	0
10	4	7	1	7	9
6	4	4	4	2	4
9	1	9	9	3	1
$\bar{x}_1 = 8$	10	$\bar{x}_2 = 6$	18	$\bar{x}_3 = 4$	12

Table 5.26 Example 5.24 ANOVA calculations

$SS_b = 40$	$df_b = k - 1 = 2$	$MS_b = \frac{SS_b}{df_b} = 20$	$F_0 = \frac{MS_b}{MS_w} = 5.71$
$SS_w = 42$	$df_w = N - k = 12$	$MS_w = \frac{SS_w}{df_w} = 3.5$	

Now, computing grand mean

$$x_G = \frac{\bar{x}_1 + \bar{x}_2 + \bar{x}_3}{3} = 6.$$

Now, computing SS_b

$$SS_b = n \sum_{i=1}^{k} \bar{x}_i - \bar{x}_G^2 = 5((8-6)^2 + (6-6)^2 + (4-6)^2) = 40.$$

Now, computing SS_w

$$SS_w = \sum_{i=1}^{k} \sum_{j=1}^{n} x_{ij} - \bar{x}_i^2 = 10 + 18 + 12 = 42.$$

The other calculations are given in Table 5.25. The value of test statistics can be obtained as shown in Table 5.26. From Table A.13 in Appendix, we have $F_{2,12,0.05} = 3.8853$. Since $F_0 > F_{2,12,0.05}$, H_0 is to be rejected.

In the above, the ANOVA technique is used to examine the effect of one independent variable (factor) on a dependent (response) variable. As in this ANOVA technique, the effect of one factor is investigated, this is called one-way ANOVA technique. An extension of this is two-way ANOVA technique, where the effects of two factors on a response variable are investigated. Many examples based on one-way and two-way ANOVA techniques are presented in Chaps. 7 and 8, respectively. Further, three-way ANOVA is discussed in Chap. 8 with many examples in details.

The interested readers may refer to Cassella and Berger (2002), Freund and Miller (2004), Freund et al. (2010), Hoel et al. (1971), Medhi (1992), Meyer (1970) and Rohatgi and Saleh (2015).

Problems

5.1 There has been a great deal of controversy in recent years over the possible dangers of living near a high-level electromagnetic field (EMF). After hearing many anecdotal tales of large increase among children living near EMF, one researcher decided to study the possible dangers. In order to do his study, he followed following steps: (a) studied maps to find the locations of electric power lines, (b) used these maps to select a fairly large community that was located in a high-level EMF area. He interviews people in the local schools, hospitals, and public health facilities in order to discover the number of children who had been affected by any type of cancer in the previous 3 years. He found that there had been 32 such cases. According to government public health committee, the average number of cases of childhood cancer over a 3-year period in such a community was 16.2, with a standard deviation of 4.7. Is the discovery of 32 cases of childhood cancers significantly large enough, in comparison with the average number of 16.2, for the researcher to conclude that there is some special factor in the community being studied that increases the chance for children to contract cancer? Or is it possible that there is nothing special about the community and that the greater number of cancers is solely due to chance?

5.2 Let $Y_1 < Y_2 < \cdots < Y_n$ be the order statistics of a random sample of size 10 from a distribution with the following PDF

$$f(x; \theta) = \frac{1}{2} e^{-|x-\theta|}, \quad -\infty < x < \infty$$

for all real θ. Find the likelihood ratio test λ for testing $H_0 : \theta = \theta_0$ against the alternative $H_1 : \theta \neq \theta_0$.

5.3 Let $X_1, X_2, \ldots, X_n$ and $Y_1, Y_2, \ldots, Y_n$ be independent random samples from the two normal distributions $N(0, \theta_1)$ and $N(0, \theta_2)$.

(a) Find the likelihood ratio test λ for testing the composite hypothesis $H_0 : \theta_1 = \theta_2$ against the composite alternative hypothesis $H_1 : \theta_1 \neq \theta_2$.
(b) The test statistic λ is a function of which F statistic that would actually be used in this test.

5.4 Let $X_1, X_2, \ldots, X_{50}$ denote a random sample of size 50 from a normal distribution $N(\theta, 100)$. Find a uniformly most powerful critical region of size $\alpha = 0.10$ for testing $H_0 : \theta = 50$ against $H_1 : \theta > 50$.

5.5 Consider a queueing system that describes the number of telephone ongoing calls in a particular telephone exchange. The mean time of a queueing system is required to be at least 180 s. Past experience indicates that the standard deviation of the talk time is 5 s. Consider a sample of 10 customers who reported the following talk time

210, 195, 191, 202, 152, 70, 105, 175, 120, 150.

Would you conclude at the 5% level of significance that the system is unacceptable? What about at the 10% level of significance.

5.6 Let $X_1, X_2, \ldots, X_n$ be a random sample from a distribution with the following PDF

$$f(x; \theta) = \begin{cases} \theta x^{\theta-1}, & 0 < x < \infty \\ 0 & \text{otherwise} \end{cases}$$

where $\theta > 0$. Find a sufficient statistics for θ and show that a uniformly most powerful test of $H_0 : \theta = 6$ against $H_1 : \theta < 6$ is based on this statistic.

5.7 If $X_1, X_2, \ldots, X_n$ is a random sample from a beta distribution with parameters $\alpha = \beta = \theta > 0$, find a best critical region for testing $H_0 : \theta = 1$ against $H_1 : \theta = 2$.

5.8 Let $X_1, X_2, \ldots, X_n$ denote a random sample of size 20 from a Poisson distribution with mean θ. Show that the critical region C defined by $\sum_{i=1}^{20} x_i \geq 4$.

5.9 Let X be a discrete type random variable with PMF

$$P(x; \theta) = \begin{cases} \theta^x (1 - \theta)^{1-x}, & x = 0, 1 \\ 0 & \text{otherwise} \end{cases}.$$

We test the simple hypothesis $H_0 : \theta = \frac{1}{4}$ against the alternative composite hypothesis $H_1 : \theta < \frac{1}{4}$ by taking a random sample of size 10 and rejecting H_0 if and only if the observed values $x_1, x_2, \ldots, x_{10}$ of the sample observations are such that $\sum_{i=1}^{10} x_i < 1$. Find the power of this test.

5.10 In a certain chemical process, it is very important that a particular solution that is to be used as a reactant has a pH of exactly 8.20. A method for determining pH that is available for solutions of this type is known to give measurements that are normally distributed with a mean equal to the actual pH and with a standard deviation of .02. Suppose ten independent measurements yielded the following pH values: 8.18, 8.17, 8.16, 8.15, 8.17, 8.21, 8.22, 8.16, 8.19, 8.18.

1. What conclusion can be drawn at the $\alpha = 0.10$ level of significance?
2. What about at the $\alpha = 0.05$ level of significance?

5.11 An automobile manufacturer claims that the average mileage of its new two-wheeler will be at least 40 km. To verify this claim 15 test runs were conducted independently under identical conditions and the mileage recorded (in km) as: 39.1, 40.2, 38.8, 40.5, 42, 45.8, 39, 41, 46.8, 43.2, 43, 38.5, 42.1, 44, 36. Test the claim of the manufacturer at $\alpha = 0.05$ level of significance.

5.12 The life of certain electrical equipment is normally distributed. A random sample of lives of twelve such equipments has a standard deviation of 1.3 years. Test the hypothesis that the standard deviation is more than 1.2 years at 10% level of significance.

5.13 Random samples of the yields from the usage of two different brands of fertilizers produced the following results: $n_1 = 10$, $\bar{X} = 90.13$, $s_1^2 = 4.02$; $n_2 = 10$, $\bar{Y} = 92.70$, $s_2^2 = 3.98$. Test at 1 and 5% level of significance whether the difference between the two sample means is significant.

5.14 Consider the strength of a synthetic fiber that is possibly affected by the percentage of cotton in the fiber. Five levels of this percentage are considered with five observations at each level. The data are shown in Table 5.27. Use the F-test, with $\alpha = 0.05$ to see if there are differences in the breaking strength due to the percentages of cotton used.

5.15 It is desired to determine whether there is less variability in the marks of probability and statistics course by IITD students than in that by IITB students. If independent random samples of size 10 of the two IIT's yield $s_1 = 0.025$ and $s_2 = 0.045$, test the hypothesis at the 0.05 level of significance.

5.16 Two analysts A and B each make +ve determinations of percent of iron content in a batch of prepared ore from a certain deposit. The sample variances for A and B turned out to be 0.4322 and 0.5006, respectively. Can we say that analyst A is more accurate than B at 5% level of significance?

5.17 Elongation measurements are made of ten pieces on steel, five of which are treated with method A (aluminum only), and the remaining five are method B (aluminum plus calcium). The results obtained are given in Table 5.28. Test the hypothesis that

1. $\sigma_A^2 = \sigma_B^2$.
2. $\mu_B - \mu_A = 10\%$.

Table 5.27   Data for Problem 5.14

15	7	7	15	11	9
20	12	17	12	18	18
25	14	18	18	19	19
30	19	25	22	19	23
35	7	10	11	15	11

Table 5.28 Data for Problem 5.17

Method A	78	29	25	23	30
Method B	34	27	30	26	23

Table 5.29 Data for Problem 5.18

1	0	0	1	3	4	0	2	1	4
2	2	0	0	5	2	1	3	0	1
1	8	0	2	0	1	9	3	3	5
1	3	2	0	7	0	0	0	1	3
3	3	1	6	3	0	1	2	1	2
1	1	0	0	2	1	3	0	0	2

Table 5.30 Data for Problem 5.19

Day	Sun	Mon	Tue	Wed	Thur	Fri	Sat
Number of Earthquakes (f_i)	156	144	170	158	172	148	152

at 2% level of significance by choosing approximate alternatives.

5.18 Suppose the weekly number of accidents over a 60-week period in Delhi is given in Table 5.29. Test the hypothesis that the number of accidents in a week has a Poisson distribution. Assume $\alpha = 0.05$.

5.19 A study was investigated to see if Southern California earthquakes of at least moderate size (having values of at least 4.4 on the Richter Scale) are more likely to occur on certain days of the week then on others. The catalogs yielded the following data on 1100 earthquakes given in Table 5.30. Test at the 5% level of significance, the hypothesis that an earthquake is equally likely to occur on any of the 7 days of the week.

5.20 A builder claims that a particular brand water heaters are installed in 70% of all homes being constructed today in the city of Delhi, India. Would you agree with this claim if a random survey of new homes in this city shows that 9 out of 20 had water heater installed? Use a 0.10 level of significance.

5.21 The proportions of blood types O, A, B and AB in the general population of a particular country are known to be in the ratio 49:38:9:4, respectively. A research team, investigating a small isolated community in the country, obtained the frequencies of blood type given in Table 5.31. Test the hypothesis that the proportions in this community do not differ significantly from those in the general population. Test at the 5% level of significance.

Table 5.31 Data for Problem 5.21

Blood type	O	A	B	AB
Frequency (f_i)	87	59	20	4

Table 5.32 Data for Problem 5.22

4	3	3	1	2	3	4	6	5	6
2	4	1	3	4	5	3	4	3	4
3	3	4	5	4	5	6	4	5	1
6	3	6	2	4	6	4	6	3	5
6	3	6	2	4	6	4	6	3	2
5	4	6	3	3	3	5	3	1	4

Table 5.33 Data for Problem 5.23

	Accident	No accident
Cellular phone	22	278
No phone	26	374

Table 5.34 Data for Problem 5.24

	Smokers	Nonsmokers
Lung cancer	62	14
No Lung cancer	9938	19986

5.22 Consider the data of Table 5.32 that correspond to 60 rolls of a die. Test the hypothesis that the die is fair ($P_i = \frac{1}{6}$, $i = 1, \ldots, 6$), at 0.5% level of significance.

5.23 A sample of 300 cars having cellular phones and one of 400 cars without phones are tracked for 1 year. Table 5.33 gives the number of cars involved in accidents over that year. Use the above to test the hypothesis that having a cellular phone in your car and being involved in an accident are independent. Use the 5% level of significance.

5.24 A randomly chosen group of 20,000 nonsmokers and one of 10,000 smokers were followed over a 10-year period. The data of Table 5.34 relate the numbers of them that developed lung cancer during the period. Test the hypothesis that smoking and lung cancer are independent. Use the 1% level of significance.

5.25 A politician claims that she will receive at least 60% o the votes in an upcoming election. The results of a simple random sample of 100 voters showed that 58 of those sampled would vote for her. Test the politician's claim at the 5% level of significance.

Table 5.35 Data for Problem 5.27

	Agency 1	Agency 2	Agency 3
For the pension plan	67	84	109
Against the pension plan	33	66	41
Total	100	150	150

Table 5.36 Data for Problem 5.29

	Lubricant 1	Lubricant 2	Lubricant 3
Acceptable	144	152	140
Not acceptable	56	48	60
Total	200	200	200

5.26 Use the 10% level of significance to perform a hypothesis test to see if there is any evidence of a difference between the Channel A viewing area and Channel B viewing area in the proportion of residents who viewed a news telecast by both the channels. A simple random sample of 175 residents in the Channel A viewing area and 225 residents in the Channel B viewing area is selected. Each resident in the sample is asked whether or not he/she viewed the news telecast. In the Channel A telecast, 49 residents viewed the telecast, while 81 residents viewed the Channel B telecast.

5.27 Can it be concluded from the following sample data of Table 5.35 that the proportion of employees favouring a new pension plan is not the same for three different agencies. Use $\alpha = 0.05$.

5.28 In a study of the effect of two treatments on the survival of patients with a certain disease, each of the 156 patients was equally likely to be given either one of the two treatments. The result of the above was that 39 of the 72 patients given the first treatment survived and 44 of the 84 patients given the second treatment survived. Test the null hypothesis that the two treatments are equally effective at $\alpha = 0.05$ level of significance.

5.29 Three kinds of lubricants are being prepared by a new process. Each lubricant is tested on a number of machines, and the result is then classified as acceptable or nonacceptable. The data in the Table 5.36 represent the outcome of such an experiment. Test the hypothesis that the probability p of a lubricant resulting in an acceptable outcome is the same for all three lubricants. Test at the 5% level of significance.

5.30 Twenty-five men between the ages of 25 and 30, who were participating in a well-known heart study carried out in New Delhi, were randomly selected. Of these,

Table 5.37 Data for Problem 5.30

Smokers	124	134	136	125	133	127	135	131	133	125	118			
Nonsmokers	130	122	128	129	118	122	116	127	135	120	122	120	115	123

Table 5.38 Data for Problem 5.31

Method 1	6.2	5.8	5.7	6.3	5.9	6.1	6.2	5.7
Method 2	6.3	5.7	5.9	6.4	5.8	6.2	6.3	5.5

11 were smokers, and 14 were not. The data given in Table 5.37 refer to readings of their systolic blood pressure. Use the data of Table 5.37 to test the hypothesis that the mean blood pressures of smokers and nonsmokers are the same at 5% level of significance.

5.31 Polychlorinated biphenyls (PCB), used in the production of large electrical transformers and capacitors, are extremely hazardous when released into the environment. Two methods have been suggested to monitor the levels of PCB in fish near a large plant. It is believed that each method will result in a normal random variable that depends on the method. Tests the hypothesis at the $\alpha = 0.10$ level of significance that both methods have the same variance, if a given fish is checked eight times by each method with the data (in parts per million) recorded given in Table 5.38.

5.32 An oil company claims that the sulfur content of its diesel fuel is at most 0.15 percent. To check this claim, the sulfur contents of 40 randomly chosen samples were determined; the resulting sample mean, and sample standard deviation was 0.162 and 0.040, respectively. Using the five percent level of significance, can we conclude that the company's claims are invalid?

5.33 Historical data indicate that 4% of the components produced at a certain manufacturing facility are defective. A particularly acrimonious labor dispute has recently been concluded, and management is curious about whether it will result in any change in this figure of 4%. If a random sample of 500 items indicated 16 defectives, is this significant evidence, at the 5% level of significance, to conclude that a change has occurred.

5.34 An auto rental firm is using 15 identical motors that are adjusted to run at fixed speeds to test three different brands of gasoline. Each brand of gasoline is assigned to exactly five of the motors. Each motor runs on ten gallons of gasoline until it is out of fuel. Table 5.39 gives the total mileage obtained by the different motors. Test the hypothesis that the average mileage obtained is not affected by the type of gas used. Use the 5% level of significance.

Table 5.39 Data for Problem 5.34

Gas 1	Gas 2	Gas 3
220	244	252
251	235	272
226	232	250
246	242	238
260	225	256

Table 5.40 Data for Problem 5.35

	n	Mean	SD
Control	15	82.52	9.24
Pets	15	73.48	9.97
Friends	15	91.325	8.34

Table 5.41 Data for Problem 5.36

Narmada	72	58	74	66	70
Tapti	76	85	82	80	77
Kaveri	77	81	71	76	80

5.35 To examine the effects of pets and friends in stressful situations, researchers recruited 45 people to participate in an experiment and data are shown in Table 5.40. Fifteen of the subjects were randomly assigned to each of the 3 groups to perform a stressful task alone (Control Group), with a good friend present, or with their dog present. Each subject mean heart rate during the task was recorded. Using ANOVA method, test the appropriate hypothesis at the $\alpha = 0.05$ level to decide if the mean heart rate differs between the groups.

5.36 Suppose that to compare the food quality of three different hostel students on the basis of the weight on 15 students as shown in Table 5.41.

The means of these three samples are 68, 80, and 77. We want to know whether the differences among them are significant or whether they can be attributed to chance.

5.37 A fisheries researcher wishes to conclude that there is a difference in the mean weights of three species of fish (A,B,C) caught in a large lake. The data are shown in Table 5.42. Using ANOVA method, test the hypothesis at $\alpha = 0.05$ level.

Table 5.42   Data for Problem 5.37

A	B	C
1.5	1.5	6
4	1	4.5
4.5	4.5	4.5
3	2	5.5

References

Cassella G, Berger RL (2002) Statistical inference. Duxbury, Pacific Grove

Freund JE, Miller I (2004) John E. Freund's mathematical statistics: with applications. Pearson, New York

Freund R, Mohr D, Wilson W (2010) Statistical methods. Academic Press, London

Hoel PG, Port SC, Stone CJ (1971) Introduction to statistical theory. Houghton Mifflin, Boston

Medhi J (1992) Statistical methods: an introductory text. New Age International, New Delhi

Meyer PL (1970) Introductory probability and statistical applications. Addison Wesley, Boston

Miller LH (1956) Table of percentage points of Kolmogorov statistics. J Am Stat Assoc 51:111–121

Owen DB (1962) Handbook of statistical tables. Addison-Wesley, Reading, Mass

Rohatgi VK, Ehsanes Saleh AKMd (2015) An introduction to probability and statistics. Wiley, New York

Chapter 6
Analysis of Correlation and Regression

6.1 Introduction

It is quite often that one is interested to quantify the dependence (positive or negative) between two or more random variables. The basic role of covariance is to identify the nature of dependence. However, the covariance is not an appropriate measure of dependence since it is dependent on the scale of observations. Hence, a measure is required which is unaffected by such scale changes. This leads to a new measure known as the correlation coefficient. Correlation analysis is the study of analyzing the strength of such dependence between the two random variables using the correlation coefficient. For instance, if X represents the age of a used mobile phone and Y represents the retail book value of the mobile phone, we would expect smaller values of X to correspond to larger values of Y and vice-versa.

Regression analysis is concerned with the problem of predicting a variable called dependent variable on the basis of information provided by certain other variables called independent variables. A function of the independent variables, say $f(X_1, \ldots, X_n)$, is called predictor of dependent variable Y that is considered.

6.2 Correlation

Often we come across the situations where we have two related random variables X and Y out of which only one say X is observed. Our interest is to predict the value of Y using the observed value of X. If the joint distribution of (X, Y) is known, then one can obtain the conditional distribution of Y given X and use it as a proxy for our prediction. The conditional expectation is a reasonable guess regarding what we might expect for Y that is $E[(Y - f(X))^2]$ over all functions $f(X)$ that depends only on X. In practice, one often restricts the choice of f to be a linear functions of

© Springer Nature Singapore Pte Ltd. 2018
D. Selvamuthu and D. Das, *Introduction to Statistical Methods,
Design of Experiments and Statistical Quality Control*,
https://doi.org/10.1007/978-981-13-1736-1_6

the form $a + bX$ where a and b are constants. The minimization of the expression $E[(Y - a - bX)^2]$, over all values of a and b, can be explicitly carried out as follows:

$$E[(Y - a - bX)^2] = E(Y^2 + a^2 + b^2X^2 - 2aY + 2abX - 2bXY)$$
$$= E(Y^2) + a^2 + b^2E(X^2) - 2aE(Y) + 2abE(X) - 2bE(XY).$$

Differentiating partially w.r.t. a and b we get,

$$\frac{\partial}{\partial a}E[(Y - a - bX)^2] = 2a - 2E(Y) + 2bE(X)$$

$$\frac{\partial}{\partial b}E[(Y - a - bX)^2] = 2bE(X^2) + 2aE(X) - 2E(XY).$$

Equating them to zero gives us the expressions

$$2b[E(X)]^2 + 2aE(X) - 2E(X)E(Y) = 0$$
$$2b[E(X)]^2 + 2aE(X) - 2E(XY) = 0$$
$$\Rightarrow 2bE(X^2) - [E(X)]^2 = 2[E(XY) - E(X)E(Y)]$$
$$\Rightarrow \hat{b} = \frac{CoV(XY)}{VarX}$$
$$\therefore \hat{a} = E(Y) - bE(X).$$

One can decompose $Y - E(Y)$ as follows:

$$Y - E(Y) = Y - \hat{a} - \hat{b}E(X)$$
$$= \hat{b}(X - E(X)) + Y - \hat{a} - \hat{b}X.$$

Because the cross term vanishes, we get

$$Var(Y) = \frac{(Cov(X, Y))^2}{Var(X)Var(Y)}Var(Y) + \left[1 - \frac{(Cov(X, Y))^2}{Var(X)Var(Y)}\right]Var(Y).$$

Definition 6.1 The Pearson's (or linear) correlation coefficient of X and Y, denoted by $Cor(X, Y)$, or ρ_{XY}, is defined as

$$\rho_{XY} = Cor(X, Y) = \frac{Cov(X, Y)}{\rho_X \rho_Y} = \frac{\sum(X_i - \overline{X})(Y_i - \overline{Y})}{\sqrt{\sum(X_i - \overline{X})^2 \sum(Y_i - \overline{Y})^2}} = \frac{S_{xy}}{\sqrt{S_{xx}}\sqrt{S_{yy}}}$$

where $S_{xx} = \sum\limits_{i=1}^{n}(X_i - \overline{X})^2$; $S_{yy} = \sum\limits_{i=1}^{n}(Y_i - \overline{Y})^2$; $S_{xy} = \sum\limits_{i=1}^{n}(X_i - \overline{X})(Y_i - \overline{Y})$, and ρ_X and ρ_Y are the standard deviations of X and Y, respectively.

Table 6.1 Values of X and Y

X	30	32	34	36	38	40
Y	1.6	1.7	2.5	2.8	3.2	3.5

Rewriting the earlier relation, we have

$$Var(Y) = \rho^2 Var(Y) + (1 - \rho^2)Var(Y).$$

The first term, i.e., ρ^2, represents the amount of the variance reduced due to "pre-dictable" component and the second term, i.e., $(1 - \rho^2)$, represents the residual variance. The sign of the correlation coefficient determines whether the dependence between the two variables is positive or negative, whereas the magnitude of the correlation coefficient gives the strength of the dependence.

There is another measure known as the coefficient of determination which is obtained as the square of the correlation coefficient (ρ^2). This statistic quantifies the proportion of the variance of one variable (in a statistical sense, not a casual sense) by the other.

Example 6.1 Consider the data given in Table 6.1. Find the correlation coefficient and coefficient of determination.

Solution:

$$S_{xx} = 70, S_{yy} = 3.015, S_{xy} = 14.3.$$

Hence, the value of correlation coefficient and coefficient of determination, i.e., ρ and ρ^2, are given by

$$\rho = \frac{14.3}{\sqrt{70 \times 3.015}} = 0.984$$

and

$$\rho^2 = 0.984^2 = 0.968.$$

From Schwartz's inequality, it follows that $-1 \leq \rho \leq 1$. Further $\rho = 1$ indicates that there is a perfect positive correlation between the variables and all the points lie exactly on a straight line of positive slope. On the other hand, $\rho = -1$ signifies that there is a perfect negative correlation between the variables and point lies on a line with negative slope, and $\rho = 0$ tells us that there is no linear dependence between the two variables under study. The sketches in Figs. 6.1, 6.2 and 6.3 indicate these and in between cases.

Note that, linear correlation is not a good measure of the dependence of two variables. Spearman's correlation coefficient gives us a better measure of the dependence. It is defined as follows.

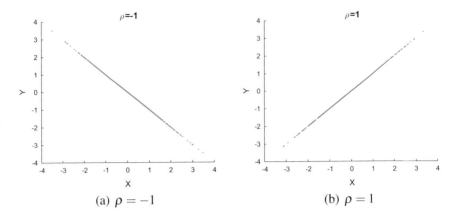

Fig. 6.1 Scatter plot when $\rho = -1, 1$

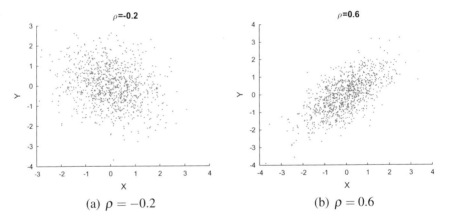

Fig. 6.2 Scatter plot when $\rho = -0.2, 0.6$

Definition 6.2 The Spearman's (or monotone) correlation coefficient of X and Y, denoted by $(\rho^s)_{XY}$, is

$$(\rho^s)_{XY} = \rho U_1 U_2.$$

where if X and Y are continuous random variables, then $U_1 = F_X(x)$ *and* $U_2 = F_Y(y)$ are uniform distributed random variables between the intervals 0 and 1.

The rationale behind the definition of the uniform correlation is based on the fact that if two uniform random variables are monotonically dependent (i.e., either positively or negatively dependent), then they are linearly dependent. Thus, by transforming X and Y into uniform random variables, their monotone dependence becomes linear dependence.

Fig. 6.3 Scatter plot when
$\rho = 0$

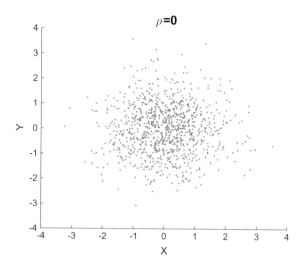

Note that if the variables were qualitative in nature, that is, nominal or ordinal, then it would be advisable to use a nonparametric method of determining the correlation coefficient, namely Spearman's correlation coefficient.

6.2.1 Causality

Causality or causation indicates that the one event is the outcome of the occurrence of some other event; i.e., there is a cause and effect relationship between the two events. However, correlation is just a statistical measure that measures the linear relationship between the two variables and does not mean that the change in one variable causes the change in the other variable. Hence, correlation and casuality are two different concepts and correlation does not imply causality. For example, we might find that there is a positive correlation between the time spent driving on road and the number of accidents but this does not mean that spending more time on road causes accident. Because in that case, in order to avoid accidents one may drive fast so that time spent on road is less.

To illustrate how two variables are related, the values of X and Y are pictured by drawing the scatter diagram, sketching the graph of the combination of the two variables. The sample correlation coefficient r is the estimator of population correlation coefficient ρ. If we have n independent observations from a bivariate normal distribution with means μ_X, μ_Y, variances σ_X^2, σ_Y^2 and correlation coefficient ρ, one might want to test the hypothesis $\rho = 0$. In this context, the null and alternative hypothesis is $H_0 : \rho = 0$, against $H_1 : \rho \neq 0$.

The hypothesis can be tested using a t-statistic

$$T = \frac{r}{SE_r} \tag{6.1}$$

and the measure ρ is estimated by the sample correlation coefficient r where

$$r = \frac{S_{xy}}{\sqrt{S_{xx}}\sqrt{S_{yy}}}. \tag{6.2}$$

and SE_r represents the standard error of the correlation coefficient

$$SE_r = \sqrt{\frac{1-r^2}{n-2}}. \tag{6.3}$$

This T-statistic has $n-2$ degrees of freedom.

Example 6.2 (*Parameter Estimation*) The weight of lion cub per week after its birth is given in Table 6.2. Estimate the linear regression model with weight as the dependent variable. What would be the cub's weight after 42 weeks?

Solution:

For the given data, the values of $\Sigma x = 210$, $\Sigma y = 15.3$, $n = 6$, $S_{xy} = 14.3$ and $S_{xx} = 70$. Using these values, we get the value of slope parameter

$$\hat{b} = \frac{S_{xy}}{S_{xx}} = \frac{14.3}{70} = 0.2043.$$

The mean values are

$$\bar{x} = \frac{\Sigma x}{n} = \frac{210}{6} = 35,$$
$$\bar{y} = \frac{\Sigma y}{n} = \frac{15.3}{6} = 2.55.$$

Hence, we have

$$\hat{a} = \bar{y} - \hat{\beta}\bar{x} = 2.55 - (0.2043)(35) = -4.60.$$

Table 6.2 Example 6.2 - Weight of lion cub per week after its birth

Weeks	30	32	34	36	38	40
Fetal weight	1.6	1.7	2.5	2.8	3.2	3.5

Using the estimated values of regression coefficients, we get

$$\hat{y} = \hat{a} + \hat{b}x = -4.60 + (0.2043)(42) = 3.98.$$

A test of the more general hypothesis $\rho = \rho_0$ against suitable alternative is easily conducted from the sample information. Suppose X and Y follow the bivariate normal distribution, the quantity $\frac{1}{2} \ln(\frac{1+r}{1-r})$ is the value of a random variable that follows approximately the normal distribution with mean $\frac{1}{2} \ln(\frac{1+\rho}{1-\rho})$ and variance $\frac{1}{n-3}$. Thus, the test procedure is to compute

$$Z_0 = \frac{\sqrt{n-3}}{2} \left[\ln \left(\frac{(1+r)(1-\rho_0)}{(1-r)(1+\rho_0)} \right) \right] \tag{6.4}$$

and compare with the critical points of the standard normal distribution. It is also possible to obtain an approximate test of size α by using

$$W = \frac{1}{2} \ln \left(\frac{1+\rho}{1-\rho} \right).$$

Note that, W has an approximate normal distribution with mean $\frac{1}{2} \ln \left(\frac{1+\rho}{1-\rho} \right)$ and variance $\frac{1}{n-3}$. We can also test the hypothesis like $H_0 : \rho = \rho_0$, against $H_1 : \rho \neq \rho_0$ where ρ_0 is not necessarily zero. In that case, W is $\frac{1}{2} \ln \left(\frac{1+\rho_0}{1-\rho_0} \right)$.

Example 6.3 The data in Table 6.3 were obtained in a study of the relationship between the weight and chest size of infants at birth. Determine the sample correlation coefficient r and then test the null hypothesis $H_o : \rho = 0$ against the alternative hypothesis $H_1 : \rho \neq 0$ at a significance level 0.01.

Solution:

We can find from the given data that

$$\bar{x} = 3.46, \quad \text{and} \quad \bar{y} = 29.51.$$

Table 6.3 Data for weight and chest size of infants at birth

x (weights in kg)	2.76	2.17	5.53	4.31	2.30	3.70
y (chest size in cm)	29.5	26.3	36.6	27.8	28.3	28.6

$(x-\bar{x})(y-\bar{y})$	$(x-\bar{x})^2$	$(y-\bar{y})^2$
0.007	0.49	0
4.141	1.664	10.304
14.676	4.285	50.286
-1.453	0.722	2.924
1.404	1.346	1.464
-0.218	0.058	0.828

$$S_{xy} = 18.557,\ S_{xx} = 8.565,\ S_{yy} = 65.788.$$

$$\therefore \rho = \frac{S_{xy}}{\sqrt{S_{xx}S_{yy}}} = \frac{18.557}{\sqrt{(8.565)(65.788)}} = 0.782.$$

Hence, $t = \sqrt{n-2}\dfrac{r}{\sqrt{1-r^2}} = 2.509$. Since, $2.504 < t_{0.005,4} = 4.604$, we fail to reject the null hypothesis.

The hypothesis test for the Spearman correlation coefficient is basically the same as the test for the Pearson correlation coefficient. The standard deviation of R is approximated by $\sqrt{\frac{1-R^2}{n-2}}$ and when n is 10 or more, R is approximated by a student t-distribution with $n - 2$ degrees of freedom. When the hypothesis is $H_0 : \rho = 0$, against $H_1 : \rho \neq 0$. The standardized t-statistic can be written as

$$t = R\sqrt{\frac{n-2}{1-R^2}}.$$

6.2.2 Rank Correlation

We know that ranking is the process of assigning ordered labels "first," "second," "third," etc., or 1, 2, 3, etc to various observations of a particular variable. In statistics, rank correlation is a measure of this ordinal relationship, i.e., the association between the rankings of various ordinal variables or different rankings of a particular variable. Thus, a rank correlation coefficient gives the degree of this association between rankings. One such correlation coefficient is Spearman's rank difference correlation coefficient and is denoted by R. In order to calculate R, we arrange the data in ranks computing the difference in rank "d" for each other. R is given by the formula

$$R = 1 - 6\frac{\sum\limits_{i=1}^{n}(d_i^2)}{n(n^2-1)}. \tag{6.5}$$

Example 6.4 Consider the data given in Table 6.4. Find the Spearman rank correlation coefficient.

Table 6.4 Values of X and Y

X	35	23	47	17	10	43	9	6	28
Y	30	33	45	23	8	49	12	4	31

Table 6.5 Values of X and Y

X	35	23	47	17	10	43	9	6	28
Y	30	33	45	23	8	49	12	4	31
Rank of X	3	5	1	6	7	2	8	9	4
Rank of Y	5	3	2	6	8	1	7	9	4
d_i	2	2	1	0	1	1	1	0	0

Solution:

The calculations are given in the following Table 6.5.

Hence, the value of Spearman's rank correlation coefficient is given by

$$R = 1 - 6 \times \frac{12}{9 \times 80} = 0.9.$$

Note that, if we are provided with only ranks without the values of x and y we can still find Spearman's rank difference correlation R by taking the difference of the ranks and proceeding as shown above.

This produces a correlation coefficient which has a maximum value of 1, indicating a perfect positive association between the ranks, and a minimum value of -1, indicating a perfect negative association between ranks. A value of 0 indicates no association between the ranks for the observed values of X and Y.

6.3 Multiple Correlation

Let us consider one dependent variable and more than one independent variables. The degree of relationship existing between the dependent variable and all the independent variables together is called multiple correlation.

Consider three variables X_1, X_2, and X_3 such that X_1 is a dependent variable and X_2, X_3 are independent variables. The linear regression equation of X_1 on independent variables is of the form

$$X_1 = a_{1.23} + a_{12.3}X_2 + a_{13.2}X_3 \tag{6.6}$$

where $a_{1.23}, a_{12.3}, a_{13.2}$ are constants to be determined. Here, $a_{1j.k}$ denotes the slope of line between X_1 and X_j when X_k is held constant. We can observe that variation in X_1 is partially due to X_2 and partially due to X_3; we call $a_{1j.k}$ as partial regression coefficients of X_1 on X_j keeping X_k constant.

Following the same approach of least square approximation as we used for simple linear regression, we can obtain the coefficients of partial regression. Let N observations on each X_i are obtained. Summing Equation (6.6) on both sides, we have

$$\sum X_1 = a_{1.23}N + a_{12.3} \sum X_2 + a_{13.2} \sum X_3. \tag{6.7}$$

Multiplying Equation (6.6) by X_2 and X_3 successively and taking summation over N values, we have

$$\sum X_1 X_2 = a_{1.23} \sum X_2 + a_{12.3} \sum X_2^2 + a_{13.2} \sum X_2 X_3 \tag{6.8}$$

$$\sum X_1 X_3 = a_{1.23} \sum X_3 + a_{12.3} \sum X_2 X_3 + a_{13.2} \sum X_3^2. \tag{6.9}$$

Solving equations (6.7), (6.8) and (6.9) simultaneously, we can obtain the coefficients.

Standard Error of Estimate

Following the same approach as in simple linear regression, the standard error of estimate of X_1 on X_2 and X_3 can be defined as

$$\sigma_{1.23} = \sqrt{\frac{\sum (X_1 - X_{1,est})^2}{N}}$$

where $X_{1,est}$ is estimated value of X_1 obtained from regression Equation (6.6).

Coefficient of Multiple Correlation

The coefficient of multiple correlation is given by

$$R_{1.23} = \sqrt{1 - \frac{\sigma_{1.23}^2}{\sigma_1^2}} \tag{6.10}$$

where σ_1 represents the standard deviation of the variable X_1. $R_{1.23}^2$ is called the coefficient of multiple determination. The value of $R_{1.23}$ lies between 0 and 1 with value near 1 representing the good linear relationship between the variables and value near 0 giving the worst linear relationship. The value 0 gives no linear relationship that does not imply independence because nonlinear relationship may be present.

Similarly, there are many situations where we have to find the correlation between two variables after adjusting the effect of one or more independent variables. This leads us to the concept of partial correlation.

6.3.1 Partial Correlation

In some practical situations, we need to study the correlation between one dependent variable and one particular independent variable keeping the effect of all other independent variables constant. This can be done using coefficient of partial correlation. It is given by

$$r_{12.3} = \frac{r_{12} - r_{13}r_{23}}{\sqrt{(1 - r_{13}^2)(1 - r_{23}^2)}} \tag{6.11}$$

where $r_{12.3}$ represents the partial correlation coefficient between X_1 and X_2 keeping X_3 constant and r_{ij} represents the correlation coefficient between X_i and X_j.

Relationship Between Partial and Multiple Correlation Coefficient

From Eqs. (6.10) and (6.11), we can obtain the following relationship between partial and multiple correlation coefficients:

$$1 - R_{1.23}^2 = (1 - r_{12}^2)(1 - r_{13.2})^2.$$

Example 6.5 The given table shows the weights (in Kg), the heights (in inches), and the age of 12 boys.

Weight (X_1)	64	71	53	67	55	58	77	57	56	57	76	68
Height (X_2)	57	59	49	62	51	50	55	48	52	42	61	57
Age (X_3)	8	10	6	11	8	7	10	9	10	6	12	9

(a) Find the least square regression equation of X_1 on X_2 and X_3.
(b) Determine the estimated values of X_1 from the given values of X_2 and X_3.
(c) Estimate the weight of a boy who is 9 years old and 54 inches tall.
(d) Compute the standard error of estimate of X_1 on X_2 and X_3.
(e) Compute the coefficient of linear multiple correlation of X_1 on X_2 and X_3.
(f) Also, compute the coefficients of linear partial correlation $r_{12.3}$, $r_{13.2}$ and $r_{23.1}$.

Solution:

(a) The linear regression equation of X_1 on X_2 and X_3 is

$$X_1 = b_{1.23} + b_{12.3}X_2 + b_{13.2}X_3.$$

The normal equations of least square regression equation are

$$\sum X_1 = b_{1.23}N + b_{12.3}\sum X_2 + b_{13.2}\sum X_3$$

$$\sum X_1X_2 = b_{1.23}\sum X_2 + b_{12.3}\sum X_2^2 + b_{13.2}\sum X_2X_3$$

$$\sum X_1X_3 = b_{1.23}\sum X_3 + b_{12.3}\sum X_2X_3 + b_{13.2}\sum X_3^2.$$

Computing and substituting the values from table, we get

$$12b_{1.23} + 643b_{12.3} + 106b_{13.2} = 753$$
$$643b_{1.23} + 34843b_{12.3} + 5779b_{13.2} = 40830$$
$$106b_{1.23} + 5779b_{12.3} + 976b_{13.2} = 6796.$$

Solving the above system of equations, we get

$$b_{1.23} = 3.6512, \ b_{12.3} = 0.8546, \ b_{13.2} = 1.5063.$$

The required regression equation is

$$X_1 = 3.6512 + 0.8546X_2 + 1.5063X_3. \tag{6.12}$$

(b) The estimated values of X_1 can be obtained by substituting the corresponding values of X_2 and X_3 in Eq. (6.12). The obtained values are given as

$X_{1,est}$	64.414	69.136	54.564	73.206	59.286	56.925	65.717	58.229	63.153	48.582	73.857	65.920
X_1	64	71	53	67	55	58	77	57	56	51	76	68

(c) Substituting $X_2 = 54$ and $X_3 = 9$ in Eq. (6.12), the estimated weight of the boy is 63.356 Kg.

(d) The standard error of estimated is given by

$$S_{1.23} = \sqrt{\frac{\sum(X_1 - X_{1,est})^2}{N}}$$

$$= \sqrt{\frac{(64 - 64.414)^2 + \cdots + (68 - 65.920)^2}{12}} = 4.6\text{Kg}.$$

The population standard error of estimate is given by

$$\hat{S}_{1.23} = \sqrt{\frac{N}{N-3}}S_{1.23} = 5.3 \text{ Kg}.$$

(e) The coefficient of linear multiple correlation of X_1 on X_2 and X_3 is

$$R_{1.23} = \sqrt{1 - \frac{S_{1.23}^2}{S_1^2}} = \sqrt{1 - \frac{4.64472^2}{8.60352^2}} = 0.8418.$$

(f) The coefficients of linear partial correlation are

$$r_{12.3} = \frac{r_{12} - r_{13}r_{23}}{\sqrt{(1 - r_{13}^2)(1 - r_{23}^2)}} = 0.5334.$$

Similarly,

$$r_{13.2} = 0.3346 \quad \text{and} \quad r_{23.1} = 0.4580.$$

6.4 Regression

Sir Francis Galton coined the term "regression" in 1800s to describe a biological phenomenon. The main purpose of regression is to explore the dependence of one variable on another. The technique of regression, in particular, linear regression, is the most popular statistical tool. There are all forms of regression, namely linear, nonlinear, simple, multiple, parametric, nonparametric, etc. A regression model is said to be linear when it is linear in parameters. For instance,

$$y = \beta_0 + \beta_1 x + \varepsilon,$$

$$y = \beta_0 + \beta_1 x_1 + \beta_2 x_2 \varepsilon$$

are the linear models. In this chapter, we will look at the simplest case of linear regression with one predictor variable.

In simple linear regression, we have a relationship of the form

$$Y_i = \beta_0 + \beta_1 X_i + \varepsilon_i \tag{6.13}$$

where Y_i is a random variable and X_i is another observable variable.

The quantities β_0 and β_1, the intercept and slope of the regression, which are unknown parameters, are assumed to be fixed, and ε_i is necessarily a random variable. It is also common to suppose that $E(\varepsilon_i) = 0$.

From (6.13), we have

$$E(Y_i) = \beta_0 + \beta_1 X_i. \tag{6.14}$$

Note that, the function that gives $E(Y)$ is a function of x and it is called the population regression function. Thus, Equation (6.14) defines the population regression function for simple linear regression. The fact that our inferences about the relationship between Y_i and X_i assuming knowledge of X_i, Eq. (6.14) can be written as

$$E(Y_i/X_i) = \beta_0 + \beta_1 X_i. \tag{6.15}$$

The Y_i is called the response or dependent variable, whereas X_i is called predictor or independent variable. Recall that the word regression in connection with conditional expectations. That is, the regression of Y on X was defined as $E(Y/X)$, the conditional expectation of Y given $X = x$. In general, the word regression is used in statistics to signify a relationship between variables. When we refer to linear regression, we mean that the conditional expectation of Y given $X = x$ is a linear function of x. Thus, Eq. (6.15) concludes the regression is linear whenever x_i is fixed and known, and it is a realization of the observable random variable X_i. Note that, if we assume that the pair (X_i, Y_i) has a bivariate normal distribution, it immediately follows that the regression of Y on X is linear.

In the previous sections, we have seen that correlation is a measure of the linear relationship between two variables, whereas simple linear regression is used to predict a linear relationship between one dependent variable and one independent variable. But in many practical situations, we have more than one independent variable. For example, the credit quality of a borrower depends on many factors; the sale of a product depends on many independent factors like advertisement, price, other products, quality. Thus, we have to model the relationship between these and this leads us to the concept of multiple correlation.

6.4.1 Least Squares Method

The least squares method is a method to obtain the unknown parameters such that the mean square error is minimized. It should be considered only as a method of "fitting a line" to a set of data, not as a method of statistical inference. In this case, we observe data consisting of n pairs of observations, $(x_1, y_1), (x_2, y_2), \ldots, (x_n, y_n)$. For instance, consider n pairs of data points listed in Table 6.2 of Example 6.2 plotted as a scatter plot in Fig. 6.4. Here, we estimate β_0 and β_1 without any statistical assumptions on (x_i, y_i).

The residual sum of squares (RSS) is defined for the line $y = c + dx$ as

$$RSS = \sum_{i=1}^{n} (y_i - (c + dx_i))^2.$$

The least squares estimates of β_0 and β_1 are defined to be those values a and b such that the line $a + bx$ minimizes RSS; i.e., the least squares estimates a and b satisfy

$$\min_{c,d} \sum_{i=1}^{n} (y_i - (c + dx_i))^2 = \sum_{i=1}^{n} (y_i - (a + bx_i))^2.$$

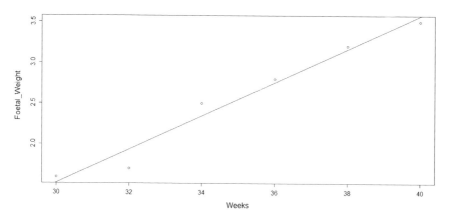

Fig. 6.4 Regression plot

Simple calculations yield the estimates of β_0, β_1, denoted by a and b, respectively, and are given by

$$\hat{\beta}_0 = \bar{y} - b\bar{x}, \text{ and } \hat{\beta}_1 = \frac{S_{xy}}{S_{xx}}.$$

If x is the predictor variable, y is the response variable and we think of predicting y from x, then the vertical distance measured in RSS is reasonable.

Example 6.6 It is expected that Y is linearly related to X. Determine the least square regression line for $E(Y)$ on the basis of ten observations given in Table 6.6
Solution:

Here,

$$\bar{x} = \frac{\Sigma_{i=1}^{10} x_i}{10} = 67.5; \quad \bar{y} = \frac{\Sigma_{i=1}^{10} y_i}{10} = 55.5$$

$$\Sigma_{i=1}^{10}(x_i - \bar{x})^2 = 2062.5; \quad \Sigma_{i=1}^{10}(x_i - \bar{x})(y_i - \bar{y}) = 1182.5.$$

Thus,

$$b = \frac{1182.5}{2062.5} = 0.57; \quad a = \bar{y} - b\bar{x} = 17.03.$$

Table 6.6 Data for Y and X

i	1	2	3	4	5	6	7	8	9	10
x	45	50	55	60	65	70	75	80	85	90
y	43	45	48	57	55	57	59	63	66	68

The estimated regression line is $y = 17.03 + 0.57x$.

In many situations, there is a simple response variation y, also called the dependent variable, which depends on the value of a set of input, also called independent variables $x_1, x_2, \ldots, x_r$. The simplest type of relationship between the dependent variable y and the input variables $x_1, x_2, \ldots, x_r$ is a linear relationship. That is, for some constants $\beta_0, \beta_1, \beta_2, \ldots, \beta_r$, the equation

$$y = \beta_0 + \beta_1 x_1 + \beta_2 x_2 + \cdots + \beta_r x_r \qquad (6.16)$$

will hold. However, in practice, such precision would be valid subject to random error. That means, the explicit relationship can be written as

$$y = \beta_0 + \beta_1 x_1 + \beta_2 x_2 + \cdots + \beta_r x_r + \varepsilon \qquad (6.17)$$

where, ε, representing the random error is assumed to be a random variable having mean 0. Equation (6.16) is called a linear regression equation. The regression coefficients $\beta_0, \beta_1, \beta_2, \ldots, \beta_r$ are to be estimated from a set of data.

A regression equation containing a single independent variable, i.e., one in which $r = 1$, is called a single regression equation, whereas one containing many independent variables is called a multiple regression equation. In Chap. 7, the second-order model (quadratic model) and the third-order model (cubic model) are discussed through examples.

6.4.2 Unbiased Estimator Method

A simple linear regression model supposes a linear relationship between the mean response and the value of a single independent variable. It can be expressed as

$$y = \beta_0 + \beta_1 x + \varepsilon. \qquad (6.18)$$

In this method, the random errors ε_i are assumed to be independent normal random variables having mean 0 and variance σ^2. Thus, we suppose that if y_i is the response corresponding to the input values x_i, then $y_1, y_2, \ldots, y_n$ are independent and y_i are normally distributed random variables with mean $\beta_0 + \beta_1 x_i$ and variance σ^2.

Note that, the variance of the random error is assured to be a constant and does not depend on the input value. This value σ^2 has to be estimated from the data. Using the least squares method, the estimator $\hat{\beta}_1$ of β_1 can be expressed as

$$\hat{\beta}_1 = \frac{\sum_{i=1}^{n}(x_i - \bar{x})y_i}{\sum_{i=1}^{n}x_i^2 - n\bar{x}^2}.$$ (6.19)

The mean of $\hat{\beta}_1$ is as follows:

$$E(\hat{\beta}_1) = \frac{\sum_{i=1}^{n}(x_i - \bar{x})E(y_i)}{\sum_{i=1}^{n}x_i^2 - n\bar{x}^2}.$$ (6.20)

Substituting $E(y_i) = \beta_0 + \beta_1 x_i$ and $\sum_i x_i - x = 0$, we get

$$E(\hat{\beta}_1) = \beta_1.$$

Thus, $\hat{\beta}_1$ is an unbiased estimator of β_1. The variance of $\hat{\beta}_1$ is given by

$$Var(\hat{\beta}_1) = \frac{Var\left(\sum_{i=1}^{n}(x_i - \bar{x})y_i\right)}{\left(\sum_{i=1}^{n}x_i^2 - n\bar{x}^2\right)^2}.$$

Using $Var(y_i) = \sigma^2$, we get

$$Var(\hat{\beta}_1) = \frac{\alpha^2}{\sum_{i=1}^{n}x_i^2 - n\bar{x}^2}.$$ (6.21)

Hence, the estimator of β_0 is given by

$$\hat{\beta}_0 = \sum_{i=1}^{n}\frac{y_i}{n} - \hat{\beta}_1\bar{x}.$$

This gives that $\hat{\beta}_0$ can be expressed as a linear combination of the independent normal random variables y_i, $i = 1, 2, \ldots, n$ and is thus normally distributed. Also, the mean of $\hat{\beta}_0$ is as follows:

$$E(\hat{\beta}_0) = \sum_{i=1}^{n} \frac{E(y_i)}{n} - \bar{x}E(\hat{\beta}_1) = \beta_0. \tag{6.22}$$

Thus, $\hat{\beta}_0$ is also an unbiased estimator. The variance of $\hat{\beta}_0$ is computed (left as an exercise) as follows:

$$Var(\hat{\beta}_0) = \frac{\sigma^2 \sum_{i=1}^{n} x_i^2}{n \left(\sum_{i=1}^{n} x_i^2 - n\bar{x}^2 \right)}. \tag{6.23}$$

In this method of unbiased estimator of regression parameter, the RSS

$$\text{RSS} = \sum_{i=1}^{n} (y_i - \bar{y})^2 - \hat{\beta}_1^2 \sum_{i=1}^{n} (x_i - \bar{x})^2 \tag{6.24}$$

can be utilized to estimate the unknown error variance σ^2. Indeed, it can be shown that $\frac{RSS}{\sigma^2}$ has a chi-square distribution with $n - 2$ degrees of freedom. Hence,

$$E\left(\frac{RSS}{\sigma^2}\right) = n - 2 \text{ or } E\left(\frac{RSS}{n-2}\right) = \sigma^2. \tag{6.25}$$

Thus, $\frac{RSS}{n-2}$ is an unbiased estimator of σ^2. Therefore,

$$\hat{\sigma}^2 = \frac{RSS}{n-2}.$$

Further, it can be shown that RSS is independent of the pair $\hat{\beta}_0$ and $\hat{\beta}_1$ (left as an exercise). The fact that RSS is independent of $\hat{\beta}_0$ and $\hat{\beta}_1$ is quite similar to the fundamental result that is normally distributed sampling $\bar{X}$ and S^2 are independent.

Moreover, when the y_i's are normal distributed random variables, the least square estimators are also the maximum likelihood estimators, consequently, the maximum likelihood estimators of β_0 and β_1 are precisely the values of β_0 and β_1 that minimize $\sum_{i=1}^{n} (y_i - \beta_0 - \beta_1 x_i)^2$. That is, they are least squares estimators.

Example 6.7 (*Parameter estimation under unbiased estimator method*) Prove that maximum likelihood estimators of β_0 and β_1 are the same as those obtained from ordinary least squares (OLS)?

Solution:

Let us assume that there are m sample points each represented by (x^i, y^i), $i = 1, \ldots, m$. It is easy to show that the regression model can be transformed into

$\theta^T x^i + \varepsilon^i$ where θ is a column vector containing α and β. This can be done by transforming the vector of the independent variables appropriately. Now, the likelihood is given by (using the information that each y^i is *i.i.d.*):

$$L(\theta; x) = f(Y|X; \theta) = \Pi_{i=1}^{m} f(y^i|x^i; \theta) \qquad (6.26)$$

where

$$f(y^i|x^i; \theta) = \frac{1}{\sqrt{2\pi}\sigma} exp\left[-\frac{(y^i - \theta^T x^i)^2}{2\sigma^2}\right].$$

Hence,

$$L(\theta; x) = f(Y|X; \theta) = \Pi_{i=1}^{m} \frac{1}{\sqrt{2\pi}\sigma} exp\left[-\frac{(y^i - \theta^T x^i)^2}{2\sigma^2}\right],$$

$$\ln(L) = \sum_{i=1}^{m}\left(\ln\left(\frac{1}{\sqrt{2\pi}\sigma}\right) - \frac{(y^i - \theta^T x^i)^2}{2\sigma^2}\right). \qquad (6.27)$$

The first term in the above equation is independent of θ. Thus, when we differentiate $\ln(L)$ with respect to θ only the second term comes into picture. Since, we want to maximize log-likelihood, we are minimizing $\frac{(y^i - \theta^T x^i)^2}{2\sigma^2}$, which is same as OLS. Hence, we will get the same estimates of the regression coefficients.

6.4.3 Hypothesis Testing Regarding Regression Parameters

Consider an important hypothesis regarding the simple linear regression model.

$$y = \beta_0 + \beta_1 x + \varepsilon.$$

The null hypothesis is that $\beta_1 = 0$. It is equivalent to stating that the mean response does not depend on the input, or equivalently, that there is no regression on the input variable.

$$H_0 : \beta_1 = 0, \quad \text{against} \quad H_1 : \beta_1 \neq 0.$$

Now,

$$\frac{\hat{\beta}_1 - \beta_1}{\sqrt{\frac{\sigma^2}{S_{xx}}}} = \sqrt{S_{xx}}\frac{\hat{\beta}_1 - \beta_1}{\sigma} \sim N(0, 1) \qquad (6.28)$$

and is independent of

$$\frac{RSS}{\sigma^2} \sim \chi^2(n-2) \qquad (6.29)$$

where

$$S_{xx} = \sum_{i=1}^{n} (x_i - \bar{x})^2 = \sum_{i=1}^{n} x_i^2 - n\bar{x}^2.$$

Thus, from the definition of t-distribution it follows that

$$\frac{\sqrt{S_{xx}} \frac{(\hat{\beta}_1 - \beta_1)}{\sigma}}{\sqrt{\frac{RSS}{\sigma^2(n-2)}}} = \sqrt{\frac{(n-2)S_{xx}}{RSS}} (\hat{\beta}_1 - \beta_1) \sim t_{n-2}. \qquad (6.30)$$

That is, $\sqrt{\frac{(n-2)S_{xx}}{RSS}} (\hat{\beta}_1 - \beta_1)$ has a t-distribution with $n - 2$ degrees of freedom. Therefore, if H_0 is true,

$$\sqrt{\frac{(n-2)S_{xx}}{RSS}} (\hat{\beta}_1) \sim t_{n-2} \qquad (6.31)$$

which gives rise to the following test of H_0. A hypothesis test of size α is to reject H_0 if

$$\sqrt{\frac{(n-2)S_{xx}}{RSS}} |\hat{\beta}_1| > t_{\frac{\alpha}{2},n-2} \qquad (6.32)$$

and accept H_0 otherwise.

This test can be performed by first computing the value of the test statistic $\sqrt{\frac{(n-2)S_{xx}}{RSS}} |\hat{\beta}_1|$—call its value v—and then rejecting H_0 if the test significance level is at least as large as p value which is equal to

$$pvalue = P(|T_{n-2}| > v) = 2P(|T_{n-2}| > v|) \qquad (6.33)$$

where $T_{(n-2)}$ is a t-distributed random variable with $n - 2$ degrees of freedom.

Remark 6.1 1. Although the t and F tests are equivalent, the t test has some advantages. It may be used to test a hypothesis for any given values of β_1, not just for $\beta_1 = 0$.

2. In many applications, a regression coefficient is useful only if the sign of the coefficient agrees with the underlying theory of the model. Thus, it may be used for a one-sided test.

Example 6.8 (Hypothesis testing for significance of regression coefficients] The estimated slope coefficients for the WTI regression is 0.64 with a standard error of 0.26. Assume that the sample consists of 36 observations, test for the significance of the slope coefficient at 5% level of significance?

Solution:

The null and alternative hypothesis are $H_0 : \beta_1 = 0$ versus $H_1 : \beta_1 \neq 0$. The test statistics for the given data is

$$t = \frac{\hat{\beta}_1 - \beta_1}{s_{\hat{\beta}_1}} = \frac{0.64 - 0}{0.26} = 2.46. \tag{6.34}$$

This is a two-tailed t test and the critical two-tailed t value with $n - 2 = 34$ degrees of freedom are ± 2.03. Since test statistics is greater than the critical value, we reject the null and conclude the slope coefficient is different from zero.

Example 6.9 Use the results given in Example 6.6 and determine the unbiased estimate for σ^2.

Solution:

We have obtained previously that $\sum_{i=1}^{n}(x_i - \bar{x})^2 = 2062.5$, $\hat{\beta}_1 = 0.57$. Also,

$\sum_{i=1}^{n}(y_i - \bar{y})^2 = 680.5$.

Thus, we obtain, $\hat{\sigma}^2 = 1.30$.

Example 6.10 [Hypothesis testing with greater than a value] For the given data of claims and payments on settlement for crop insurance, the amounts, in units of 100 rupees, are shown in Table 6.7. Estimate the coefficients of the regression model and conduct a test for slope parameter, testing whether $\beta_1 \geq 1$ at 5% level of significance?

Solution:

For the given data, $n = 10$, $\Sigma x = 35.4$, $\Sigma y = 32.87$, $S_{xx} = 8.444$, $S_{yy} = 7.1588$, $S_{xy} = 7.4502$. Thus, the coefficients of the regression model are

$$\hat{\beta}_1 = \frac{7.4502}{8.444} = 0.88231,$$
$$\hat{\beta}_0 = 3.287 - (0.88231)(3.54) = 0.164.$$

The null and alternative hypothesis is $H_0 : \beta_1 \geq 1$ versus $H_1 : \beta_1 < 1$, which is equivalent to testing $H_0 : \beta_1 = 1$ versus $H_1 : \beta_1 < 1$, as mentioned in theory. The value of $\hat{\sigma}^2 = \frac{RSS}{n-2} = 0.0732$, and the standard error is given by:

$$se(\hat{\beta}_1) = \sqrt{\frac{0.0732}{8.444}} = 0.0931. \tag{6.35}$$

Table 6.7 Data for Example 6.10

Claim x	2.10	2.40	2.50	3.20	3.60	3.80	4.10	4.20	4.50	5.00
Claim y	2.18	2.06	2.54	2.61	3.67	3.25	4.02	3.71	4.38	4.45

Thus, the test statistics is:

$$t_0 = \frac{\hat{\beta}_1 - 1}{se(\hat{\beta}_1)} = \frac{-0.11769}{0.0931} = -1.264. \tag{6.36}$$

From Table A.10 in Appendix, the one-tailed critical value of $t_{0.05,8} = 1.8595$. Since $t_0 > t_{0.05,8}$, we fail to reject $H_0 : \beta_1 \geq 1$.

6.4.4 Confidence Interval for β_1

A confidence interval estimator for β_1 is easily obtained from Eq. 6.30. Indeed, it follows from Eq. (6.30) that for any α, $0 < \alpha < 1$.

$$P(-t_{\frac{\alpha}{2},n-2} < \sqrt{\frac{(n-2)S_{xx}}{RSS}} (\hat{\beta}_1 - \beta_1) < t_{\frac{\alpha}{2},n-2}) = 1 - \alpha$$

or equivalently,

$$P\left(\hat{\beta}_1 - \sqrt{\frac{RSS}{(n-2)S_{xx}}} t_{\frac{\alpha}{2},n-2} < \beta < \hat{\beta}_1 + \sqrt{\frac{RSS}{(n-2)S_{xx}}} t_{\frac{\alpha}{2},n-2}\right) = 1 - \alpha.$$

A $100(1-\alpha)\%$ confidence interval estimator of β_1 is

$$\left(\hat{\beta}_1 - \sqrt{\frac{RSS}{(n-2)S_{xx}}} t_{\frac{\alpha}{2},n-2}, \hat{\beta}_1 + \sqrt{\frac{RSS}{(n-2)S_{xx}}} t_{\frac{\alpha}{2},n-2}\right) = 1 - \alpha.$$

As a remark, we observe that the result

$$\frac{\hat{\beta}_1 - \beta_1}{\frac{\sigma^2}{S_{xx}}} \sim N(0, 1)$$

cannot be directly be applied to draw inferences about β because it involves σ^2 which is an unknown. To solve this problem, we replace σ^2 by its estimator $\frac{RSS}{n-2}$ which changes the distribution of the above statistics to the t-distribution with $n - 2$ degree of freedom from standard normal distribution.

Example 6.11 (Confidence Interval for parameters of regression with single regressor) The estimated slope coefficient from regression of WTI oil market on S&P500 is 0.64 with a standard error equal to 0.26. Assuming that the sample had 32 observations, calculate the 95% confidence interval for the slope coefficient and determine if it is significantly different from zero?

Solution:

The confidence interval for slope coefficient β_1 is:

$$\hat{\beta}_1 \pm (t_c s_{\hat{\beta}_1}). \tag{6.37}$$

Here $\hat{\beta}_1$ is the estimated value of the slope coefficient, $s_{\hat{\beta}_1}$ is its standard error and t_c is the two-tailed critical value. In this case, the critical two-tail t values are ± 2.042 (from the t table with $n - 2 = 30$ degrees of freedom). We can compute the 95% confidence interval as:

$$0.64 \pm (2.04)(0.26) = (0.11, 1.17). \tag{6.38}$$

Because this confidence interval does not include zero, we can conclude that the slope coefficient is significantly different from zero.

6.4.5 Regression to the Mean

Francis Galton, while describing the laws of inheritance, coined the term regression. He believed that these laws of inheritance made the population extremes to regress toward the mean. In other words, the children of the individuals who have extreme values of a particular feature would tend to have less extreme values of this particular feature as compared to their parents.

If a linear relationship between the feature of the child (y) and that of the parent x is assumed, then a regression to the mean will occur if the regression parameter β_1 lies in the interval $(0, 1)$. In other words, if we have

$$E(y) = \beta_0 + \beta_1 x$$

and $0 < \beta_1 < 1$, then $E(y)$ will be greater than x when x is small and smaller than x when x is small.

Example 6.12 To illustrate Galton's theory of regression to the mean, a British statistician plotted the heights of ten randomly chosen against the heights of their fathers A scatter diagram representing the data is presented in Fig. 6.5

Father's Height	60	62	64	65	66	67	68	70	72	74
Son's Height		63.6	65.2	66	65.5	66.9	67.1	67.4	68.3	70.1	70

It can be observed from Fig. 6.5 that taller fathers tend to have taller sons. Also, it appears that the sons of fathers, who are either extremely small or extremely tall, tend to be more "average" as compared to the height of their fathers, that is a "regression

Fig. 6.5 Scatter plot

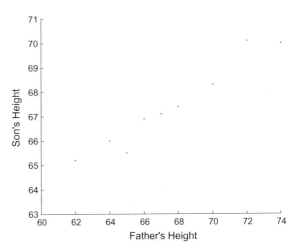

toward the mean." Now, we will test whether the above data provide strong evidences to prove that there is a regression toward the mean; we will use this data set to test

$$H_0 : \beta_1 \geq 1 \ \text{ against } \ H_1 : \beta_1 < 1$$

which is equivalent to a test

$$H_0 : \beta_1 = 1 \ \text{ against } \ H_1 : \beta_1 < 1.$$

If not, it follows from equation (19) that when $\beta_1 = 1$, the test statistic

$$T = \sqrt{\frac{8S_{XX}}{RSS}} (\beta_1 - 1)$$

has a t-distribution with 8 degrees of freedom.

The significance level α - test will reject H_0 when the value of T is sufficiently small. The test is to reject H_0 if $\sqrt{\frac{8S_{xx}}{RSS}} (\beta_1 - 1) < -t_{\alpha,8}$ we have

$$\sqrt{\frac{8S_{xx}}{RSS}} (\beta_1 - 1) = 30.2794(0.4646 - 1) = -16.213.$$

Since $t_{0.01,8} = 2.896$, we see that

$$T < -t_{0.01,8}$$

and hence we can reject the null hypothesis, $\beta_1 = 1$ at 99% confidence level. Also, the *pvalue*

$$pvalue = P(t_8 \leq -16.213)$$

is approximately zero which shows that the null hypothesis that $\beta_1 \geq 1$ can be rejected at almost any significance leve; thus, proving that there is a regression toward the mean.

Example 6.13 (*Confidence Interval for a predicted value*) Given the regression equation:

$$\hat{WTI} = -2.3\% + (0.64)(S\&P500). \tag{6.39}$$

Calculate a 95% prediction interval on the predicted value of WTI crude oil market. Assume the standard error of forecast is 3.67 and forecasted value of S&P500 excess return is 10%?

Solution: The predicted value for WTI is:

$$\hat{WTI} = -2.3\% + (0.64)(10\%) = 4.1\%. \tag{6.40}$$

The 5% two-tailed critical t value with 34 degrees of freedom is 2.03. The prediction interval at the 95% confidence level is:

$$\hat{WTI} \pm (t_c s_f) = [4.1\% \pm ((2.03)(3.67\%)] = 4.1\% \pm 7.5\%. \tag{6.41}$$

where t_c is two-tailed critical value and s_f is standard error of forecast. This range can be interpreted as, given a forecasted value for S&P500 excess returns of 10%, we can be 95% confident that the WTI excess returns will be between −3.4 and 11.6%.

Example 6.14 (*Confidence interval for mean predicted value*) For the data given in Example 6.2, estimate the mean weight of a cub at 33 weeks. Construct a 95% confidence interval for the same, given that the estimated value of $\hat{\sigma}^2 = 0.0234$?

Solution:

Recall that the estimated value for the coefficients is $\hat{\beta}_0 = -4.60$, $\hat{\beta}_1 = 0.2043$, $\bar{x} = 35$ and $S_{xx} = 70$. The least square regression line is $\hat{y} = -4.60 + 0.2043x$. Given that $x_o = 33$, so we have $\hat{\mu}_o = -4.60 + (0.2043)33 = 2.142$. The mean weight of the cub at 33 weeks is expected to be 2.142 kg. Now the standard error is given by

$$se(\hat{\mu}_o) = \sqrt{\left(\frac{1}{n} + \frac{(x_o - \bar{x})^2}{S_{xx}}\right)\hat{\sigma}^2},$$

$$se(\hat{\mu}_o) = \sqrt{\left(\frac{1}{6} + \frac{(33 - 35)^2}{70}\right)0.0234} = 0.072. \tag{6.42}$$

Also,

$$t_{0.05,4} = 2.132.$$

Hence 95% confidence interval is

$$2.142 \pm (2.132)(0.072) = 2.142 \pm 0.1535 \equiv (1.9885, 2.2955).$$

6.4.6 Inferences Covering β_0

The determination of confidence intervals and hypothesis tests for β_0 is accomplished in exactly same manner as was evaluated for β_1. Specifically, (left as an exercise)

$$\sqrt{\frac{n(n-2)S_{xx}}{\sum x_i^2 RSS}} (\hat{\beta}_0 - \beta_0) \sim t_{n-2}$$

which leads to the following confidence interval estimator of α. The $100(1-\alpha)$ confidence interval for β_0 is the interval

$$\hat{\beta}_0 \pm \sqrt{\frac{\sum x_i^2 RSS}{n(n-2)S_{xx}}} t_{\frac{\gamma}{2}, n-2}.$$

6.4.7 Inferences Concerning the Mean Response of $\beta_0 + \beta_1 x_0$

It is often of interest to use the data pairs (x_i, y_i), $i = 1, 2, \ldots, n$, to estimate $\beta_0 + \beta_1 x_0$, the mean response for a given input level x_0. If it is a point estimator that is desired, then the neutral estimator is $A + Bx_0$, which is an unbiased estimator since

$$E(A + Bx_0) = E(A) + x_0 E(B) = \beta_0 + \beta_1 x_0.$$

Since the y_i's are independent normal random variable, and

$$A + Bx_0 = \sum_{i=1}^{n} y_i \left[\frac{1}{n} - \frac{1}{S_{xx}} (x_i - \bar{x})(\bar{x} - x_0) \right].$$

We can show that (left as an exercise)

$$A + Bx_0 \sim N\left(\beta_0 + \beta_1 x_0, \sigma^2 \left(\frac{1}{n} + \frac{(x_i - \bar{x})^2}{S_{xx}} \right) \right).$$

In addition, because $A + Bx_0$ is independent of $\frac{RSS}{\sigma^2}$ and

$$\frac{RSS}{\sigma^2} \sim \chi^2_{n-2}.$$

Hence, it follows that

$$\frac{A + Bx_0 - (\beta_0 + \beta_1 x_0)}{\sqrt{\frac{1}{n} + \frac{(x_0 - \bar{x})^2}{S_{xx}}} \sqrt{\frac{RSS}{n-2}}} \sim t_{n-2}.$$

Then, with $100(1 - \alpha)\%$ confidence, $\beta_0 + \beta_1 x_0$ will lie within

$$A + Bx_0 \pm \sqrt{\frac{1}{n} + \frac{(x_0 - \bar{x})^2}{S_{xx}}} \sqrt{\frac{RSS}{n-2}} t_{\frac{\alpha}{2}, n-2}.$$

6.5 Logistic Regression

In many practical situations, where one is interested in studying relationship between input and output variables as in regression, the output variable is discrete rather than a continuous variable. In particular, in many practical situations, we have a binary output variable. Further, some of the input variables may or may not be continuous variables. The question is how can one model and analyze such situations? In this section, we will address such situations where the output variable is a binary variable, i.e., success or failure. We will assume that these experiments can be carried at various levels and a performance of an experiment at level x results in a success with probability $p(x)$. In conclusion, the problem is to model the conditional probability $P(Y = 1/X = x)$ as a function of x when the output variable Y is a binary variable with values 0 and 1.

The experiments comes from a logistic regression model if $p(x)$ takes a form of inverse logistic function, i.e.,

$$\log \frac{p(x)}{1 - p(x)} = a + bx$$

which is equivalent to

$$p(x) = \frac{1}{[e^{-1(a+bx)} + 1]}.$$

We call $p(x)$ the logistics regression function. Based on the different possible values of b, we have the following possibilities,

1. When $b = 0$, $p(x)$ is constant function.
2. If $b > 0$, then $p(x)$ is an increasing function and it converges to 1 when $x \to \infty$.
3. if $b < 0$, then $p(x)$ is a decreasing function and it converges to 0 as $x \to \infty$.

The unknown parameters that need to be estimated in a logistic regression model are a and b. The estimation of unknown parameters can be performed by using the maximum likelihood estimation technique in the following manner. Suppose that there are following levels $x_1, \ldots, x_k$ at which an experiment can be performed. Then, for each data point in training data, we have observed class y_i corresponding to each feature x_i. The probability of that class is either p, if $y_i = 1$, or $1 - p$, if $y_i = 0$. The likelihood is then

$$L(a, b) = \prod_{i=1}^{n} (p(x_i))^{y_i} (1 - p(x_i))^{1-y_i}.$$

Taking logarithm on both sides, we get the log-likelihood function as

$$\log L(a, b) = \sum_{i=1}^{n} (y_i \log(p(x_i)) + (1 - y_i) \log(1 - p(x_i)))$$

$$= \sum_{i=1}^{n} \log(1 - p(x_i)) + \sum_{i=1}^{n} y_i \log\left(\frac{p(x_i)}{1 - p(x_i)}\right)$$

$$= \sum_{i=1}^{n} \log(1 - p(x_i)) + \sum_{i=1}^{n} y_i(a + bx_i).$$

The maximum likelihood estimates of the unknowns a and b can be obtained by finding the values of a and b that maximizes the log-likelihood function given above. But one needs to perform this numerically as closed form solutions are not available as the log-likelihood function is nonlinear in nature.

If the function $p(x)$ takes the following form for some constants $a > 0$ and $b > 0$

$$p(x) = \Phi(a + bx) = \frac{1}{\sqrt{2\pi}} \int_{-\infty}^{a+bx} e^{-\frac{y^2}{2}} dy.$$

Such models are known as *probit* models. It can be observed that $p(x)$ is nothing but $\Phi(a + bx)$ where $\Phi(\cdot)$ is CDF of standard normal random variable.

The interested readers to know more about correlation and regression may refer to Panik (2012) and Rohatgi and Saleh (2015).

Problems

6.1 A random sample of size 8 from a bivariate normal distribution yields a value of the correlation coefficient of 0.75. Would we accept or reject at the 5% significance level, the hypothesis that $\rho = 0$.

6.2 The correlation between scores on a traditional aptitude test and scores on a final test is known to be approximately 0.7. A new aptitude test has been developed and is tried on a random sample of 100 students, resulting in a correlation of 0.67. Does this result imply that the new test is better?

Table 6.8 Data for Problem 6.6

x	4	2	5	3	2	3	4	3	5	2
y	3.12	3.00	4.5	4.75	3	3.5	3.75	4.12	4.54	3.1

6.3 A sample of size 100 from a normal population had an observed correlation of 0.6. Is the shortfall from the claimed correlation of at least 0.75 significant at 5% level of significance? What would a confidence interval for the correlation coefficient be at 95% level of significance?

6.4 A sample of size 27 from a bivariate normal population had an observed correlation of 0.2. Can you discard the claim that components are independent? Use 5% level of significance.

6.5 Show that for any collection $(X_1, X_2, \ldots, X_n)$ of random variables the covariance matrix $\sum = (Cov(X_i, X_j))$, which is symmetric, is always positive definite.

6.6 Students' scores in the probability course examination, x, and on the semester CGPA, y, are given in Table 6.8.

(a) Calculate the least square regression line for the data of Table 6.8.
(b) Plot the points and the least square regression line on the same graph.
(c) Find point estimates for β_0, β_1 and σ^2.
(d) Find 95% confidence interval for the α and β under the usual assumptions.

6.7 Consider the weight X_1, age X_2 and height X_3 of 12 students of a school given in Table 6.9. Find the least square linear regression line of X_1 on X_2 and X_3. Also find the coefficient of determination.

6.8 Prove that

$$Var(\hat{\beta_0}) = \frac{\sigma^2 \sum_{i=1}^{n} x_i^2}{n\left(\sum_{i=1}^{n} x_i^2 - n\bar{x}^2\right)}.$$

6.9 Prove that residual sum of squares (RSS) is independent of pairs $\hat{\beta_0}$ and $\hat{\beta_1}$.

Table 6.9 Height and weight data of 12 students of a school

Weight (pounds)	74	81	63	77	65	68	87	67	66	61	86	78
Age (years)	9	11	7	12	9	8	11	10	11	7	13	10
Height (inches)	52	54	44	57	46	45	50	43	47	37	56	52

References

Panik MJ (2012) Statistical inference: a short course. Wiley, New York

Rohatgi VK, Ehsanes Saleh AKMd (2015) An introduction to probability and statistics. Wiley, New York

Chapter 7
Single-Factor Experimental Design

7.1 Introduction

Often, we wish to investigate the effect of a factor (independent variable) on a response (dependent variable). We then carry out an experiment where the levels of the factor are varied. Such experiments are known as single-factor experiment. There are many designs available to carry out such experiment. The most popular ones are completely randomized design, randomized block design, Latin square design and balanced incomplete block design. In this chapter, we will discuss these four designs along with the statistical analysis of the data obtained by following such designs of experiments.

7.2 Completely Randomized Design

This is a very important single-factor experimental design. This design includes two basic principles of design of experiments, that are, randomization and replication. They are discussed below.

Randomization refers to random allocation of the experimental materials to different objects of comparison (treatments) in an experiment. Let us illustrate this with the help of an example. Suppose an agricultural scientist wishes to know if the quantity of a fertilizer affects the yield of a crop. He divides a large agricultural plot into $3^2 = 9$ subplots and allocates the nine subplots to three different quantities $(A_1, A_2, \text{and } A_3)$ of the fertilizer (treatments) randomly. This is shown in Fig. 7.1. Sometimes, randomization also refers to performing the individual runs or trials of the experiment randomly. Let us illustrate this with the help of an example. Suppose a product development engineer wishes to examine if the percentage of jute fibers determine the tensile strength of jute–polypropylene composites. She carries out nine runs with three levels (5, 10, and15%) of jute fiber percentage and three replicates.

© Springer Nature Singapore Pte Ltd. 2018
D. Selvamuthu and D. Das, *Introduction to Statistical Methods,*
Design of Experiments and Statistical Quality Control,
https://doi.org/10.1007/978-981-13-1736-1_7

Fig. 7.1 Random allocation
of treatments to plots

A_1	A_2	A_3
A_2	A_3	A_2
A_2	A_1	A_1

Table 7.1 Experimental run
numbers

Jute fiber percentage	Experimental run number		
5	1	2	3
10	4	5	6
15	7	8	9

The experimental run numbers are shown in Table 7.1. The tensile strength of thus
prepared nine specimens is carried out randomly such that the specimens prepared
by experimental run numbers 5, 1, 7, 2, 4, 6, 8, 3, 9 are tested as per the following
test sequence: 1, 2, 3, 4, 5, 6, 7, 8, 9. Randomization can be done by using tables of
random numbers or computer programs based on random number generators. Randomization is very important as far as the statistical methods of data analysis are
concerned. In many statistical methods of data analysis, it is assumed that the observations (errors) are independently distributed random variables and randomization
makes this assumption valid. Also, randomizing the treatments over the experimental
materials averages out the effect of extraneous factors over which we have no control,
including rise or fall of environmental conditions, drift in calibration of instruments
and equipments, fertility of soil. For example, suppose in the above example there is
a sudden change in relative humidity of the testing laboratory. If all specimens prepared at 10% jute fiber percentage would have been tested at higher level of humidity
in the testing laboratory, there would have been a systematic bias in the experimental
results which consequently invalidates the results. Randomly testing the specimens
alleviates this problem.

Replication means repetition of the treatments under investigation. In the earlier
example of composite, replication would mean preparing a composite specimen by
keeping the amount of jute fiber as 5%. Thus, if three specimens of composite are
prepared each by keeping the amount of jute fiber as 5%, we say that three replicates
are obtained. As known, replication has two important properties. First, it permits us
to obtain an estimate of experimental error. The estimate of error becomes a basic unit
of measurement for determining whether the observed differences are statistically
significant. Second, if the sample mean is used to estimate the effect of a factor in the
experiment, replication permits the experimenter to obtain a more precise estimate
of the effect. As known, the variance of sample mean decreases with the increase in
the number of replicates. This is shown below.

$$\sigma_{\bar{y}}^2 = \frac{\sigma^2}{n}$$

where $\sigma_{\bar{y}}^2$ denotes the variance of sample mean $\bar{y}$, σ^2 indicates the variance of individual observations, and n refers to the number of replicate. Note that the replicates are different from the repeated measurements. Suppose a large sample of composite is prepared by keeping the amount of jute fibers as 5%, then the large sample is divided into four small samples and finally the tensile strength of the four small samples is obtained. Here, the measurements on the four samples are not replicates but repeated measurements. In this case, the repeated measurements reflect the variability within runs, while the replicates reflect the variability between and (potentially) within runs. Let us take one more example. Suppose a cotton fiber is measured for its length three times. These measurements are not replicates, they are a form of repeated measurements. In this case, the observed variability in the three repeated measurements indicates the inherent variability in measurement or gauge of the length tester.

7.2.1 A Practical Problem

A product development engineer was interested in investigating the filtration efficiency of fibrous filter media that were suitable for HVAC filtration application. He knew that the filtration efficiency was affected by the shape of cross-section of fibers that were used to prepare the filter media. He then decided to prepare fibrous filter media by mixing fibers of deep-groove cross-section and circular cross-section in different weight proportions. He chose five levels of weight percentage of deep-grooved fibers (0, 25, 50, 75, 100) and prepared five specimens at each of the five levels of weight percentage of deep-grooved fibers. He thus prepared 25 specimens as shown in Table 7.2.

The 25 specimens were prepared in a random manner. This was done by using a random number table. By using this table, a random number between 1 and 25 was selected. Suppose a number 11 was selected, then, specimen number 11 was prepared first. By using the table, the engineer selected another random number between 1 and 25. Suppose this number was 16. Then, specimen number 16 was prepared. This process was repeated until 25 specimens were assigned run numbers randomly. The only restriction on randomization was that if the same number was drawn again, it was discarded. In this way, the runs were made as shown in Table 7.2. The randomization was necessary to average out the effect of any extraneous factor over which the experimenter had no control, for example, sudden rise or fall in environmental conditions, drift in processing equipments, etc. Also, randomization was required so far the statistical methods of data analysis were concerned. As known, many statistical methods of data analysis demand that the experimental errors are independently distributed random variables and randomization makes this assumption valid.

Table 7.2 Specimen numbers versus run numbers

Specimen no.	Weight percentage of fibers		Run no.	Specimen no.	Weight percentage of fibers		Run no.
	Deep-grooved	Circular			Deep-grooved	Circular	
1	0	100	19	16	75	25	2
2	0	100	20	17	75	25	12
3	0	100	22	18	75	25	3
4	0	100	9	19	75	25	10
5	0	100	15	20	75	25	16
6	25	75	4	21	100	0	13
7	25	75	6	22	100	0	11
8	25	75	17	23	100	0	25
9	25	75	24	24	100	0	18
10	25	75	7	25	100	0	21
11	50	50	1				
12	50	50	8				
13	50	50	5				
14	50	50	23				
15	50	50	14				

7.2.2 Data Visualization

Table 7.3 reports the filtration efficiency of the filter media. It is always better to plot the experimental data graphically. Figure 7.2a displays the scatter diagram of filtration efficiency against the weight percentage of deep-grooved fibers. The solid dots denote the individual observation and the hollow circles indicate the average filtration efficiencies. Figure 7.2b shows the box plot of filtration efficiency, to know more about this plot, please refer to Sect. 3.2.4 in Chap. 3 of this book. Both graphs

Table 7.3 Experimental results of filtration efficiency

Weight of deep-grooved fibers (%)	Filtration efficiency (%)				
	I	II	III	IV	V
0	44	47	47	45	46
25	59	57	61	53	58
50	61	59	63	58	60
75	69	66	69	67	65
100	71	74	74	72	71

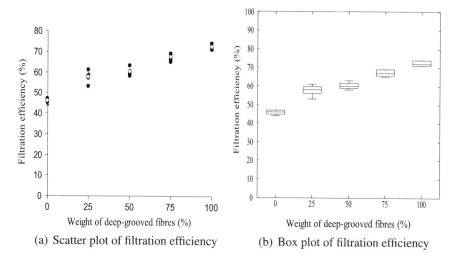

(a) Scatter plot of filtration efficiency   (b) Box plot of filtration efficiency

Fig. 7.2 Plots for experimental data in practical problem given in Sect. 7.2.1

indicate that the filtration efficiency increases with the increase in weight proportion of deep-grooved fibers in the filter media. It is strongly suspected that the weight proportion of deep-grooved fibers affects the filtration efficiency.

7.2.3 Descriptive Model

Suppose we wish to compare a treatment at different levels of a single factor x on a response variable y. The observations are taken in a random order. The experimental dataset would look as shown in Table 7.4.

The total $y_{i.}$ and average $\bar{y}_{i.}$ are obtained as follows

Table 7.4 Data for single-factor experiment

Treatment (Level)	Observations				Total	Average
1	y_{11}	y_{12}	$\ldots$	y_{1n}	$y_{1.}$	$\bar{y}_{1.}$
2	y_{21}	y_{22}	$\ldots$	y_{2n}	$y_{2.}$	$\bar{y}_{2.}$
$\vdots$	$\vdots$	$\vdots$	$\vdots$	$\vdots$	$\vdots$	$\vdots$
a	y_{m1}	y_{m2}	$\ldots$	y_{mn}	$y_{m.}$	$\bar{y}_{m.}$
					$y_{..}$	$\bar{y}_{..}$

$$y_{i.} = \sum_{j=1}^{n} y_{ij}, \quad i = 1, 2, \ldots, m \quad \text{and} \quad \bar{y}_{i.} = \frac{y_{i.}}{n}$$

The grand total $y_{..}$ and grand average $\bar{y}_{..}$ are obtained as follows

$$y_{..} = \sum_{i=1}^{m} \sum_{j=1}^{n} y_{ij} \quad \text{and} \quad \bar{y}_{..} = \frac{y_{..}}{mn} = \frac{y_{..}}{N},$$

where $N = mn$.

The observations of the experiment can be described by the following linear statistical model

$$y_{ij} = \mu + \tau_i + \varepsilon_{ij}; \quad i = 1, 2, \ldots, m; \ j = 1, 2, \ldots, n$$

where y_{ij} is a random variable denoting the ijth observation, μ is a parameter common to all levels called the overall mean, τ_i is a parameter associated with the ith treatment (level) called the ith treatment (level) effect, and ε_{ij} is a random error component.

The model can also be written as

$$y_{ij} = \mu_i + \varepsilon_{ij}; \quad i = 1, 2, \ldots, m; \ j = 1, 2, \ldots, n$$

where $\mu_i = \mu + \tau_i$ is the mean of ith level.

The model can describe two situations depending on how the treatments are chosen. Suppose that the model involves specifically chosen treatments (levels) of a factor, then its conclusion is limited to these levels of the factor only. It means that conclusions cannot be extended to similar treatments (levels) that are not explicitly considered. In such cases, the levels of the factor are said to be fixed and the model is called fixed effect model. On the other hand, suppose that the model involves levels of a factor chosen from a larger population of levels of that factor, then its conclusion can be extended to all levels of the factor in the population, whether or not they are explicitly considered in the experiment. Then, these levels of the factor are said to be variables and the model is called random effect model. Needless to say that the analysis of the model changes if the type of model changes. Here, we will discuss the analysis of a fixed effect model. This means that our conclusions will be valid only for the treatments or levels chosen for the study.

The followings are the reasonable estimates of the model parameters

$$\hat{\mu} = \bar{y}_{..}$$
$$\hat{\tau}_i = \bar{y}_{i.} - \bar{y}_{..}, \quad i = 1, 2, \ldots, k$$
$$\hat{\mu}_i = \hat{\mu} + \hat{\tau}_i = \bar{y}_{i.}, \quad i = 1, 2, \ldots, k$$

The estimate of the observations is $\hat{y}_{ij} = \hat{\mu} + \hat{\tau}_i = \bar{y}_{..} + \bar{y}_{i.} - \bar{y}_{..} = \bar{y}_{i.}$.
Hence, the residual is $e_{ij} = y_{ij} - \hat{y}_{ij}$.

These formulas are used to estimate the observations and the residuals. Table 7.5 compares the estimated observations with the actual ones and also reports the residuals. The estimated observations are placed within parentheses, and the residuals are shown at the bottom of each cell.

It is necessary to check whether the model is adequate. The adequacy of a model is investigated by examining the residuals. If the model is adequate then the residuals should be structureless; that is, they should not follow any trend or pattern. Figure 7.3a plots the residuals against the run numbers. It can be observed that the residuals do not follow any specific trend or pattern. Plotting the residuals against the time order of experiments is helpful in detecting the correlation between the residuals. A tendency to have runs of positive and negative residuals indicates positive correlation. Such a pattern would indicate that the residuals are not independent. The violation of independence assumption on the residuals is potentially a serious problem, and proper randomization of the experiment is an important step in obtaining independence. It can be seen from Fig. 7.3a that there is no reason to suspect any violation of independence assumption on the residuals. There is one more check for the residuals to display structureless character. The residuals should not be related to any other variable including the predicted response. A simple check for this is to plot the residuals versus the fitted responses, and it should not reveal any obvious pattern. Figure 7.3b plots the residuals versus the fitted responses. It can be observed that no obvious pattern is seen. Also, it is important to check if the residuals can be regarded as taken from a normal distribution or not. Figure 7.4 plots the normal probability plot of the residuals. However, it can be observed that this plot is not grossly nonnormal.

Table 7.5 Data and residuals for filtration efficiency

Weight of deep-grooved fibers (%)	Filtration efficiency (%)				
	I	II	III	IV	V
0	44 (45.8)	47 (45.8)	47 (45.8)	45 (45.8)	46 (45.8)
	−1.8	1.2	1.2	−0.8	0.2
25	59 (57.6)	57 (57.6)	61 (57.6)	53 (57.6)	58 (57.6)
	1.4	−0.6	3.4	−4.6	0.4
50	61 (60.2)	59 (60.2)	63 (60.2)	58 (60.2)	60 (60.2)
	0.8	−1.2	2.8	−2.2	−0.2
75	69 (67.2)	66 (67.2)	69 (67.2)	67 (67.2)	65 (67.2)
	1.8	−1.2	1.8	−0.2	−2.2
100	71 (72.4)	74 (72.4)	74 (72.4)	72 (72.4)	71 (72.4)
	−1.4	1.6	1.6	−0.4	−1.4

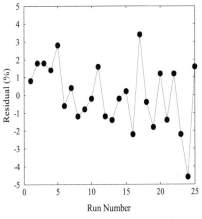

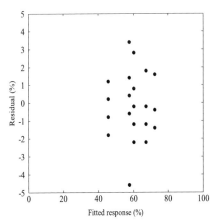

(a) Plot of residuals over time for filtration efficiency

(b) Plot of residuals versus fitted responses for filtration efficiency

Fig. 7.3 Plots for residuals

Fig. 7.4 Normal probability plot of residuals for filtration efficiency

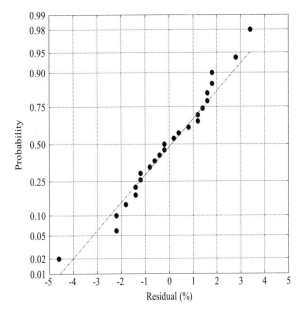

7.2.4 Test of Hypothesis

The detailed discussion on testing of hypothesis is given in Chap. 5. Here, we are giving a brief introduction. Suppose we are interested in testing the equality of treatment means of a single factor x on a response variable y. Often, we select a sample from a population, carry out experiment, and see that the sample treatment means are different. Then, an obvious question arises regarding whether the population treatment means are different or not. In order to answer such a question, a statistical test of hypothesis is carried out.

The statistical test of hypothesis is conducted in five steps—Step 1: State the statistical hypothesis, Step 2: Select the level of significance to be used, Step 3: Find out the value of the test statistic, Step 4: Specify the critical region to be used, Step 5: Take a decision.

In the first step, it is necessary to state the hypothesis. The null hypothesis states that there is no difference among the population treatment means. This null hypothesis is tested for possible rejection under the assumption that the null hypothesis is true. If this null hypothesis is proved to be false, then the alternative hypothesis is accepted. The alternative hypothesis is complementary to null hypothesis. It states that there is a difference between the population treatment means for at least a pair of populations.

In the second step, the level of significance, usually denoted by α, is stated in terms of some small probability value such as 0.10 (one in ten) or 0.05 (one in twenty) which is equal to the probability that the test statistic falling in the critical region, thus indicating falsity of the null hypothesis. The selection of the level of significance is critical. This can be understood with a view to the errors associated with the testing of hypothesis.

There are two types of errors involved in test of hypothesis. They are described in Table 7.6. The error committed by rejecting the null hypothesis when it is true is called Type I error, and the error committed by accepting the null hypothesis when it is false is called Type II error. In situations where Type I error is possible, the level of significance α represents the probability of such an error. The higher is the value of level of significance, the higher is the probability of Type I error. Here, $\alpha = 0$ means complete elimination of occurrence of Type I error. Of course, it implies that no critical region exists; hence, the null hypothesis is retained always. In this case, in fact, there is no need to analyze or even collect any data at all. Obviously, while such a procedure would completely eliminate the possibility of making a Type I error, it

Table 7.6 Errors in test of hypothesis

Possibilities	Course of Action	
Null hypothesis is true.	Accept (Desired correct action)	Reject (Undesired erroneous action)
Null hypothesis is false.	Accept (Undesired erroneous action)	Reject (Desired correct action)

does not provide a guarantee against error, for every time that the null hypothesis stated is false, a Type II error would necessarily occur. Similarly, by letting $\alpha = 1$, it would be possible to eliminate entirely the occurrence of Type II error at the cost of committing a Type I error for every true null hypothesis tested. Thus, the choice of a level of significance represents a compromise effect at controlling the two types of errors that may occur in testing statistical hypothesis. We see that for a given choice of α, there is always a probability for Type II error. Let us denote this probability by β. This depends on the value α chosen. The higher is the value of α, the lower is the value of β.

The third step is related to finding the value of test statistic. The phrase test statistic is simply used here to refer to the statistic employed in effecting the test of hypothesis. Here, the test statistic is

$$F_0 = \frac{MS_{\text{Treatments}}}{MS_{\text{Error}}}$$

where $\quad MS_{\text{Treatments}} = \dfrac{SS_{\text{Treatments}}}{(m-1)} = \dfrac{\sum\limits_{i=1}^{m}\sum\limits_{j=1}^{n}(\bar{y}_i - \bar{y}_{..})^2}{k-1}$

and $\quad MS_{\text{Error}} = \dfrac{SS_{\text{Error}}}{m(n-1)} = \dfrac{\sum\limits_{i=1}^{m}\sum\limits_{j=1}^{n}(\bar{y}_{ij} - \bar{y}_{i.})^2}{m(n-1)}$

Generally, analysis of variance (ANOVA) table is constructed to calculate the value of test statistic.

The fourth step is concerned with specifying a critical region. A critical region is a portion of the scale of possible values of the statistic so chosen that if the particular obtained value of the statistic falls within it, rejection of the hypothesis is indicated. Here, we should reject H_0 when $F_0 > F_{\alpha, a-1, a(n-1)}$.

The fifth and final step is the decision making step. In this step, we refer the value of the test statistic as obtained in Step 4 to the critical region adopted. If the value falls in this region, reject the hypothesis. Otherwise, retain or accept the hypothesis as a tenable (not disproved) possibility.

Let us carry out a test of hypothesis in case of filtration efficiency data given in Table 7.3.

Step 1: Statement of Hypothesis

Here, the null hypothesis states that there is no difference among the population treatment means of filtration efficiency. This means

$$H_0: \mu_1 = \mu_2 = \cdots = \mu_m \quad \text{or, equivalently,} \quad H_0: \tau_1 = \tau_2 = \cdots = \tau_m = 0$$

Table 7.7 Calculations for treatment totals and treatment means for filtration efficiency

Weight of deep-grooved fibers (%)	Filtration efficiency (%)						
	I	II	III	IV	V	Total	Mean
0	44	47	47	45	46	229	45.8
25	59	57	61	53	58	288	57.6
50	61	59	63	58	60	301	60.2
75	69	66	69	67	65	336	67.2
100	71	74	74	72	71	362	72.4
Filtration efficiency (%)						1516	60.64

The alternative hypothesis states that there is difference between the population treatment means for at least a pair of populations. This means

$H_1 : \mu_i \neq \mu_j$ for at least one i and j or, equivalently, $H_1 : \tau_i \neq 0$ for at least one i

Step 2: Selection of Level of Significance

Suppose the level of significance is chosen as 0.05. It means that the probability of rejecting the null hypothesis when it is true is less than or equal to 0.05.

Step 3: Computation of Test Statistic

The test statistic is computed by carrying out an analysis of variance of the data. The details of analysis of variance are given in Sect. 5.3 of this book. The basic calculations for treatment totals and treatment means are shown in Table 7.7.

The calculations for ANOVA are shown below.

$$SS_{Total} = \sum_{i=1}^{m}\sum_{j=1}^{n}(\bar{y}_{ij} - \bar{y})^2$$

$$
\begin{aligned}
= &(44 - 60.64)^2 + (47 - 60.64)^2 + (47 - 60.64)^2 + (45 - 60.64)^2 + (46 - 60.64)^2 + \\
&(59 - 60.64)^2 + (57 - 60.64)^2 + (61 - 60.64)^2 + (53 - 60.64)^2 + (58 - 60.64)^2 + \\
&(61 - 60.64)^2 + (59 - 60.64)^2 + (63 - 60.64)^2 + (58 - 60.64)^2 + (60 - 60.64)^2 + \\
&(69 - 60.64)^2 + (66 - 60.64)^2 + (69 - 60.64)^2 + (67 - 60.64)^2 + (65 - 60.64)^2 + \\
&(71 - 60.64)^2 + (74 - 60.64)^2 + (74 - 60.64)^2 + (72 - 60.64)^2 + (71 - 60.64)^2
\end{aligned}
$$

$$= 2133.76$$

$$SS_{Treatments} = n\sum_{i=1}^{m}(\bar{y}_i - \bar{y})^2$$

$$= 5[(45.8 - 60.64)^2 + (57.6 - 60.64)^2 + (60.2 - 60.64)^2 + (67.2 - 60.64)^2 + (72.4 - 60.64)^2]$$

$$= 2054.96$$

Table 7.8 ANOVA Table for filtration efficiency

Source of variation	Sum of squares	Degree of freedom	Mean square	F_0
Fiber weight percentage	2054.96	4	513.74	130.39
Error	78.8	20	3.94	
Total	2133.76	24		

Fig. 7.5 Display of critical region

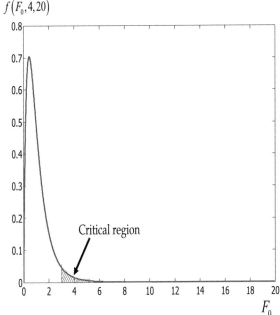

$$SS_{\text{Error}} = SS_{\text{Total}} - SS_{\text{Treatments}} = 2133.76 - 2054.96 = 78.8$$

The calculations are summarized in Table 7.8.

Step 4: Specification of Critical Region

The null hypothesis is rejected if $F_0 > F_{\alpha,m-1,m(n-1)} = F_{0.05,5-1,5(5-1)} = 2.87$ (Table A.11). The critical region is shown in Fig. 7.5.

Step 5: Take a Decision

As the value of the test statistic falls in the critical region adopted, this calls for rejection of null hypothesis. It is therefore concluded that there is difference between population treatment means, at least for a pair of populations.

7.2.5 Multiple Comparison Among Treatment Means (Tukey's Test)

Tukey's test compares pairs of treatment means. This test declares two means to be significantly different if the absolute value of their difference exceeds

$$T_\alpha = q_\alpha(a,f)\sqrt{\frac{MS_{Error}}{n}}.$$

Here, $q_\alpha(a,f)$ is called **studentized range statistic** corresponding to the level of significance of α, a is the number of factor level, f is the degree of freedom associated with the MS_{Error}, and n is the number of treatment levels. The numerical value of the studentized range statistic can be obtained from Table A.17.

Here $T_{0.05} = q_\alpha(a,f)\sqrt{\frac{MS_{Error}}{n}} = q_{0.05}(5,20)\sqrt{\frac{3.94}{5}} = 3.76$.

The five treatment averages are $\bar{y}_1 = 45.8$, $\bar{y}_2 = 57.6$, $\bar{y}_3 = 60.2$, $\bar{y}_4 = 67.2$, $\bar{y}_5 = 72.4$. The differences in averages are

$$|\bar{y}_1 - \bar{y}_2| = 11.8^* \quad |\bar{y}_2 - \bar{y}_3| = 2.6 \quad |\bar{y}_3 - \bar{y}_4| = 7^* \quad |\bar{y}_4 - \bar{y}_5| = 5.2^*$$
$$|\bar{y}_1 - \bar{y}_3| = 14.4^* \quad |\bar{y}_2 - \bar{y}_4| = 9.6^* \quad |\bar{y}_3 - \bar{y}_5| = 12.2^*$$
$$|\bar{y}_1 - \bar{y}_4| = 21.4^* \quad |\bar{y}_2 - \bar{y}_5| = 14.8^*$$
$$|\bar{y}_1 - \bar{y}_5| = 26.6^*$$

The starred values indicate the pairs of means that are significantly different. Sometimes it is useful to draw a graph, as shown below, underlining pairs of means that do not differ significantly.

$$\bar{y}_1 = 45.8, \quad \bar{y}_2 = 57.6, \quad \bar{y}_3 = 60.2, \quad \bar{y}_4 = 67.2, \quad \bar{y}_5 = 72.4$$

Another popular test for comparing the pairs of treatment means is Fisher's test. This test declares two means to be significantly different if the absolute value of their difference exceeds the least significant difference (*LSD*)

$$LSD = t_{\frac{\alpha}{2},N-a}\sqrt{\frac{2MS_{Error}}{n}}$$

Here, $LSD = t_{\frac{\alpha}{2},N-a}\sqrt{\frac{2MS_{Error}}{n}} = t_{\frac{0.05}{2},25-5}\sqrt{\frac{2 \times 3.94}{5}} = 2.62$

(From Table A.10, $t_{0.025,20} = 2.086$) The five treatment averages are $\bar{y}_1 = 45.8$, $\bar{y}_2 = 57.6$, $\bar{y}_3 = 60.2$, $\bar{y}_4 = 67.2$, $\bar{y}_5 = 72.4$. The differences in averages are

$$|\bar{y}_1 - \bar{y}_2| = 11.8^* \quad |\bar{y}_2 - \bar{y}_3| = 2.6 \quad |\bar{y}_3 - \bar{y}_4| = 7^* \quad |\bar{y}_4 - \bar{y}_5| = 5.2^*$$
$$|\bar{y}_1 - \bar{y}_3| = 14.4^* \quad |\bar{y}_2 - \bar{y}_4| = 9.6^* \quad |\bar{y}_3 - \bar{y}_5| = 12.2^*$$
$$|\bar{y}_1 - \bar{y}_4| = 21.4^* \quad |\bar{y}_2 - \bar{y}_5| = 14.8^*$$
$$|\bar{y}_1 - \bar{y}_5| = 26.6^*$$

The starred values indicate the pairs of means that are significantly different.

$$\bar{y}_1 = 45.8, \quad \bar{y}_2 = 57.6, \quad \bar{y}_3 = 60.2, \quad \bar{y}_4 = 67.2, \quad \bar{y}_5 = 72.4$$

Here, the results obtained using Fisher's test are similar to that using Tukey's test.

Example 7.1 The brushite cement is known to be a potential bone replacement material. In order to examine the effect of poly(methyl methacrylate) in determining the compressive strength of the cement, a completely randomized experiment is conducted. Four levels of poly(methyl methacrylate) concentration are taken and each treatment is replicated four times. The results are shown in Table 7.9.

Solution: We are required to know if the concentration of poly(methyl methacrylate) affects the compressive strength of the cement. Let us carry out the test of hypothesis, described in Sect. 7.2.4.

Step 1: Statement of Hypothesis Here, the null hypothesis states that there is no difference among the population mean compressive strengths of the brushite cements. The alternative hypothesis states that there is a difference between the population treatment means for at least a pair of populations.

Step 2: Selection of Level of Significance Suppose the level of significance is chosen as 0.05.

Step 3: Computation of Test Statistic The test statistic can be computed by carrying out an analysis of variance of the data. The calculations for ANOVA are summarized in Table 7.10.

Step 4: Specification of Critical Region The null hypothesis is rejected if $F_0 > F_{0.05,4,15} = 3.06$ (Table A.11).

Table 7.9 Calculations for treatment totals and treatment means for compressive strength of brushite cement

Concentration of polymethyl methacrylate (%)	Compressive strength (MPa)			
	I	II	III	IV
30	20	21	19	22
35	27	26	29	28
40	32	34	30	33
45	42	40	44	43
50	28	32	30	26

Table 7.10 ANOVA table for compressive strength of brushite cement

Source of variation	Sum of squares	Degree of freedom	Mean square	F_0
Concentration of poly(methyl methacrylate)	1008.70	4	252.175	79.63
Error	47.50	15	3.167	
Total	1056.20	19		

Step 5: Take a Decision As the value of the test statistic falls in the critical region adopted, this calls for rejection of null hypothesis. It is therefore concluded that there is difference between population treatment means, at least for a pair of populations.

Example 7.2 Consider Example 7.1. Let us find out which of the means are significantly different by applying Tukey's test, as described in Sect. 7.2.5. Here, the mean compressive strengths are given below.

$$\bar{y}_1 = 20.5, \; \bar{y}_2 = 27.5, \; \bar{y}_3 = 32.25, \; \bar{y}_4 = 42.25, \; \bar{y}_5 = 29.$$

The numerical value of the studentized range statistic is $T_{0.05} = q_{0.05}(5, 15) = 4.37$. The differences in mean compressive strengths are obtained in the following manner.

$$|\bar{y}_1 - \bar{y}_2| = 7^* \quad |\bar{y}_2 - \bar{y}_3| = 4.75^* \quad |\bar{y}_3 - \bar{y}_4| = 10^* \quad |\bar{y}_4 - \bar{y}_5| = 13.25^*$$
$$|\bar{y}_1 - \bar{y}_3| = 11.75^* \quad |\bar{y}_2 - \bar{y}_4| = 14.75^* \quad |\bar{y}_3 - \bar{y}_5| = 3.25$$
$$|\bar{y}_1 - \bar{y}_4| = 21.75^* \quad |\bar{y}_2 - \bar{y}_5| = 1.5$$
$$|\bar{y}_1 - \bar{y}_5| = 8.5^*$$

The starred values indicate the pairs of means that are significantly different. Clearly, the only pairs of means that are not significantly different are 2 and 5, and 3 and 5, and the treatment 4 (45% concentration) produces significantly greater compressive strength than the other treatments.

Example 7.3 Let us now apply Fisher's test, as described in Sect. 7.2.5, to compare among the means of compressive strength. The least significant difference (*LSD*) can be found as.

$$LSD = t_{0.025, 15}\sqrt{\frac{2 \times 3.167}{4}} = 2.131 \times 1.2584 = 2.68$$

(From Table A.10, $t_{0.025, 15} = 2.1314$) In the following, the differences in mean compressive strengths are shown and the pairs of means that are significantly different are indicated by starred values.

$$|\bar{y}_1 - \bar{y}_2| = 7^* \quad |\bar{y}_2 - \bar{y}_3| = 4.75^* \quad |\bar{y}_3 - \bar{y}_4| = 10^* \quad |\bar{y}_4 - \bar{y}_5| = 13.25^*$$
$$|\bar{y}_1 - \bar{y}_3| = 11.75^* \quad |\bar{y}_2 - \bar{y}_4| = 14.75^* \quad |\bar{y}_3 - \bar{y}_5| = 3.25^*$$
$$|\bar{y}_1 - \bar{y}_4| = 21.75^* \quad |\bar{y}_2 - \bar{y}_5| = 1.5$$
$$|\bar{y}_1 - \bar{y}_5| = 8.5^*$$

Clearly, the only one pair of means that is not significantly different is 2 and 5, and the treatment 4 (45% concentration) produces significantly greater compressive strength than the other treatments.

Example 7.4 Let us find out a regression model to the data shown in Example 7.1. Figure 7.6 displays a scatter diagram of compressive strength of brushite cement versus concentration of poly(methyl methacrylate). The open circles represent the experimental results of compressive strength of the cement obtained at different concentrations of poly(methyl methacrylate). The solid circles denote the mean compressive strength of the cement at each value of concentration of poly(methyl methacrylate). Clearly, the relationship between compressive strength and concentration of poly (methyl methacrylate) is not linear. As a first approximation, an attempt can be made to fit the data to a quadratic model of the following form

$$y = \beta_0 + \beta_1 x + \beta_2 x^2 + \varepsilon$$

where y denotes the compressive strength, x indicates concentration of poly(methyl methacrylate), β_0, β_1, β_2 are regression coefficients, and ε refers to error. The least square fit of the quadratic model to the data yields

$$\hat{y} = -151.2 + 8.692x - 0.1007x^2.$$

The method of estimation of regression coefficients is described in Sect. 9.3.2 of this book. The behavior of the quadratic model is displayed in Fig. 7.6. As shown, it overestimates the compressive strength at 40% concentration of poly(methyl methacrylate) and largely underestimates the compressive strength at 45% concentration of poly(methyl methacrylate). Overall, it does not appear to represent the data satisfactorily. The R^2 statistic and the adjusted-R^2 statistic are found to be 0.7517 and 0.5034, respectively. Afterwards, an attempt can be made to fit the data to a cubic model of the following form

$$y = \beta_0 + \beta_1 x + \beta_2 x^2 + \beta_3 x^3 + \varepsilon$$

where β_3, a new regression coefficient, is added to the cubic term in x. The least square fit of the cubic model to the data yields

$$\hat{y} = 697.2 - 57.32x + 1.579x^2 - 0.014x^3$$

Fig. 7.6 Scatter diagram for compressive strength of brushite cement

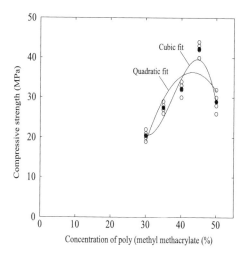

Figure 7.6 displays the behavior of the cubic model. The cubic model fits better than the quadratic model at 40 and 45% concentrations of poly(methyl methacrylate). The R^2 statistic and the adjusted-R^2 statistic for the cubic model are found to be 0.9266 and 0.7063, respectively. Overall, the cubic model appears to be superior to the quadratic model.

7.3 Randomized Block Design

In any experiment, there are factors that influence the outcomes although they may not be of interest to study. Such factors are known as nuisance factors. If the nuisance factors are unknown and uncontrollable then a design technique called randomization is used to guard against them. However, if they are known but uncontrollable then their effect might be compensated by a technique called analysis of covariance. If the nuisance factors are known and controllable then a design technique called blocking can be employed to systematically eliminate its effect on the statistical comparison among treatments. Blocking is an important technique that has been used extensively while conducting design of experiments. Randomized block design is a very popular design that utilizes the principle of blocking. Let us now discuss this principle.

Blocking is used in situations where it is impossible to carry out all of the runs in an experiment under homogeneous conditions. For example, a single batch of raw material may not be enough large to perform all of the runs in an experiment. Also, there are situations in which we desire to deliberately vary the experimental conditions including machines, operators, environments. This is done to ensure that the process or product is so robust that it is unaffected by such conditions. The principle of experimentation used in such situations involves blocking, where each set of nonhomogeneous conditions defines a block. Suppose in our earlier example of electroconductive yarns (Sect. 1.3.2 of Chap. 1), the four runs in Replicate I were performed by using the electrically conducting monomer supplied by Supplier X and the four runs in Replicate II were conducted by using the electrically conducting monomer supplied by Supplier Y. Then, the data would have been analyzed by using the technique called blocking, where each block corresponded to one replicate. Further, statistical analysis could be used to quantify the block effect. In fact, one can find that the block effect is relatively small.

7.3.1 A Practical Problem

A process engineer wishes to investigate the effect of water removal process on the moisture content of textile fabrics. He selects three different water removal processes, namely hydro, suction, and mangle, and chooses four different types of fabrics, namely worsted, barathea, twill, and melton. As the fabrics chosen are of different types (nonhomogeneous), the completely randomized design seems to be inappropriate. Also, the engineer wants to be sure that any effect causing due to different types of fabrics should not have any influence on the comparison of the three processes (treatments). He therefore wishes to use the technique of blocking and accordingly divides the fabrics into four different blocks, depending on their types. Further, he randomizes the order of runs within each block. Obviously, the blocks put a restriction on randomization. This is how typically the experiment is carried out in accordance with a randomized block design.

7.3.2 Data Visualization

Table 7.11 reports the results of the experiment (Leaf 1987). Here, the data represent the percentage of moisture content remaining in the fabrics after removal of water.

Table 7.11 Experimental results of moisture content

Processes (Treatments)	Fabric types (Blocks)			
	Worsted	Barathea	Twill	Melton
Hydro	46	35	47	42
Suction	54	57	79	86
Mangle	56	56	61	65

7.3.3 Descriptive Model

Suppose we have m number of treatments that are needed to be compared and n number of blocks such that there is one observation per treatment in each block, and the order in which the treatments are run within each block is determined randomly. The experimental dataset looks like as shown in Table 7.12.

The treatment totals $y_{i.}$ and treatment means $\bar{y}_{i.}$ are obtained as follows

$$y_{i.} = \sum_{j=1}^{n} y_{ij} \quad \bar{y}_{i.} = \frac{y_{i.}}{n}, \quad i = 1, 2, \ldots, m.$$

The block totals $y_{.j}$ and means $\bar{y}_{.j}$ are obtained as follows

$$y_{.j} = \sum_{i=1}^{m} y_{ij}, \quad \bar{y}_{.j} = \frac{y_{.j}}{m}, \quad j = 1, 2, \ldots, n$$

The grand total and means are obtained as shown below

$$y_{..} = \sum_{i=1}^{m}\sum_{j=1}^{n} y_{ij} = \sum_{i=1}^{m} y_{i.} = \sum_{j=1}^{n} y_{.j}.$$

$$\bar{y}_{..} = \frac{y_{..}}{mn}.$$

The observations of the experiment can be described by the following linear statistical model

$$y_{ij} = \mu + \tau_i + \beta_j + \varepsilon_{ij}; \quad i = 1, 2, \ldots, m; \; j = 1, 2, \ldots, n.$$

Table 7.12 Data for randomized block experiment

Treatment (Level)	Observations				Total	Average
	Block 1	Block 2	...	Block b		
1	y_{11}	y_{12}	...	y_{1b}	$y_{1.}$	$\bar{y}_{1.}$
2	y_{21}	y_{22}	...	y_{2b}	$y_{2.}$	$\bar{y}_{2.}$
⋮	⋮	⋮	⋮	⋮	⋮	⋮
a	y_{a1}	y_{a2}	...	y_{ab}	$y_{a.}$	$\bar{y}_{a.}$
Total	$y_{.1}$	$y_{.2}$	...	$y_{.b}$	$y_{..}$	
Average	$\bar{y}_{.1}$	$\bar{y}_{.2}$	...	$\bar{y}_{.b}$		$\bar{y}_{..}$

Here, y_{ij} is a random variable denoting the ijth observation. μ is a parameter common to all levels called the overall mean, τ_i is a parameter associated with the ith level called the ith level effect, β_j is a parameter associated with jth block called jth block effect, ε_{ij} is a random error component, m denotes no. of treatments and n indicates no. of blocks. It follows traditional fixed effect model. This model can also be written as

$$y_{ij} = \mu_{ij} + \varepsilon_{ij}; \quad i = 1, 2, \ldots, m; \; j = 1, 2, \ldots, n$$

where $\mu_{ij} = \mu + \tau_i + \beta_j$ is the mean of ijth observations.

The followings are the reasonable estimates of the model parameters

$$\hat{\mu} = \bar{y}_{..}$$
$$\hat{\tau}_i = \bar{y}_{i.} - \bar{y}_{..}, \quad i = 1, 2, \ldots, m$$
$$\hat{\beta}_j = \bar{y}_{.j} - \bar{y}_{..}, \quad j = 1, 2, \ldots, n$$
$$\hat{\mu}_{ij} = \hat{\mu} + \hat{\tau}_i + \hat{\beta}_j = \bar{y}_{..} + (\bar{y}_{i.} - \bar{y}_{..}) + (\bar{y}_{.j} - \bar{y}_{..}) = \bar{y}_{i.} + \bar{y}_{.j} - \bar{y}_{..}$$

The estimate of the observations is $\hat{y}_{ij} = \hat{\mu} + \hat{\tau}_i + \hat{\beta}_j = \bar{y}_{i.} + \bar{y}_{.j} - \bar{y}_{..}$. Hence, the residual is $e_{ij} = y_{ij} - \hat{y}_{ij}$.

These formulas are used to estimate the observations and residuals. Table 7.13 compares the estimated observations with the experimental ones and also reports the residuals. Here, in each cell, three values are reported for a given process and for a given fabric type. The leftmost value is obtained from experiment, the middle one is the fitted value, and the rightmost value represents the residual.

It is necessary to check whether the model is adequate. The adequacy of a model is investigated by examining the residuals. If the model is adequate then the residuals should be structureless, that is, they should not follow any trend or pattern. Figure 7.7a plots the residuals against treatments. It can be observed that the residuals do not follow any specific trend or pattern with respect to treatments as well as blocks (Fig. 7.7b). Figure 7.7c does not display any trend or pattern when the residuals are plotted against the fitted responses. Further, it is important to check if the residuals

Table 7.13 Data and Residuals for moisture content

Processes (Treatments)	Fabric types (Blocks)											
	Worsted			Barathea			Twill			Melton		
Hydro	46	37.5	8.5	35	34.8	0.2	47	47.8	−0.8	42	49.8	−7.8
Suction	54	64	−10	57	61.3	−4.3	79	74.3	4.7	86	76.3	9.7
Mangle	56	54.5	1.5	56	51.8	4.2	61	64.8	−3.8	65	66.8	−1.8

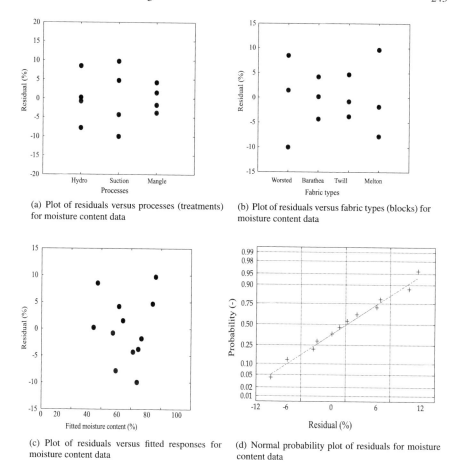

(a) Plot of residuals versus processes (treatments) for moisture content data

(b) Plot of residuals versus fabric types (blocks) for moisture content data

(c) Plot of residuals versus fitted responses for moisture content data

(d) Normal probability plot of residuals for moisture content data

Fig. 7.7 Plots for residuals

can be regarded as taken from a normal distribution or not. Figure 7.7d plots the normal probability plot of the residuals. It can be concluded that the residuals are taken from a normal distribution.

7.3.4 Test of Hypothesis

As mentioned in Chap. 7, the test of hypothesis is hereby carried out in the following manner.

Step 1: Statement of Hypothesis

We are interested in testing the equality of population treatment means of moisture content in fabrics. Then, the null hypothesis is that there is no difference among the

Table 7.14 Calculations for treatment and block totals and means for moisture content data

Processes (Treatments)	Fabric types (Blocks)				Total	Mean
	Worsted	Barathea	Twill	Melton		
Hydro	46	35	47	42	170	42.5
Suction	54	57	79	86	276	69
Mangle	56	56	61	65	238	59.5
Total	156	148	187	193	684	
Mean	52	49.3	62.3	64.3		57

population treatment means. This is stated as follows

$$H_0 : \mu_1 = \mu_2 = \cdots = \mu_m \text{ or, equivalently, } H_0 : \tau_1 = \tau_2 = \cdots = \tau_m = 0$$

The alternative hypothesis states that there is a difference between the population treatment means for at least a pair of populations. This can written as

$$H_1 : \mu_i \neq \mu_j \text{ for at least one } i \text{ and } j \text{ or, equivalently, } H_1 : \tau_i \neq 0 \text{ for at least one } i$$

Step 2: Selection of Level of Significance
Suppose the level of significance is chosen as 0.05. It means that the probability of rejecting the null hypothesis when it is true is less than or equal to 0.05.

Step 3: Computation of Test Statistic
The test statistic is computed by carrying out an analysis of variance of the data. The basic calculations for treatment totals and treatment means are shown in Table 7.14.
The calculations for ANOVA are shown below

$$SS_{Total} = \sum_{i=1}^{m}\sum_{j=1}^{n}(\bar{y}_{ij} - \bar{y})^2 = \sum_{i=1}^{3}\sum_{j=1}^{4}(\bar{y}_{ij} - \bar{y})^2 = (46 - 57)^2 + (35 - 57)^2 + (47 - 57)^2 + (42 - 57)^2$$
$$+ (54 - 57)^2 + (57 - 57)^2 + (79 - 57)^2 + (86 - 57)^2 + (56 - 57)^2 + (56 - 57)^2 + (61 - 57)^2 + (65 - 57)^2$$
$$= 2346$$

$$SS_{Treatments} = n\sum_{i=1}^{m}(\bar{y}_{i.} - \bar{y})^2 = 4[(42.5 - 57)^2 + (69 - 57)^2 + (59.5 - 57)^2] = 1442$$

$$SS_{Blocks} = m\sum_{j=1}^{n}(\bar{y}_{.j} - \bar{y})^2 = 3[(52 - 57)^2 + (49.3 - 57)^2 + (62.3 - 57)^2 + (64.3 - 57)^2] = 498$$

$$SS_{Error} = SS_{Total} - SS_{Treatments} - SS_{Blocks} = 406$$

The calculations are summarized in Table 7.15.

Step 4: Specification of Critical Region
The null hypothesis is rejected if $F_0 > F_{\alpha,a-1,(a-1)(b-1)} = F_{0.05,2,6} = 5.14$ (Table A.11). The critical region is shown in Fig. 7.8.

Table 7.15 ANOVA Table for moisture content data

Source of variation	Sum of squares	Degree of freedom	Mean squares	F_0
Treatments	1442	2	721	10.66
Blocks	498	3	166	
Errors	406	6	67.66	
Total	2346	11		

Fig. 7.8 Display of critical region for moisture content data

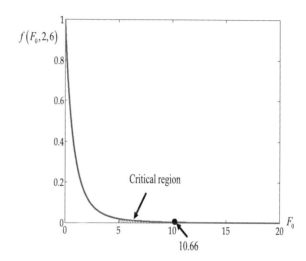

Step 5: Take a Decision

As the value of the test statistic falls in the critical region adopted, this calls for rejection of null hypothesis. It is therefore concluded that there is a difference between population treatment means, at least for a pair of populations. The moisture content of the fabrics obtained from different processes is significantly different at 0.05 level of significance.

7.3.5 Multiple Comparison Among Treatment Means

Tukey's test compares pairs of treatment means. This test declares two means to be significantly different if the absolute value of their difference exceeds

$$T_\alpha = q_\alpha(m, f)\sqrt{\frac{MS_{\text{Error}}}{n}}$$

where $q_\alpha(m, f)$ is called studentized range statistic corresponding to level of significance of α, m is the number of factor level, and f is the degrees of freedom associated with the MS_{Error}, and n is the number of treatment levels. The numerical value of the studentized range statistic can be obtained from a standard table. Here, using Table A.17, $q_{0.05}(2, 6) = 3.46$, we have

$$T_{0.05} = q_\alpha(m, f)\sqrt{\frac{MS_{Error}}{n}} = q_{0.05}(2, 6)\sqrt{\frac{67.66}{4}} = 14.23.$$

The three treatment averages are $\bar{y}_{1.} = 42.5$, $\bar{y}_{2.} = 69$, $\bar{y}_{3.} = 59.5$. The absolute differences in treatment averages are

$$|\bar{y}_{1.} - \bar{y}_{2.}| = 26.5^* \quad |\bar{y}_{1.} - \bar{y}_{3.}| = 17^* \quad |\bar{y}_{2.} - \bar{y}_{3.}| = 9.5$$

The starred values indicate the pair of means which is significantly different. Sometimes it is useful to draw a graph, as shown below, underlining pairs of means that do not differ significantly.

$$\bar{y}_{1.} = 42.5, \quad \bar{y}_{3.} = 59.5, \quad \bar{y}_{2.} = 69.$$

As shown, the hydro process results in lowest moisture content in the fabric and this is significantly different than those given by suction and mangle processes.

Another popular test for comparing the pairs of treatment means is Fisher's test. This test was also described in the earlier chapter. The least significant difference (LSD) is calculated as follows (using Table A.10, $t_{0.025, 6} = 2.4469$),

$$LSD = t_{\frac{0.05}{2}, 6}\sqrt{\frac{MS_{Error}}{n}} = 10.06$$

The three treatment averages are $\bar{y}_{1.} = 42.5$, $\bar{y}_{2.} = 69$, $\bar{y}_{3.} = 59.5$. The differences in averages are

$$|\bar{y}_{1.} - \bar{y}_{2.}| = 26.5^* \quad |\bar{y}_{1.} - \bar{y}_{3.}| = 17^* \quad |\bar{y}_{2.} - \bar{y}_{3.}| = 9.5$$

The starred values indicate the pairs of means that are significantly different. Sometimes it is useful to draw a graph, as shown below, underlining pairs of means that do not differ significantly.

$$\bar{y}_{1.} = 42.5, \quad \bar{y}_{3.} = 59.5, \quad \bar{y}_{2.} = 69.$$

It is seen that in this case, the results of Fisher's test are the same as those of Tukey's test.

Example 7.5 Four different algorithms (*A*, *B*, *C*, and *D*) are used to solve three problems of research in accordance with a randomized block design such that each problem is considered as one block. The results of this experiment are given in terms of efficiency of algorithms (min) in Table 7.16. It is required to know whether the algorithms are same in terms of their efficiency or not at a level of significance of 0.05. Let us carry out the test of hypothesis, described in Sect. 7.2.4.

Step 1: Statement of Hypothesis: Here, the null hypothesis states that there is no difference among the population mean efficiencies of the algorithms (treatment). The alternative hypothesis states that there is a difference between the population mean efficiencies of the algorithms for at least a pair of populations.

Step 2: Selection of Level of Significance: Suppose the level of significance is chosen as 0.05.

Step 3: Computation of Test Statistic: The test statistic can be computed by carrying out an analysis of variance of the data. The calculations for ANOVA are summarized in Table 7.17.

Step 4: Specification of Critical Region: The null hypothesis is rejected if $F_0 > F_{0.05,3,6} = 4.76$ (Table A.11).

Step 5: Take a Decision: As the value of the test statistic falls in the critical region adopted, this calls for rejection of null hypothesis. It is therefore concluded that there is a significant difference between the population mean efficiencies of the algorithms for at least a pair of populations at 0.05 level of significance.

Example 7.6 Let us see what would have happened if the experimenter in Example 7.5 was not aware of randomized block design. Suppose he would have assigned the algorithms to each problem randomly and the same results (Table 7.16)

Table 7.16 Data for algorithm efficiency

Algorithm A	Problem (Blocks)		
	20	23	35
B	5	12.8	20
C	2	3.2	4
D	5	20.5	40

Table 7.17 ANOVA table for algorithm efficiency

Source of variation	Sum of squares	Degree of freedom	Mean square	F_0
Treatments (Algorithm)	938.4	3	312.8	6.50
Blocks (Problem)	567.1	2	283.56	
Errors	288.6	6	48.10	
Total	1794.1	11		

Table 7.18 Incorrect analysis of algorithm efficiency based on a completely randomized design

Source of variation	Sum of squares	Degree of freedom	Mean square	F_0
Algorithm	938.4	3	312.8	6.50
Error	855.7	8	107	
Total	1794.1	11		

would have been obtained by chance. The incorrect analysis of the data based on a completely randomized experiment is shown in Table 7.18. Because $F_0 > F_{0.05,3,8} = 4.07$ (Table A.11), the test statistic does not fall in the critical region adopted, hence this does not call for rejection of null hypothesis. It is therefore concluded that there is no significant difference between the population mean efficiencies of the algorithms at 0.05 level of significance. It can be thus observed that the randomized block design reduces the amount of experimental error such that the differences among the four algorithms are detected. This is a very important feature of randomized block design. If the experimenter fails to block a factor, then the experimental error can be inflated so much that the important differences among the treatment means may remain undetected.

7.4 Latin Square Design

Latin square design is used to eliminate two nuisance sources of variability by systematically blocking in two directions. The rows and columns represent two restrictions on randomization. In general, a $p \times p$ Latin square is a square containing p rows and p columns. In this Latin square, each of the resulting p^2 cells contains one of the p letters that corresponds to the treatment, and each letter appears only once in each row and column. Some Latin square designs are shown in Fig. 7.9. Here, the Latin letters A, B, C, D, and E denote the treatments.

Fig. 7.9 Display of Latin square designs

A	B	C
B	C	A
C	A	B

3×3

A	B	C	D
B	C	D	A
C	D	A	B
D	A	B	C

4×4

A	B	C	D	E
B	C	D	E	A
C	D	E	A	B
D	E	A	B	C
E	A	B	C	D

5×5

7.4.1 A Practical Problem

Suppose a material supplied by three different vendors is required to be analyzed for its weight in grams. It is known from past experience that the weighing machines used and the operators employed for measurements have influence on the weights. The experimenter therefore decides to run an experiment such that it takes into account of the variability caused by the machine and the operator. A 3×3 Latin square design is chosen where each material supplied by three different vendors is weighed by using three different weighing machines and three different operators.

7.4.2 Data Visualization

As stated earlier, the material is analyzed for weight in grams from three different vendors (a, b, c) by three different operators (I, II, III) and using three different weighing machines (1, 2, 3) in accordance with a 3×3 Latin square design. The results of experiment are shown in Table 7.19. Here, A, B, and C denote the weights obtained by three different vendors.

7.4.3 Descriptive Model

Let us describe the observations using the following linear statistical model

$$y_{ijk} = \mu + \alpha_i + \tau_j + \beta_k + \varepsilon_{ijk}; \quad i = 1, 2, \ldots, p; \; j = 1, 2, \ldots, p; \; k = 1, 2, \ldots, p$$

Here, y_{ijk} is a random variable denoting the ijkth observation. μ is a parameter common to all levels called the overall mean, α_i is a parameter associated with the ith row called the ith row effect, and τ_j is a parameter associated with the jth treatment called the jth treatment effect, β_k is a parameter associated with kth column called kth column effect, and ε_{ijk} is a random error component. The above expression can also be written as $y_{ijk} = \mu_{ijk} + \varepsilon_{ijk}$ where $\mu_{ijk} = \mu + \alpha_i + \tau_j + \beta_k$.

Note that this is a fixed effect model. The followings are the reasonable estimates of the model parameters

Table 7.19 Experimental results of material weight

Operators	Weighing machines		
	1	2	3
I	$A = 16$	$B = 10$	$C = 11$
II	$B = 15$	$C = 9$	$A = 14$
III	$C = 13$	$A = 11$	$B = 13$

Table 7.20 Data and residuals for material weight

Operators	Weighing machines								
	1			2			3		
I	A = 16	A = 15.79	0.21	B = 10	B = 10.12	−0.12	C = 11	C = 11.12	−0.12
II	B = 15	B = 15.13	−0.13	C = 9	C = 8.79	0.21	A = 14	A = 14.13	−0.13
III	C = 13	C = 13.12	−0.12	A = 11	A = 11.12	−0.12	B = 13	B = 12.79	0.21

$$\hat{\mu} = \bar{y}_{...} \quad \text{where} \quad y_{...} = \sum_{i=1}^{p}\sum_{j=1}^{p}\sum_{k=1}^{b} y_{ijk} \quad \text{and} \quad \bar{y}_{...} = \frac{y_{...}}{p^3}$$

$$\hat{\alpha}_i = \bar{y}_{i..} - \bar{y}_{...} \quad \text{where} \quad y_{i..} = \sum_{j=1}^{p}\sum_{k=1}^{p} y_{ijk}, \quad i = 1, 2, \ldots, p \quad \text{and} \quad \bar{y}_{i..} = \frac{y_{i..}}{p}$$

$$\hat{\tau}_j = \bar{y}_{.j.} - \bar{y}_{...} \quad \text{where} \quad y_{.j.} = \sum_{i=1}^{p}\sum_{k=1}^{p} y_{ijk}, \quad j = 1, 2, \ldots, p \quad \text{and} \quad \bar{y}_{.j.} = \frac{y_{.j.}}{p}$$

$$\hat{\beta}_k = \bar{y}_{..k} - \bar{y}_{...} \quad \text{where} \quad y_{..k} = \sum_{i=1}^{p}\sum_{j=1}^{p} y_{ijk}, \quad k = 1, 2, \ldots, p \quad \text{and} \quad \bar{y}_{..k} = \frac{y_{..k}}{p}$$

$$\hat{\mu}_{ijk} = \hat{\mu} + \hat{\alpha}_i + \hat{\tau}_j + \hat{\beta}_k = \bar{y}_{i..} + \bar{y}_{.j.} + \bar{y}_{..k} - 2\bar{y}_{...}$$

The estimate of the observations is $\hat{y}_{ijk} = \bar{y}_{i..} + \bar{y}_{.j.} + \bar{y}_{..k} - 2\bar{y}_{...}$. Hence, the residual is $e_{ijk} = y_{ijk} - \hat{y}_{ijk}$.

These formulae are used to estimate the observations and the residuals. Table 7.20 compares the estimated observations with the actual ones and reports the residuals. Here, in each cell, three values are reported for a given operator and for a given weighing machine. The leftmost value is obtained from experiment, the middle one is the fitted value, and the rightmost value represents residual.

As in any design problem, it is necessary to check whether the model is adequate or not. The methods of doing this are described in Sect. 8.2.3. The reader is instructed to carry out a similar analysis and conclude on the adequacy of the model.

7.4.4 Test of Hypothesis

The test of hypothesis is carried out as stated below.

Step 1: Statement of Hypothesis
We are interested here in testing the equality of vendor means. Then, the null hypothesis states that the population vendor means are equal.

$$H_0 : \mu_{.1.} = \mu_{.2.} = \cdots = \mu_{.p.}.$$

The alternative hypothesis states that there is a difference between the population treatment means for at least a pair of populations.

$$H_1 : \mu_{.i.} \neq \mu_{.j.} \text{ for at least one } i \text{ and } j.$$

Step 2: Selection of Level of Significance
Suppose the level of significance is chosen as 0.05. It means that the probability of rejecting the null hypothesis when it is true is less than or equal to 0.05.

Step 3: Computation of Test Statistic
The test statistic is computed by carrying out an analysis of variance of the data. The calculations for the sum of squares are shown here.

$$SS_{Total} = \sum_{i=1}^{3}\sum_{j=1}^{3}\sum_{k=1}^{3} y_{ij}^2 - \frac{y_{...}^2}{3 \times 3} = (16^2 + 10^2 + 11^2 + 15^2 + 9^2 + 14^2 + 13^2 + 11^2 + 13^2) - \left(\frac{112^2}{9}\right)$$

$$= 1438 - 1393.78 = 44.22$$

$$SS_{Vendors} = \frac{1}{3}\sum_{j=1}^{3} y_{.j.}^2 - \frac{y_{...}^2}{3 \times 3} = \frac{1}{3}[(16 + 14 + 11)^2 + (10 + 15 + 13)^2 + (11 + 9 + 13)^2] - \frac{112^2}{9}$$

$$= 1404.67 - 1393.78 = 10.89$$

$$SS_{Operators} = \frac{1}{3}\sum_{i=1}^{3} y_{i..}^2 - \frac{y_{...}^2}{3 \times 3} = \frac{1}{3}[37^2 + 38^2 + 37^2] - \frac{112^2}{9} = 1394 - 1393.78 = .22$$

$$SS_{Weighing\ machines} = \frac{1}{3}\sum_{k=1}^{3} y_{..k}^2 - \frac{y_{...}^2}{3 \times 3} = \frac{1}{3}[44^2 + 30^2 + 38^2] - \frac{112^2}{9} = 1426.67 - 1393.78 = 32.89$$

$$SS_{Error} = SS_{Total} - SS_{Vendors} - SS_{Inspectors} - SS_{Scales} = 44.22 - 10.89 - .22 - 32.89 = .22$$

The calculations are summarized in Table 7.21.

Step 5: Specification of Critical Region
The null hypothesis is rejected if $F_0 > F_{\alpha, p-1, (p-1)(p-2)} = F_{0.05, 2, 2} = 19.00$ (Table A.11). The critical region is shown in Fig. 7.10.

Step 6: Decision
As the value of the test statistic falls in the critical region adopted, this calls for rejection of null hypothesis. It is therefore concluded that there is a difference between

Table 7.21 ANOVA table for material weight data

Source of variation	Sum of squares	Degree of freedom	Mean square	F Value
Vendors	10.89	2	5.445	49.5
Operators	0.22	2	0.11	
Weighing machines	32.89	2	16.445	
Errors	.22	2	.11	
Total	44.22	8		

Fig. 7.10 Critical region

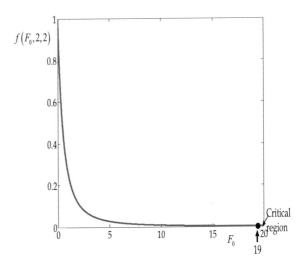

the weight of the materials supplied by the vendors, at least for a pair of weights of the materials supplied by the vendors.

7.4.5 Multiple Comparison Among Treatment Means

Tukey's test compares between the pairs of mean values. This test is described in the earlier section. The numerical value of the studentized range statistic is obtained as follows.

Here, the studentized range statistic is mentioned below. Using Table A.17, $q_{0.05}(2, 2) = 6.09$, we have

$$T_\alpha = q_\alpha(a, f)\sqrt{\frac{MS_{\text{Error}}}{n}} = q_{0.05}(2, 2)\sqrt{\frac{0.11}{3}} = 1.166.$$

The three averages are $\bar{y}_{1.} = 13.67$, $\bar{y}_{.2.} = 12.67$, $\bar{y}_{.3.} = 11$. The absolute differences in averages are

$$|\bar{y}_{1.} - \bar{y}_{.2.}| = 1.00^* \quad |\bar{y}_{1.} - \bar{y}_{.3.}| = 2.67^* \quad |\bar{y}_{.2.} - \bar{y}_{.3.}| = 1.67^*.$$

The starred values indicate the pairs of means that are significantly different. Sometimes, it is useful to draw a graph, as shown below, underlining pairs of means that do not differ significantly.

$$\bar{y}_{.3.} = 11, \quad \underline{\bar{y}_{.2.} = 12.67, \quad \bar{y}_{1.} = 13.67.}$$

It can be observed that the material supplied by vendor A is on an average the heaviest, followed by those supplied by vendors B and C, respectively. Further, the differences between the materials supplied by vendors A and C and B and C are statistically significant.

There is another test, Fisher's test, that compares whether the difference between the averages is statistically significant or not. In this test, the least significance difference (LSD) is obtained as follows (Using Table A.10, $t_{0.05,2} = 4.3027$).

$$LSD = t_{\frac{0.05}{2},2}\sqrt{\frac{2MS_{\text{Error}}}{n}} = t_{\frac{0.05}{2},2}\sqrt{\frac{2 \times .11}{3}} = 1.165.$$

The three averages are $\bar{y}_{.1.} = 13.67$, $\bar{y}_{.2.} = 12.67$, $\bar{y}_{.3.} = 11$. Then, the differences in averages are

$$|\bar{y}_{.1.} - \bar{y}_{.2.}| = 1.00 \quad |\bar{y}_{.1.} - \bar{y}_{.3.}| = 2.67^* \quad |\bar{y}_{.2.} - \bar{y}_{.3.}| = 1.67^*.$$

The starred values indicate the pairs of means that are significantly different.

Example 7.7 An agricultural scientist wishes to examine the effect of five varieties of barley $(A, B, C, D, \text{and } E)$ on their yield (kg per plot). An experiment is conducted according to a Latin Square Design so that the effects of plot and season are systematically controlled. Yield of barley (kg per plot) is shown in Table 7.22. Suppose we wish to know if the five varieties of barley exhibit same yield or not at a level of significance of 0.05. Let us carry out the test of hypothesis, described in Sect. 8.3.4.

Step 1: Statement of Hypothesis: Here, the null hypothesis states that there is no difference among the population mean yields of barley produced from five different varieties (treatment). The alternative hypothesis states that there is a difference between the population mean yields of barley for at least a pair of populations.

Step 2: Selection of Level of Significance: Suppose the level of significance is chosen as 0.05.

Step 3: Computation of Test Statistic: The test statistic can be computed by carrying out an analysis of variance of the data. The calculations for ANOVA are summarized in Table 7.23.

Step 4: Specification of Critical Region: The null hypothesis is rejected if $F_0 > F_{0.05,3,6} = 4.76$ (Table A.11).

Table 7.22 Experimental data barley yield as per Latin Square design

Plot	Season			
	I	II	III	IV
1	$B = 40$	$C = 26$	$D = 26$	$A = 52$
2	$D = 15$	$A = 50$	$B = 38$	$C = 38$
3	$C = 30$	$B = 45$	$A = 48$	$D = 24$
4	$A = 55$	$D = 20$	$C = 35$	$B = 42$

Table 7.23 ANOVA for barley yield

Source of variation	Sum of squares	Degree of freedom	Mean square	F_0
Variety of barley	1963	3	654.33	25.83
Plot	16.40	3	5.5	
Season	40.50	3	13.5	
Errors	152.00	6	25.33	
Total	2172.00	15		

Step 5: Take a Decision: As the value of the test statistic falls in the critical region adopted, this calls for rejection of null hypothesis. It is therefore concluded that there is a significant difference between the population mean yields of barley for at least a pair of populations at 0.05 level of significance.

Example 7.8 Consider Example 7.7. Let us find out which of the mean barley yields are significantly different by applying Tukey's test, as described in Sect. 8.4.5. Here, the mean barley yields are as follows.

$$\bar{y}_{.1.} = 51.25, \; \bar{y}_{.2.} = 41.25, \; \bar{y}_{.3.} = 32.25, \; \bar{y}_{.4.} = 4.08.$$

The numerical value of the studentized range statistic is $T_{0.05} = q_{0.05}(4, 15) = 4.08$ (Table A.17). The differences in mean barley yields are

$$|\bar{y}_{.1.} - \bar{y}_{.2.}| = 10^*, \;\; |\bar{y}_{.1.} - \bar{y}_{.3.}| = 19^*, \;\; |\bar{y}_{.1.} - \bar{y}_{.4.}| = 30^*$$
$$|\bar{y}_{.2.} - \bar{y}_{.3.}| = 9^*, \;\; |\bar{y}_{.2.} - \bar{y}_{.4.}| = 19^*$$
$$|\bar{y}_{.3.} - \bar{y}_{.4.}| = 11^*.$$

The starred values indicate the pairs of means that are significantly different. Clearly, all the pairs of means are significantly different in terms of yielding barley at 0.05 level of significance.

7.5 Balanced Incomplete Block Design

Sometimes, in a randomized block design, it is not possible to carry out runs with all treatment combinations in each block. This type of situation occurs because of shortage of apparatus or unavailability of sufficient materials or lack of time or limited facilities. In such situation, it is possible to use a block design in which every treatment may not be present in every block. Then, the block is said to be incomplete, and a design constituted of such blocks is called incomplete block design. However, when all treatment combinations are equally important as it ensures equal precision of the estimates of all pairs of treatment effects, the treatment combinations used in each block should be selected in a balanced manner such that any pair of treatments appear together an equal number of times. An incomplete block design constituted in such a balanced manner is called balanced incomplete block design (BIBD). In a nutshell, it is an incomplete block design in which any two treatments appear together an equal number of times.

7.5.1 A Practical Problem

A process development engineer was interested to identify an appropriate method of pigment dispersion in a coloring paint. In a pilot plant, he thought to prepare a particular mix of a pigment, apply it to a panel by five application methods (screening, brushing, spraying, rolling, and jetting), and measure the percentage reflectance of pigment. As in the pilot plant, it was possible to produce only four runs in a day, the engineer chose a balanced incomplete block design for experimentation.

7.5.2 Experimental Data

The engineer decided to choose five days and follow five application methods. He carried out runs in such a manner that any two processes appear together for four number of times. The experimental results are displayed in Table 7.24.

7.5.3 Descriptive Model

Suppose we have m number of treatments and n number of blocks such that each block contains k number of treatments and each treatment occurs r times. So, in this design, the total number of observations is $N = mr = nk$. In a balanced incomplete block design, if $m = n$ then the design becomes symmetric. In this particular numerical

Table 7.24 Experimental results of percentage reflectance of pigment

Process	Day				
	1	2	3	4	5
Screening	66	64	64	68	
Brushing	76		74	76	76
Spraying		82	86	84	82
Rolling	62	64	66		66
Jetting	54	56		56	58

problem, $m = 5, n = 5, k = 4$, and $r = 4$. As, here, $m = n = 4$, hence the design is said to be symmetric.

The statistical model for the balanced incomplete block design (BIBD) is

$$y_{ij} = \mu + \tau_i + \beta_j + \varepsilon_{ij}$$

where y_{ij} is a random variable indicating the ith observation in the jth block, μ is a parameter common to all treatments called the overall mean, τ_i is a parameter associated with the ith treatment called ith treatment effect, β_j is another parameter related to the jth block called jth block effect, and ε_{ij} is random error.

7.5.4 Test of Hypothesis

As stated in Chap. 7, the test of hypothesis is carried out in the following manner.

Step 1: Statement of Hypothesis
We are interested in testing the equality of population treatment means of percentage reflectance of pigment. Our null hypothesis states that there is no difference among the population treatment means. Hence, we write

$$H_0 : \mu_1 = \mu_2 = \cdots = \mu_m \text{ or equivalently } H_0 : \tau_1 = \tau_2 = \cdots = \tau_m$$

The alternative hypothesis stats that there is a difference between population treatment means for at least a pair of populations. So, we write

$$H_1 : \mu_i \neq \mu_j \text{ for at least one } i \text{ and } j, \text{ or, equivalently, } H_1 : \tau_i \neq 0 \text{ for at least one } i.$$

Step 2: Selection of Level of Significance
Suppose the level of significance is chosen as 0.05. It means that the probability of rejecting the null hypothesis when it is true is less than or equal to 0.05.

Step 3: Computation of Test Statistic
The test statistic is computed by carrying out an analysis of variance of the data. The basic calculations for treatment totals and treatment means are shown in Table 7.25.

The calculations for ANOVA are shown below. The total variability can be calculated as

$$SS_{Total} = \sum_{i=1}^{m} \sum_{j=1}^{n} y_{ij}^2 - \frac{y_{..}^2}{N} = (66)^2 + (64)^2 + \cdots + (56)^2 + (58)^2 - \frac{(1380)^2}{4 \times 5} = 1860.$$

This total variability may be partitioned into

$$SS_{Total} = SS_{Treatments(adjusted)} + SS_{Blocks} + SS_{Error}.$$

Table 7.25 Calculations for treatments and blocks for pigment dispersion data

Process	Day					y_i
	1	2	3	4	5	
Screening	66	64	64	68		262
Brushing	76		74	76	76	302
Spraying		82	86	84	82	334
Rolling	62	64	66		66	258
Jetting	54	56		56	58	224
y_j	258	266	290	284	282	1380

where the sum of squares for treatments is adjusted to separate the treatment effect and the block effect. This adjustment is required because each treatment is represented in a different set of r blocks. Thus, the differences between unadjusted treatment totals $y_1., y_2., \ldots, y_m.$ are also affected by the differences between blocks.

The adjusted treatment sum of squares is expressed as follows

$$SS_{Treatments(adjusted)} = \frac{k \sum_{i=1}^{m} Q_i^2}{\frac{mr(k-1)}{(a-1)}}$$

where Q_i is the adjusted total for the ith treatment. This can be calculated in a manner shown below

$$Q_i = y_{i.} - \frac{1}{k} \sum_{j=1}^{n} \alpha_{ij} y_{.j}, \quad i = 1, 2, \ldots, m$$

with $\alpha_{ij} = 1$ if treatment i appears in block j and $\alpha_{ij} = 0$ otherwise. The summation of adjusted treatment totals will be equal to zero. Needless to say that $SS_{Treatments(adjusted)}$ has $a - 1$ degrees of freedom.

In order to compute the adjusted treatment sum of squares, we need to calculate the adjusted total for each treatment. This is done as shown below.

$$Q_1 = y_{1.} - \frac{1}{4} \sum_{j=1}^{n} \alpha_{1j} y_{.j} = 262 - \frac{1}{4}(258 + 266 + 290 + 284) = -12.5$$

$$Q_2 = y_{2.} - \frac{1}{4} \sum_{j=1}^{n} \alpha_{2j} y_{.j} = 302 - \frac{1}{4}(258 + 290 + 284 + 282) = 23.5$$

$$Q_3 = y_{3.} - \frac{1}{4} \sum_{j=1}^{n} \alpha_{3j} y_{.j} = 334 - \frac{1}{4}(266 + 290 + 284 + 282) = 53.5$$

$$Q_4 = y_{4.} - \frac{1}{4} \sum_{j=1}^{n} \alpha_{4j} y_{.j} = 258 - \frac{1}{4}(258 + 266 + 290 + 282) = -16$$

$$Q_5 = y_{5.} - \frac{1}{4} \sum_{j=1}^{n} \alpha_{5j} y_{.j} = 224 - \frac{1}{4}(258 + 266 + 284 + 282) = -48.5$$

Then, the adjusted treatment sum of squares can be computed as stated hereunder

$$SS_{Treatments(adjusted)} = \frac{k \sum_{i=1}^{m} Q_i^2}{\frac{mr(k-1)}{(a-1)}} = \frac{4[(12.5)^2 + (23.5)^2 + (53.5)^2 + (16)^2 + (48.5)^2]}{\frac{5 \times 4 \times (4-1)}{(5-1)}} = 1647.73$$

The block sum of squares is

$$SS_{Blocks} = \frac{1}{k} \sum_{i=1}^{m} y_{.j}^2 - \frac{y_{..}^2}{N} = \frac{1}{4}[(258)^2 + (266)^2 + (290)^2 + (284)^2 + (282)^2] - \frac{(1380)^2}{4 \times 5} = 180$$

The error sum of squares is

$$SS_{Error} = SS_{Total} - SS_{Treatments(adjusted)} - S_{Blocks} = 1860 - 1647.73 - 180 = 32.27$$

The calculations are summarized in Table 7.26.

Step 4: Specification of Critical Region
The null hypothesis is rejected if $F_0 > F_{\alpha,a-1,N-(a-1)(b-1)} = F_{0.05,4,11} = 3.36$ (Table A.11).

Step 5: Take a Decision
As the numerical value (140.59) of the test statistic is greater than 3.36, the former falls in the critical region adopted, thus calling for rejection of null hypothesis. It is therefore concluded that the pigment application methods are significantly different at a level of significance of 0.05.

Table 7.26 ANOVA Table for pigment dispersion data

Sources of variation	Sum of squares	Degrees of freedom	Mean square	F_0
Treatments (adjusted for blocks)	1647.73	4	411.93	140.59
hline Blocks	180	4	–	
Error	32.27	11	2.93	
Total		19		

Problems

7.1 A completely randomized experiment was carried out to examine the effect of an inductor current sense resistor on the power factor of a PFC circuit. Four resistor values were chosen and three replicates of each treatment were carried out. The results are displayed in Table 7.27.

(a) Draw a scatter plot and a box plot of power factor of the PFC circuit.
(b) Fit a descriptive model to the above-mentioned data and comment on the adequacy of the model.
(c) Test the hypothesis that the four sensor resistor values result in same power factor. Use $\alpha = 0.05$.
(d) Use Tukey's test to compare the means of power factors obtained at different sensor resistor values.
(e) Use Fisher's test to compare the means of power factors obtained at different sensor resistor values. Is the conclusion same as in 8.1(d)?
(f) Fit a suitable regression model to the data.

7.2 The austenite stainless steels are used for making various engineering components in power plants and automobile industries. In order to examine the effect of cyclic loading on fatigue properties of austenite stainless steel, a completely randomized experiment was conducted. Four levels of number of cycles to failure were chosen and the corresponding maximum stress was observed. This was replicated three times. The results of experiment are shown in Table 7.28.

Table 7.27 Data for Problem 7.1

Sensor resistor value (Ohm)	Power factor (-)		
	I	II	III
0.05	0.88	0.87	0.89
0.10	0.92	0.94	0.93
0.20	0.95	0.94	0.95
0.25	0.98	0.99	0.98

Table 7.28 Data for Problem 7.2

Number of cycles to failures	Maximum stress (MPa)		
	I	II	III
100	500	530	480
1000	350	320	380
10000	280	240	300
100000	200	220	230
1000000	150	180	160

(a) Does the number of cycles to failure affect the maximum stress of the austenite stainless steel? Use $\alpha = 0.05$.

(b) Use Tukey's test to compare the means of maximum stress obtained at the different numbers of cycles to failure.

(c) Fit a suitable regression model to the data.

7.3 In order to examine the effect of temperature on the output voltage of a thermocouple, a completely randomized experiment was carried out. Four different temperatures (250, 500, 750, and 1000°C) were chosen and the output voltages were measured. This was replicated three times. The results are shown in Table 7.29.

(a) Does the temperature affect the output voltage of the thermocouple? Use $\alpha = 0.05$.

(b) Use Tukey's test to compare the means of output voltages obtained at different temperatures.

(c) Fit a suitable regression model to the data.

7.4 Four catalysts that may affect the yield of a chemical process are investigated. A completely randomized design is followed where each process using a specific catalyst was replicated four times. The yields obtained are shown in Table 7.30.
   Do the four catalyst have same effect on yield? Use $\alpha = 0.05$.

7.5 A completely randomized experiment was carried out to examine the effect of diet on coagulation time for blood of animals. For this, 24 animals were randomly allocated to four different diets A, B, C, and D and the blood samples of the animals were tested. The blood coagulation times of the animals are shown in Table 7.31.

Table 7.29 Data for Problem 7.3

Temperature (°C)	Output voltage (mV)		
	I	II	III
250	9.8	10.2	10.1
500	19.3	20.4	19.8
750	29.5	30.6	29.9
1000	39.6	40.6	39.8

Table 7.30 Data for Problem 7.4

Catalyst	Yield (%)			
	I	II	III	IV
A	58.2	57.4	56.8	57.6
B	63.8	64.2	66.4	65.9
C	80.6	80.4	82.8	84.2
D	96.4	95.8	97.4	98.6

Table 7.31 Data for Problem 7.5

A	B	C	D
60	65	69	58
63	67	66	63
62	70	73	61
61	68	68	63
63	66	66	62
62	65	67	65

Table 7.32 Data for Problem 7.6

Cooking method	Concentration of glucosinolates (μ g/g)		
	I	II	III
Boiling	1080	1060	1100
Micro oven	1240	1260	1250
Basket steaming	1540	1550	1530
Oven steaming	2060	2040	2020

(a) Draw a scatter plot and a Box plot of blood coagulation time.
(b) Fit a descriptive model to the above-mentioned data and comment on the adequacy of the model.
(c) Test the hypothesis that the four diets result in same blood coagulation time. Use $\alpha = 0.05$.
(d) Use Tukey's test to compare the mean coagulation times.
(e) Use Fisher's test to compare the mean coagulation times. Is the conclusion same as in 8.5(d)?

7.6 In order to study the effect of domestic cooking on phytochemicals of fresh cauliflower, a completely randomized experiment was conducted. Cauliflower was processed by four different cooking methods, and the resulting glucosinolates were measured. Each treatment was replicated three times. The experimental data are shown in Table 7.32. Does the cooking method affect the concentration of glucosinolates in cooked cauliflower? Use $\alpha = 0.05$.

7.7 Three brands of batteries were investigated for their life in clocks by performing a completely randomized design of experiment. Four batteries of each brand were tested, and the results on the life (weeks) of batteries were obtained and are shown in Table 7.33.
    Are the lives of the four brands of batteries different? Use $\alpha = 0.05$?

7.8 A quality control manager wishes to test the effect of four test methods on the percentage rejection of produced items. He performed a completely randomized

Table 7.33 Data for
Problem 7.7

Brand 1	Brand 2	Brand 3	Brand 4
21	18	28	36
24	22	30	34
22	20	32	32
25	16	30	34

Table 7.34 Data for
Problem 7.8

Method 1	Method 2	Method 3	Method 4
5	12	15	16
9	10	17	12
7	16	12	14
8	18	14	18

Table 7.35 Data for
Problem 7.9

Brand 1	Brand 2	Brand 3	Brand 4	Brand 5
70	80	91	94	98
74	82	90	96	99
76	86	92	92	97
72	84	93	94	96

design with four replicates under each test methods and obtained the following data
shown in Table 7.34.

Are the percentage rejection resulting from four different test methods different?
Use $\alpha = 0.05$.

7.9 A completely randomized design of experiment was carried out to compare five
brands of air-conditioning filters in terms of their filtration efficiency. Four filters of
each brand were tested, and the results were obtained on the filtration efficiency (%)
of the filters and are shown in Table 7.35.

Are the filtration efficiencies of the five brands of filters different? Use $\alpha = 0.05$.

7.10 An experiment was conducted to examine the effect of lecture timing on the
marks (out of 100) obtained by students in a common first-year undergraduate course
of Mathematics. A randomized block design was chosen with three blocks in such
a way that the students majoring in Computer Science, Electrical Engineering, and
Mechanical Engineering constituted Block 1, Block 2, and Block 3, respectively.
The marks obtained by the students out of 100 are shown in Table 7.36.

Analyze the data and draw appropriate conclusion. Use $\alpha = 0.05$.

7.11 Consider, in the above problem, the experimenter was not aware of randomized
block design and would have assigned the lecture timing each of the students majoring
in different branches randomly. Further, consider that the experimenter would have
obtained the same results as in the above problem by chance. Would the experimenter
have concluded the same as in the above problem? Use $\alpha = 0.05$.

Table 7.36 Data for Problem 7.10

Lecture timing	Blocks		
	1	2	3
8:00 am – 8:50 am	96	90	85
11:00 am – 11:50 am	85	88	82
2:00 pm – 2:50 pm	80	76	78

Table 7.37 Data for Problem 7.12

Test method	Materials (Blocks)			
	I	II	III	IV
A	10.2	19.7	23.4	19.6
B	12.5	15.4	28.7	23.4
C	13.7	18.7	24.5	26.7

Table 7.38 Data for Problem 7.13

Region	Season		
	I	II	III
1	C = 265	B = 410	A = 220
2	A = 280	C = 300	B = 384
3	B = 360	A = 240	C = 251

7.12 Three different test methods (A, B, and C) are compared to examine the strength of four different materials (I, II, III, and IV) in accordance with a randomized block design of experiment such that each material acts like a block. The results of experiments (strength in N) are given in Table 7.37.

Analyze the data and draw appropriate conclusion. Use $\alpha = 0.05$.

7.13 A sales manager was interested to compare the sales of three products (A, B, and C). A Latin Square Design of experiment was conducted to systematically control the effects of region and season on the sales of the products. The data on sales revenue (in thousand dollars) are given in Table 7.38.

Analyze the data and draw appropriate conclusion. Use $\alpha = 0.05$.

7.14 An oil company wishes to test the effect of four different blends of gasoline (A, B, C, and D) on the fuel efficiency of cars. It is thought to run the experiment according to a Latin Square Design so that the effects of drivers and car models may be systematically controlled. The fuel efficiency, measured in km/h after driving the cars over a standard course, is shown in Table 7.39.

Analyze the data and draw appropriate conclusion. Use $\alpha = 0.05$.

7.15 A chemical engineer wishes to test the effect of five catalysts (A, B, C, D, and E) on the reaction time of a chemical process. An experiment is conducted according to a Latin Square Design so that the effects of batches and experimenters

Table 7.39 Data for
Problem 7.14

Driver	Car model			
	I	II	III	IV
1	$B = 20.2$	$C = 28.3$	$D = 18.6$	$A = 16.2$
2	$D = 14.7$	$A = 14.3$	$B = 23.5$	$C = 27.9$
3	$C = 25.8$	$B = 22.7$	$A = 15.8$	$D = 16.5$
4	$A = 12.6$	$D = 17.8$	$C = 29.5$	$B = 23.5$

Table 7.40 Data for
Problem 7.15

Experi-menter	Reaction time				
	I	II	III	IV	V
1	$A = 8.5$	$B = 15.8$	$D = 17.5$	$C = 9.6$	$E = 12.0$
2	$C = 10.2$	$A = 7.6$	$B = 15.6$	$E = 12.6$	$D = 18.2$
3	$E = 12.4$	$D = 17.9$	$A = 8.0$	$B = 16.0$	$C = 10.0$
4	$B = 16.4$	$E = 11.8$	$C = 10.6$	$D = 18.4$	$A = 8.2$
5	$D = 18.6$	$C = 9.8$	$E = 11.6$	$A = 7.9$	$B = 16.2$

Table 7.41 Data for
Problem 7.16

Oil weight (g)	Car				
	1	2	3	4	5
5	90.45	91.23	90.75	89.47	
10	92.41		93.5	92.04	93.4
15		94.56	94.99	95.14	95.07
20	96.89	97.03	97.01		96.87
25	99.74	99.7		99.4	99.58

may be systematically controlled. The experimental data, expressed in h, is shown in Table 7.40.

Analyze the data and draw appropriate conclusion. Use $\alpha = 0.05$.

7.16 A process engineer is interested to study the effect of oil weight onto the filtration efficiency of engine intake air filters of automotives. In the road test he wishes to use cars as blocks, however, because of time constraint, he used a balanced incomplete block design. The results of experiment are shown in Table 7.41. Analyze the data using $\alpha = 0.05$.

Reference

Leaf GAV (1987) Practical statistics for the textile industry. Manchester, Part II, The Textile Institute, p 70

Chapter 8
Multifactor Experimental Designs

8.1 Introduction

In order to study the effects of two or more factors on a response variable, factorial designs are usually used. By following these designs, all possible combinations of the levels of the factors are investigated. The factorial designs are ideal designs for studying the interaction effect between factors. By interaction effect, we mean that a factor behaves differently in the presence of other factors such that its trend of influence changes when the levels of other factors change. This has already been discussed in Chap. 1. In this chapter, we will learn more about factorial design and analysis of experimental data obtained by following such designs.

8.2 Two-Factor Factorial Design

The simplest type of factorial design involves only two factors, where each factor has same level or different levels. Experiments are conducted in such a manner that all possible combinations of levels of factors are taken into account and there are replicates at each combination. As known, the replicates allow the experimenter to obtain an estimate of experimental error.

8.2.1 A Practical Problem

It is known that the strength of joints in a parachute webbing is vital to its performance. A fabric engineer wishes to examine whether the length of overlap of two pieces of webbing being joined and the stitch density (no. of stitches per cm) play any role in determining the strength of the ultimate webbing. He then selects two levels (high and low) of length of overlap and three levels (low, medium, and high) of stitch

© Springer Nature Singapore Pte Ltd. 2018
D. Selvamuthu and D. Das, *Introduction to Statistical Methods,*
Design of Experiments and Statistical Quality Control,
https://doi.org/10.1007/978-981-13-1736-1_8

Table 8.1 Experimental data of joint strength in parachute webbing

Length of overlap	Stitch density		
	Low	Medium	High
Short	32	41	68
	28	38	71
Long	37	61	77
	42	64	79

density and conducts a factorial experiment with two replicates. The experimental results are displayed in Table 8.1 (Leaf 1987).

8.2.2   Descriptive Model

In general, a two-factor factorial experiment is displayed as shown in Table 8.2. Here Y_{ijk} denotes the observed response when factor A is at ith level ($i = 1, 2, \ldots, a$) and factor B is at jth level ($j = 1, 2, \ldots, b$) for kth replicate ($k = 1, 2, \ldots, n$).

The observations of the experiment can be described by the following linear statistical model

$$y_{ijk} = \mu + \tau_i + \beta_j + (\tau\beta)_{ij} + \varepsilon_{ijk}; \quad i = 1, 2, \ldots, a; \ j = 1, 2, \ldots, b; \ k = 1, 2, \ldots, n$$

where y_{ijk} is a random variable denoting the ijkth observation, μ is a parameter common to all levels called the overall mean, τ_i is a parameter associated with the ith level effect of factor A, β_j is a parameter associated with jth level effect of factor B, and ε_{ijk} is a random error component. This corresponds to fixed effect model. It can be also written as

$$y_{ijk} = \mu_{ij} + \varepsilon_{ijk}; \quad i = 1, 2, \ldots, a; \ j = 1, 2, \ldots, b; \ k = 1, 2, \ldots, n$$

Table 8.2 Scheme of a two-factor factorial experiment data

		Observations			
		1	2	$\ldots$	b
Factor A	1	$y_{111}, y_{112}, \ldots, y_{11n}$	$y_{111}, y_{112}, \ldots, y_{11n}$	$\vdots$	$y_{1b1}, y_{1b2}, \ldots, y_{1bn}$
	2	$y_{211}, y_{212}, \ldots, y_{21n}$	$y_{221}, y_{222}, \ldots, y_{22n}$	$\vdots$	$y_{2b1}, y_{2b2}, \ldots, y_{2bn}$
	$\ldots$	$\vdots$	$\vdots$	$\ldots$	$\vdots$
	a	$y_{a11}, y_{a12}, \ldots, y_{a1n}$	$y_{a21}, y_{a22}, \ldots, y_{a2n}$	$\vdots$	$y_{ab1}, y_{ab2}, \ldots, y_{abn}$

where $\mu_{ij} = \mu + \tau_i + \beta_j + (\tau\beta)_{ij}$.

The followings are the reasonable estimates of the model parameters

$$\hat{\mu} = \bar{y}_{...} = \frac{1}{abn} \sum_{i=1}^{a} \sum_{j=1}^{b} \sum_{k=1}^{n} y_{ijk}$$

$$\hat{\tau}_i = \bar{y}_{i..} - \bar{y}_{...}, \quad i = 1, 2, \ldots, a \quad \text{where} \quad \bar{y}_{i..} = \frac{1}{bn} \sum_{j=1}^{b} \sum_{k=1}^{n} y_{ijk}$$

$$\hat{\beta}_j = \bar{y}_{.j.} - \bar{y}_{...}, \quad j = 1, 2, \ldots, b \quad \text{where} \quad \bar{y}_{i..} = \frac{1}{an} \sum_{i=1}^{a} \sum_{k=1}^{n} y_{ijk}$$

$$\hat{\tau\beta}_{ij} = \bar{y}_{ij.} - \bar{y}_{...} - (\bar{y}_{i..} - \bar{y}_{...}) - (\bar{y}_{.j.} - \bar{y}_{...}) = \bar{y}_{ij.} - \bar{y}_{i..} - \bar{y}_{.j.} + 2\bar{y}_{...}$$

where $\bar{y}_{ij.} = \frac{1}{n} \sum_{k=1}^{n} y_{ijk}$.

The estimate of the observations is $\hat{y}_{ijk} = \hat{\mu} + \hat{\tau}_i + \hat{\beta}_j + (\tau\beta)_{ij}$. Hence, the residual is $e_{ij} = y_{ijk} - \hat{y}_{ijk}$.

These formulas are used to estimate the observations and the residuals. The calculations are shown here. We consider the length of overlap as factor A and stitch density as factor B.

$$\bar{y}_{...} = \frac{1}{abn} \sum_{i=1}^{a} \sum_{j=1}^{b} \sum_{k=1}^{n} y_{ijk} = \frac{638}{12} = 53.17, \quad \bar{y}_{1..} = \frac{1}{bn} \sum_{j=1}^{b} \sum_{k=1}^{n} y_{1jk} = \frac{278}{6} = 46.33$$

$$\bar{y}_{2..} = \frac{1}{bn} \sum_{j=1}^{b} \sum_{k=1}^{n} y_{2jk} = \frac{360}{6} = 6, \quad \bar{y}_{.1.} = \frac{1}{an} \sum_{i=1}^{a} \sum_{k=1}^{n} y_{i1k} = \frac{139}{4} = 34.75$$

$$\bar{y}_{.2.} = \frac{1}{an} \sum_{i=1}^{a} \sum_{k=1}^{n} y_{i2k} = \frac{204}{4} = 51, \quad \bar{y}_{.3.} = \frac{1}{bn} \sum_{i=1}^{a} \sum_{k=1}^{n} y_{i3k} = \frac{295}{4} = 73.75$$

$$\bar{y}_{11.} = \frac{1}{n} \sum_{k=1}^{n} y_{11k} = \frac{60}{2} = 30, \quad \bar{y}_{12.} = \frac{1}{n} \sum_{k=1}^{n} y_{12k} = \frac{79}{2} = 39.5$$

$$\bar{y}_{13.} = \frac{1}{n} \sum_{k=1}^{n} y_{13k} = \frac{139}{2} = 69.5, \quad \bar{y}_{21.} = \frac{1}{n} \sum_{k=1}^{n} y_{21k} = \frac{79}{2} = 39.5$$

$$\bar{y}_{22.} = \frac{1}{n} \sum_{k=1}^{n} y_{22k} = \frac{125}{2} = 62.5, \quad \bar{y}_{23.} = \frac{1}{n} \sum_{k=1}^{n} y_{23k} = \frac{156}{2} = 78$$

$$\hat{\mu} = \bar{y}_{...} = 53.17, \quad \hat{\tau}_1 = \bar{y}_{1..} - \bar{y}_{...} = -6.84$$

$$\hat{\tau}_2 = \bar{y}_{2..} - \bar{y}_{...} = 6.83, \quad \hat{\beta}_1 = \bar{y}_{.1.} - \bar{y}_{...} = -18.42$$

$$\hat{\beta}_2 = \bar{y}_{.2.} - \bar{y}_{...} = -2.17, \quad \hat{\beta}_3 = \bar{y}_{.3.} - \bar{y}_{...} = 20.58$$

Table 8.3 Data and residuals for joint strength in parachute webbing

Length of overlap (A)	Stitch density (B)								
	Low			Medium			High		
Short	32	30	2	41	39.5	1.5	68	69.5	−1.5
	28	30	−2	38	39.5	−1.5	71	69.5	1.5
Long	37	39.5	−1.5	61	62.5	−1.5	77	78	−1
	42	39.5	2.5	64	62.5	1.5	79	78	1

$$(\tau\beta)_{11} = \bar{y}_{11.} - \bar{y}_{1..} - \bar{y}_{.1.} + \bar{y}_{...} = 2.09, \quad (\tau\beta)_{12} = \bar{y}_{12.} - \bar{y}_{1..} - \bar{y}_{.2.} + \bar{y}_{...} = -4.66$$
$$(\tau\beta)_{13} = \bar{y}_{13.} - \bar{y}_{1..} - \bar{y}_{.3.} + \bar{y}_{...} = 2.59, \quad (\tau\beta)_{21} = \bar{y}_{21.} - \bar{y}_{2..} - \bar{y}_{.1.} + \bar{y}_{...} = -2.08$$
$$(\tau\beta)_{22} = \bar{y}_{22.} - \bar{y}_{2..} - \bar{y}_{.2.} + \bar{y}_{...} = 4.67, \quad (\tau\beta)_{23} = \bar{y}_{23.} - \bar{y}_{2..} - \bar{y}_{.3.} + \bar{y}_{...} = -2.58$$

Table 8.3 reports the experimental results, fitted values, and residuals. Here, in each cell, three values are reported for a given process and for a given fabric type. The leftmost value is obtained from experiment, the middle one is the fitted value, and the rightmost value represents residual. It is now necessary to check whether the model is adequate or not. The methods of doing this are described in Sect. 7.2.3. The reader is instructed to carry out a similar analysis and conclude on the adequacy of the model.

8.2.3   Test of Hypothesis

The test of hypothesis is carried out as stated below. One can go through Chap. 5 to know more about testing of hypothesis.

Step 1: Statement of Hypothesis

We are interested in testing the equality of treatment means. Here, factor A and factor B are equally important. Then, the null hypothesis about the equality of row treatment effects is

$$H_0 : \tau_1 = \tau_2 = \cdots = \tau_a = 0$$

and the alternative hypothesis is

$$H_1 : \tau_i \neq 0 \text{ for at least one } i$$

Similarly, for the equality of column treatment effect, the null hypothesis is

$$H_0 : \beta_1 = \beta_2 = \cdots = \beta_b = 0$$

and the alternative hypothesis is

$$H_1 : \beta_j \neq 0 \text{ for at least one } j$$

We are also interested in determining whether row and column treatments interact. Thus, we also wish to test the null hypothesis

$$H_0 : (\tau\beta)_{ij} = 0 \; \forall \; i, j$$

against the alternative hypothesis

$$H_1 : (\tau\beta)_{ij} \neq 0 \text{ for at least one } i, j$$

Step 2: Choice of Level of Significance

The level of significance, usually denoted by α, is stated in terms of some small probability value such as 0.10 (one in ten) or 0.05 (one in twenty) or 0.01 (one in a hundred) or 0.001 (one in a thousand) which is equal to the probability that the test statistic falls in the critical region, thus indicating falsity of H_0.

Step 3: Find Out the Value of Test Statistics

The test statistics are stated below.

$$F_{0,i} = \frac{MS_i}{MS_{Error}}, \quad i \in \{A, B, AB\}$$

where

$$MS_A = \frac{SS_A}{(a-1)}, \quad MS_B = \frac{SS_B}{(b-1)}$$

$$MS_{AB} = \frac{SS_{AB}}{(a-1)(b-1)}, \quad MS_{Error} = \frac{SS_{Error}}{ab(n-1)}$$

$$SS_{Total} = \sum_{i=1}^{a}\sum_{j=1}^{b}\sum_{k=1}^{n}(y_{ijk} - \bar{y}_{...})^2$$

$$= bn\sum_{i=1}^{a}(\bar{y}_{i..} - \bar{y}_{...})^2 + an\sum_{j=1}^{b}(\bar{y}_{.j.} - \bar{y}_{...})^2 + n\sum_{i=1}^{a}\sum_{j=1}^{b}(\bar{y}_{ij.} - \bar{y}_{i..} - \bar{y}_{.j.} + \bar{y}_{...})^2$$

$$+ \sum_{i=1}^{a}\sum_{j=1}^{b}\sum_{k=1}^{n}(y_{ijk} - \bar{y}_{...})^2$$

$$= SS_A + SS_B + SS_{AB} + SS_{Error}$$

The sum of squares is used to construct Table 8.4.

Table 8.4 ANOVA table for two-factor factorial design

Source of variation	Sum of squares	Degree of freedom	Mean square	F_0
A	SS_A	$a-1$	$MS_A = \frac{SS_A}{(a-1)}$	$F_{0,A} = \frac{MS_A}{MS_{Error}}$
B	SS_B	$b-1$	$MS_B = \frac{SS_B}{(b-1)}$	$F_{0,B} = \frac{MS_B}{MS_{Error}}$
AB	SS_{AB}	$(a-1)(b-1)$	$MS_{AB} = \frac{SS_{AB}}{(a-1)(b-1)}$	$F_{0,AB} = \frac{MS_{AB}}{MS_{Error}}$
Error	SS_{Error}	$ab(n-1)$	$MS_{Error} = \frac{SS_{Error}}{ab(n-1)}$	
Total	SS_{Total}	$abn-1$		

Table 8.5 ANOVA table for joint strength in parachute webbing

Source of variation	Sum of squares	Degree of freedom	Mean square	F_0
A	560.4	1	560.4	93.4
B	3070.2	2	1535.1	255.85
AB	131.1	2	65.55	10.925
Error	36	6	6	
Total	3797.7	11		

The calculations for the given problem are shown below.

$$SS_{Total} = \sum_{i=1}^{2}\sum_{j=1}^{3}\sum_{k=1}^{2} y_{ijk}^2 - \frac{y_{...}^2}{2 \times 3 \times 2} = 37718 - \frac{638^2}{12} = 3797.7$$

$$SS_A = \frac{1}{3 \times 2}\sum_{i=1}^{2} y_{i..}^2 - \frac{y_{...}^2}{2 \times 3 \times 2} = \frac{1}{3 \times 2}[278^2 + 360^2] - \frac{638^2}{12} = 560.4$$

$$SS_B = \frac{1}{2 \times 2}\sum_{i=1}^{2} y_{.j.}^2 - \frac{y_{...}^2}{2 \times 3 \times 2} = \frac{1}{2 \times 2}[139^2 + 204^2 + 295^2] - \frac{638^2}{12} = 3070.2$$

$$SS_{Total} = \frac{1}{2}\sum_{i=1}^{2}\sum_{j=1}^{3} y_{ij.}^2 - \frac{y_{...}^2}{2 \times 3 \times 2} - SS_A - SS_B = 37682 - 33920.3 - 560.4 - 3070.2 = 131.1$$

$$SS_{Error} = 36.$$

The ANOVA is shown in Table 8.5.

Step 4: Specification of Critical Region

The null hypothesis is rejected if $F_0 > F_{\alpha,a-1,ab(n-1)} = F_{0.05,1,6} = 5.99$ for factor A, $F_0 > F_{\alpha,b-1,ab(n-1)} = F_{0.05,1,6} = 5.99$ for factor B, $F_0 > F_{\alpha,(a-1)(b-1),ab(n-1)} = F_{0.05,2,6} = 5.14$ for factor AB. (See Table A.11 for the above-mentioned F-values.)

Step 5: Take a Decision

As the values of the test statistic for factor A, factor B, and their interaction fall in the adopted critical region, we reject the null hypothesis. We thus conclude that factor A, factor B, and their interaction are significant at 0.05 level of significance.

8.2.4 Multiple Comparison Among Treatment Means

In order to compare among the treatment means, Tukey's test or Fisher's test can be used. The details of these tests are already given in Chap. 7. However, the comparison between the means of one factor (e.g., A) may be obscured by AB interaction if it is significant. Therefore, one approach that can be taken in such a situation is to fix a factor B at a specified level and apply Tukey's test and Fisher's test to the means of factor A at that level. This is illustrated below.

Suppose we are interested in detecting differences among the means of length of overlap (factor A). As the interaction is significant in this case, we make the comparison at one level of stitch density (factor B), low level. Further, we consider that the mean sum of squares due to error is the best estimate of error variance under the assumption that the experimental error variance is the same over all treatment combinations. This procedure can be extended to compare the differences between the means of length of overlap at all levels of stitch density. The treatment means at different levels of stitch density are shown in Table 8.6.

As per Tukey's test, the difference between any pair of means will be significant if it exceeds

$$T_\alpha = q_\alpha(a, f)\sqrt{\frac{MS_{\text{Error}}}{n}} = q_{0.05}(2, 6)\sqrt{\frac{6}{2}} = 5.99.$$

(From Table A.17, $q_{0.05}(2, 6) = 3.46$)

Here, all the differences in mean values at any level of stitch density are significant. As per Fisher's test, the difference between any pair of means will be significant if it exceeds the least significant difference (LSD) as calculated hereunder

$$LSD = t_{\frac{0.05}{2}, 6}\sqrt{\frac{MS_{\text{Error}}}{n}} = 4.238.$$

Table 8.6 Mean values of strength of parachute webbing

Length of overlap	Stitch density		
	Low	Medium	High
Short	30.0	39.5	69.5
Long	39.5	62.5	78.0

Here also, all the differences in mean values at any level of stitch density exceed the least significant difference. Hence, all the differences in mean values are significant.

8.3   Three-Factor Factorial Design

Sometimes, it is required to carry out experiments involving three factors with each factor having same level or different levels. This type of factorial design is known as three-factor factorial design. An example of this design is given below.

8.3.1   A Practical Problem

A factorial experiment was carried out in duplicate to analyze the effects of temperature, concentration, and time on the amount of a certain metal recovered from an ore sample of a given weight. The temperature was varied at two levels (1600 and $+1900\,°C$), concentration at two levels (30 and 35%), time at two levels (1 h and 3 h). The results of experiments are shown in Table 8.7.

8.3.2   Descriptive Model

In a general three-factor factorial experiment, where factor A is at ith level ($i = 1, 2, \ldots, a$), factor B is at jth level ($j = 1, 2, \ldots, b$), factor C is at kth level ($k = 1, 2, \ldots, c$) for lth replicate ($l = 1, 2, \ldots, n$), the response y_{ijkl} can be described by the following linear statistical model

$$y_{ijkl} = \mu + \tau_i + \beta_j + \gamma_k + (\tau\beta)_{ij} + (\beta\gamma)_{jk} + (\gamma\beta)_{ki} + (\tau\beta\gamma)_{ijk} + \varepsilon_{ijkl};$$

Table 8.7   Experimental data

Temperature							
1600 °C				1900 °C			
Concentration				Concentration			
30%		60%		30%		60%	
Time		Time		Time		Time	
1 h	3 h	1 h	3 h	1 h	3 h	1 h	3 h
80	81	69	91	65	84	74	93
62	79	73	93	63	86	80	93

$$i = 1, 2, \ldots, a; \quad j = 1, 2, \ldots, b; \quad k = 1, 2, \ldots, c; \quad l = 1, 2, \ldots, n$$

where y_{ijkl} is a random variable denoting the $ijkl$th observation, μ is a parameter common to all levels called the overall mean, τ_i is a parameter associated with the ith level effect of factor A, β_j is a parameter associated with jth level effect of factor B, γ_k is a parameter associated with the kth level of factor C, and ε_{ijkl} is a random error component. This corresponds to a fixed effect model.

The following are the reasonable estimates of the model parameters

$$\hat{\mu} = \bar{y}_{....} \quad \text{where} \quad \bar{y}_{....} = \frac{1}{abcn} \sum_{i=1}^{a} \sum_{j=1}^{b} \sum_{k=1}^{c} \sum_{l=1}^{n} y_{ijkl}$$

$$\hat{\tau}_i = \bar{y}_{i...} - \hat{\mu} \quad \text{where} \quad \bar{y}_{i...} = \frac{1}{bcn} \sum_{j=1}^{b} \sum_{k=1}^{c} \sum_{l=1}^{n} y_{ijkl}$$

$$\hat{\beta}_j = \bar{y}_{.j..} - \hat{\mu} \quad \text{where} \quad \bar{y}_{.j..} = \frac{1}{acn} \sum_{i=1}^{a} \sum_{k=1}^{c} \sum_{l=1}^{n} y_{ijkl}$$

$$\hat{\gamma}_k = \bar{y}_{..k.} - \hat{\mu} \quad \text{where} \quad \bar{y}_{..k.} = \frac{1}{abn} \sum_{i=1}^{a} \sum_{j=1}^{b} \sum_{l=1}^{n} y_{ijkl}$$

$$\widehat{(\tau\beta)}_{ij} = \bar{y}_{ij..} - \hat{\mu} - \hat{\tau}_i - \hat{\beta}_j \quad \text{where} \quad \bar{y}_{ij..} = \frac{1}{cn} \sum_{k=1}^{c} \sum_{l=1}^{n} y_{ijkl}$$

$$\widehat{(\beta\gamma)}_{jk} = \bar{y}_{.jk.} - \hat{\mu} - \hat{\beta}_j - \hat{\gamma}_k \quad \text{where} \quad \bar{y}_{.jk.} = \frac{1}{an} \sum_{i=1}^{a} \sum_{l=1}^{n} y_{ijkl}$$

$$\widehat{(\gamma\tau)}_{ik} = \bar{y}_{i.k.} - \hat{\mu} - \hat{\gamma}_k - \hat{\tau}_i \quad \text{where} \quad \bar{y}_{i.k.} = \frac{1}{bn} \sum_{j=1}^{b} \sum_{l=1}^{n} y_{ijkl}$$

$$\widehat{(\alpha\beta\gamma)}_{ijk} = \bar{y}_{ijk.} - \hat{\mu} - \hat{\tau}_i - \hat{\beta}_j - \hat{\gamma}_k - \widehat{(\tau\beta)}_{ij} - \widehat{(\beta\gamma)}_{jk} - \widehat{(\gamma\tau)}_{ki} \quad \text{where} \quad \bar{y}_{ijk.} = \frac{1}{n} \sum_{l=1}^{n} y_{ijkl}$$

The estimate of the observation is

$$\bar{y}_{ijkl} = \hat{\mu} + \hat{\tau}_i + \hat{\beta}_j + \hat{\gamma}_k + \widehat{(\tau\beta)}_{ij} + \widehat{(\beta\gamma)}_{jk} + \widehat{(\gamma\beta)}_{ki} + \widehat{(\tau\beta\gamma)}_{ijk}$$

The residual is

$$e_{ijkl} = y_{ijkl} - \hat{y}_{ijkl}$$

These formulas are used to estimate the observations and the residuals. They along with the experimental results are reported in Table 8.8. The numbers mentioned in the parentheses, curly brackets, and square brackets represent experimental values, fitted values, and residuals, respectively. It is now necessary to check if the model is

Table 8.8 Experimental data, fitted data, and residuals

Temperature							
1600 °C				1900 °C			
Concentration				Concentration			
30%		60%		30%		60%	
Time		Time		Time		Time	
1 h	3 h	1 h	3 h	1 h	3 h	1 h	3 h
(80)	(81)	(69)	(91)	(65)	(84)	(74)	(93)
{71}	{80}	{71}	{92}	{64}	{85}	{77}	{93}
[9]	[1]	[−2]	[−1]	[1]	[−1]	[−3]	[0]
(62)	(79)	(73)	(93)	(63)	(86)	(80)	(93)
{71}	{80}	{71}	{92}	{64}	{85}	{77}	{93}
[−9]	[−1]	[2]	[1]	[−1]	[1]	[3]	[0]

adequate or not. The methods of checking model adequacy are described in Sect. 8.2.3. The reader is required to carry out this and conclude on the adequacy of the model.

8.3.3 Test of Hypothesis

This is done as stated below.

Step 1: Statement of Hypothesis

We are interested in testing the equality of treatment means. Here, factor A, factor B, and factor C are equally important. Then, the null hypothesis about the equality of treatment effects of A is

$$H_0 : \tau_1 = \tau_2 = \cdots = \tau_a = 0$$

and the alternative hypothesis is $H_1 : \tau_i \neq 0$ for at least one i.

Similarly, for the equality of treatment effects of B, the null hypothesis is

$$H_0 : \beta_1 = \beta_2 = \cdots = \beta_b = 0$$

and the alternative hypothesis is $H_1 : \beta_j \neq 0$ for at least one j.

Similarly, for the equality of treatment effects of C, the null hypothesis is

$$H_0 : \gamma_1 = \gamma_2 = \cdots = \gamma_c = 0$$

and the alternative hypothesis is $H_1 : \gamma_k \neq 0$ for at least one k.

We are also interested in interaction effects of treatments. Thus, we also wish to test

$$H_0 : (\tau\beta)_{ij} = 0 \ \forall \ i, j \quad \text{against} \quad H_1 : (\tau\beta)_{ij} \neq 0 \ \text{ for at least one } i, j$$

$$H_0 : (\beta\gamma)_{jk} = 0 \ \forall \ j, k \quad \text{against} \quad H_1 : (\beta\gamma)_{jk} \neq 0 \ \text{ for at least one } j, k$$

$$H_0 : (\gamma\alpha)_{ki} = 0 \ \forall \ k, i \quad \text{against} \quad H_1 : (\gamma\alpha)_{ki} \neq 0 \ \text{ for at least one } k, i$$

$$H_0 : (\alpha\beta\gamma)_{ij} = 0 \ \forall \ i, j, k \quad \text{against} \quad H_1 : (\alpha\beta\gamma)_{ij} \neq 0 \ \text{ for at least one } i, j, k.$$

Step 2: Choice of Level of Significance

The level of significance, usually denoted by α, is stated in terms of some small probability value such as 0.10 (one in ten) or 0.05 (one in twenty) or 0.01 (one in a hundred) or 0.001 (one in a thousand) which is equal to the probability that the test statistic falls in the critical region, thus indicating falsity of H_0.

Step 3: Find Out the Value of Test Statistics

The test statistics are stated below

$$F_{0,i} = \frac{MS_i}{MS_{\text{Error}}}, \quad i \in \{A, B, C, AB, CA, BC, ABC\}$$

where

$$MS_A = \frac{SS_A}{(a-1)}, \quad MS_B = \frac{SS_B}{(b-1)}$$

$$MS_C = \frac{SS_C}{(c-1)}, \quad MS_{AB} = \frac{SS_{AB}}{(a-1)(b-1)}$$

$$MS_{BC} = \frac{SS_{BC}}{(b-1)(c-1)}, \quad MS_{CA} = \frac{SS_{CA}}{(c-1)(a-1)}$$

$$MS_{ABC} = \frac{SS_{ABC}}{(a-1)(b-1)(c-1)}$$

$$SS_{\text{Total}} = \sum_{i=1}^{a}\sum_{j=1}^{b}\sum_{k=1}^{c}\sum_{l=1}^{n} y_{ijkl}^2 - \frac{y_{....}^2}{abcn}$$

$$SS_A = \frac{1}{bcn}\sum_{i=1}^{a} y_{i...}^2 - \frac{y_{....}^2}{abcn}, \quad SS_B = \frac{1}{acn}\sum_{j=1}^{b} y_{.j..}^2 - \frac{y_{....}^2}{abcn}$$

$$SS_C = \frac{1}{abn}\sum_{k=1}^{c} y_{..k.}^2 - \frac{y_{....}^2}{abcn}, \quad SS_{AB} = \frac{1}{cn}\sum_{i=1}^{a}\sum_{j=1}^{b} y_{ij..}^2 - \frac{y_{....}^2}{abcn} - SS_A - SS_B$$

Table 8.9 ANOVA table

Source of variation	Sum of squares	Degree of freedom	Mean square	F_0
Temperature	6.25	1	6.25	0.26
Concentration	272.25	1	272.25	11.11
Time	1122.25	1	1122.25	45.81
Temperature × concentration	20.25	1	20.25	0.83
Concentration × time	12.25	1	12.25	0.5
Time × temperature	12.25	1	12.25	0.5
Temperature × concentration × time	72.25	1	72.25	2.95
Error	196	8	24.5	
Total	1713.75	15		

$$SS_{CA} = \frac{1}{bn} \sum_{i=1}^{a} \sum_{k=1}^{c} y_{i.k.}^2 - \frac{y_{....}^2}{abcn} - SS_C - SS_A,$$

$$SS_{ABC} = \frac{1}{n} \sum_{i=1}^{a} \sum_{j=1}^{b} \sum_{k=1}^{c} y_{ijk.}^2 - \frac{y_{....}^2}{abcn} - SS_A - SS_B - SS_C - SS_{AB} - SS_{BC} - SS_{CA}$$

$$SS_{Error} = SSS_{Total} - SS_A - SS_B - SS_C - SS_{AB} - SS_{BC} - SS_{CA} - SS_{ABC}$$

These formulas are used to construct Table 8.9.

Step 4: Specification of Critical Region

The null hypothesis is rejected if $F_0 > F_{\alpha, a-1, abc(n-1)} = F_{0.05, 1, 8} = 5.32$ (Table A.11) for factor A,

$$F_0 > F_{\alpha, b-1, abc(n-1)} = F_{0.05, 1, 8} = 5.32 \text{ for factor } B,$$
$$F_0 > F_{\alpha, c-1, abc(n-1)} = F_{0.05, 1, 8} = 5.32 \text{ for factor } C,$$
$$F_0 > F_{\alpha, (a-1)(b-1), abc(n-1)} = F_{0.05, 1, 8} = 5.32 \text{ for factor } AB.$$
$$F_0 > F_{\alpha, (b-1)(c-1), abc(n-1)} = F_{0.05, 1, 8} = 5.32 \text{ for factor } BC.$$
$$F_0 > F_{\alpha, (c-1)(a-1), abc(n-1)} = F_{0.05, 1, 8} = 5.32 \text{ for factor } CA.$$
$$F_0 > F_{\alpha, (a-1)(b-1)(c-1), abc(n-1)} = F_{0.05, 1, 8} = 5.32 \text{ for factor } ABC.$$

Step 5: Take a Decision

If the calculated F-value is higher than the table F-value, then the null hypothesis is rejected and the alternative hypothesis is accepted; otherwise, the null hypothesis is not rejected.

Hence, in this case, concentration, time, and interaction among temperature, concentration, and time are found to be significant at 0.05 level of significance.

8.4 2^2 Factorial Design

It is generally known that the factorial designs are more efficient than the one-factor-at-a-time experiments. Especially when the factors are interacting with each other, the factorial designs are preferred to avoid misleading conclusions. Also, by using factorial designs, it is possible to estimate the effects of a factor at several levels of the other factors, thereby making the conclusions valid over a wide range of experimental conditions. In the following sections, we will learn about factorial designs in more detail.

8.4.1   Display of 2^2 Factorial Design

The 2^2 factorial design involves two factors, say A and B, and each runs at two levels, say high and low, thus yielding four runs in one replicate. The design matrix of a 2^2 factorial design is displayed in Table 8.10. Here, the low level of a factor is indicated by $-$ sign and the high level of a factor is denoted by $+$ sign. In this table, the column under factor A is filled with $-$ sign and $+$ sign alternately and the column under factor B is filled with pairs of $-$ sign and $+$ sign alternately. As shown, there are four treatment combinations (low A and low B, high A and low B, low A and high B, high A and high B) and each treatment combination is replicated n times. Hence, there are $4 \times n$ runs in total. While performing the experiment, these runs are required to be randomized. Customarily, the summations of the observations obtained at different treatment combinations are indicated by (1), a, b, and ab, respectively. It can be seen that the high value of any factor is denoted by the corresponding lowercase letter and that the low value of a factor is indicated by the absence of the corresponding letter. Thus, a represents the treatment combination where A is at high level and B is at low level, b represents the treatment combination where B is at high level and

Table 8.10 Design matrix of 2^2 factorial design

Factor		Replicate				Total
A	B	I	II	$\ldots$	n	
$-$	$-$			$\ldots$		$\sum = (1)$
$+$	$-$			$\ldots$		$\sum = a$
$-$	$+$			$\ldots$		$\sum = b$
$+$	$+$			$\ldots$		$\sum = ab$

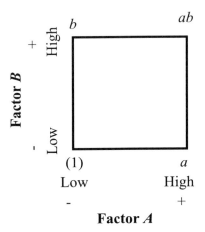

Fig. 8.1 Geometrical view of 2^2 factorial design

A is at low level, and ab denotes the treatment combination where A and B are both at high level. Sometimes, a 2^2 factorial design is represented by its geometric view as shown in Fig. 8.1.

8.4.2 Analysis of Effects in 2^2 Factorial Design

Let us now analyze the effects of the factors. By convention, the main effect of a factor is denoted by a capital Latin letter. Thus, "A" refers to the main effect of A, "B" denotes the main effect of B, and "AB" indicate the main effect of AB interaction. The main effect of A is defined by the difference between the average of the observations when A is at higher level and the average of the observations when A is at lower level. This is expressed hereunder.

$$A = \bar{y}_{A^+} - \bar{y}_{A^-} = \frac{ab + a}{2n} - \frac{b + (1)}{2n} = \frac{1}{2n}[ab + a - b - (1)]$$

Here, y denotes the observations. Similarly, the main effect of B is defined by the difference between the average of the observations when B is at higher level and the average of the observations when B is at lower level. This is shown below.

$$B = \bar{y}_{B^+} - \bar{y}_{B^-} = \frac{ab + a}{2n} - \frac{a + (1)}{2n} = \frac{1}{2n}[ab + b - a - (1)]$$

In a similar manner, the interaction effect of AB is defined by the difference between the average of the observations when the product of AB is at higher level and the average of the observations when the product of AB is at lower level.

$$AB = \bar{y}_{(AB)^+} - \bar{y}_{(AB)^-} = \frac{ab + (1)}{2n} - \frac{a + b}{2n} = \frac{1}{2n}[ab - a - b - (1)]$$

Often, in a 2^2 factorial experiment, we need to examine the magnitude and direction of the factor effects in order to determine which factors are likely to be important. The analysis of variance can generally be used to confirm this interpretation. There are many statistical software packages available that perform calculations for analysis of variance almost instantly. However, there are time-saving methods available for performing the calculations manually. One of them is based on calculation of contrasts. The contrast of a factor is defined by the total effect of the factor. For example, the contrast of factor A is expressed as follows.

$$\text{Contrast}_A = ab + a - b - (1)$$

Similarly, the contrast of factor B is expressed as follows.

$$\text{Contrast}_B = ab + b - a - (1)$$

The contrast of interaction effect AB is expressed as

$$\text{Contrast}_{AB} = ab + (1) - b - a.$$

The contrasts are used to compute the sum of squares for the factors. The sum of squares for a factor is equal to the square of the contrast divided by the number of observations in the contrast. According to this definition, the sum of squares for the factors A, B, and AB can be expressed as follows

$$SS_A = \frac{\text{Contrast}_A}{4n} = \frac{[ab + a - b - (1)]^2}{4n}$$

$$SS_B = \frac{\text{Contrast}_B}{4n} = \frac{[ab + b - a - (1)]^2}{4n}$$

$$SS_{AB} = \frac{\text{Contrast}_{AB}}{4n} = \frac{[ab + (1) - a - b]^2}{4n}$$

The total sum of squares can be found as usual in the following manner

$$SS_{\text{Total}} = \sum_{i=1}^{2}\sum_{j=1}^{2}\sum_{k=1}^{n} y_{ijk.}^2 - \frac{y_{....}^2}{4n}$$

Then, the sum of squares due to error can be found as follows

$$SS_{\text{Error}} = SS_{\text{Total}} - SS_A - SS_B - SS_{AB}$$

The complete analysis of variance can be obtained as shown in Table 9.4 considering $a = 2$ and $b = 2$.

8.4.3   A Practical Problem

Let us consider the example of electro-conductive yarn as discussed in Sect. 1.3.2 of Chap. 1. Polymerization time is denoted as factor A, and polymerization temperature is referred to as factor B. The low and high levels of polymerization time correspond to 20 and 60 min, respectively, and the low and high levels of polymerization temperature correspond to 10 and 30 °C, respectively. The experiment was performed in accordance with a 2^2 factorial design. Each of the four treatment combinations was replicated twice. Thus, in total, eight runs were carried out. The standard order of the runs is displayed under standard order column in Table 8.11. While performing the experiment, the order of these runs was randomized and the sequence of experiment was carried out as mentioned under run order column in Table 8.11. The results of experiments are also shown in Table 8.11. The geometric view of this experiment is displayed in Fig. 8.2.

Let us now analyze the main and total effects of the factors A, B, and AB. Using the expressions stated earlier, these main effects are calculated as follows.

$$A = \frac{1}{2 \times 2}[11.6 + 42.4 - 23.2 - 33.4] = -0.65$$

$$B = \frac{1}{2 \times 2}[11.6 + 23.2 - 42.4 - 33.4] = -10.25$$

$$A = \frac{1}{2 \times 2}[11.6 - 42.4 - 23.2 + 33.4] = -5.15$$

Table 8.11 Standard order, run order, and experimental results

Standard order	Run order	A	B	Response (kΩ/m)
8	1	+	+	6.4
7	2	−	+	12.4
4	3	+	+	5.2
1	4	−	−	15.8
5	5	−	−	17.6
6	6	+	−	22.1
2	7	+	−	20.3
3	8	−	+	10.8

Fig. 8.2 Geometrical view
for electro-conductive yarn
experiment

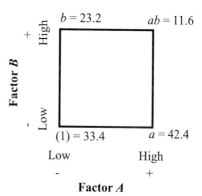

The minus sign before the values for the main effects of A, B, and AB indicates
that the change of levels of the factors A, B, and AB from higher to lower results in
increase of the response (electrical resistivity). It can be also observed that the main
effect of factor B is the highest, followed by that of factors AB and B, respectively.
The main effect of factor A is considerably smaller than that of B and AB.

Let us now analyze the total effect (contrast) of the factors A, B, and AB. Using
the expressions stated earlier, this is obtained as follows.

$$\text{Contrast}_A = 11.6 + 42.2 - 23.2 - 33.4 = -2.6$$

$$\text{Contrast}_B = 11.6 + 23.2 - 42.4 - 33.4 = -41$$

$$\text{Contrast}_{AB} = 11.6 - 42.2 - 23.2 + 33.4 = -20.6$$

The contrasts are then used to calculate the sums of squares as shown below.

$$SS_A = \frac{(-2.6)^2}{4 \times 2} = 0.845, \quad SS_B = \frac{(-41)^2}{4 \times 2} = 210.125$$

$$SS_{AB} = \frac{(-20.6)^2}{4 \times 2} = 53.045$$

The total sum of squares can be found as usual in the following manner

$$SS_{\text{Total}} = 15.8^2 + 17.6^2 + 20.3^2 + 22.1^2 + 10.8^2 + 12.4^2 + 5.2^2 + 6.4^2$$
$$- \frac{(15.8 + 17.6 + 20.3 + 22.1 + 10.8 + 12.4 + 5.2 + 6.4)^2}{4 \times 2}$$
$$= 269.255$$

Table 8.12 ANOVA table for electro-conductive yarn experiment

Source of variation	Sum of squares	Degree of freedom	Mean square	F_0
A	0.845	1	0.845	0.645
B	210.125	1	210.125	160.4
AB	53.045	1	53.045	40.49
Error	5.24	4	1.31	
Total	269.255	7		

Then, the sum of squares due to error can be found as follows

$$SS_{\text{Error}} = 269.255 - 0.845 - 210.125 - 53.045 = 5.24$$

The analysis of variance is summarized in Table 8.12. If the level of significance is taken as 0.05, then the critical region can be adopted as follows

$$F_0 > F_{\alpha,1,4(n-1)} = F_{0.05,1,4} = 7.71$$

See Table A.11 for this.

As the values of the test statistic for factor B and factor AB fall in the adopted critical region and the value of the test statistic for factor A does not fall in the adopted critical region, we conclude that factor B (polymerization temperature) and factor AB (interaction between polymerization time and polymerization temperature) are significant at 0.05 level of significance, but factor A (polymerization time) is not significant.

8.4.4 Regression Model

In the case of a 2^2 factorial design, it is easy to develop a regression model for predicting the response over a wide range of experimental conditions. The regression model is stated below

$$y = \beta_0 + \beta_1 x_1 + \beta_2 x_2 + \beta_{12} x_1 x_2 + \varepsilon$$

where y denotes response, x_1 & x_2 indicate the coded variables representing the factors, βs are the regression coefficients, and ε refers to residual. The relationship between the coded variables and the natural variables is discussed below. Take

Table 8.13 Experimental values, fitted values, and residuals for electro-conductive yarn experiment

Time (min)	Temperature (°)C	x_1 (−)	x_2 (−)	Experimental value(kΩ/m)		Fitted value (kΩ/m)	Residual (kΩ/m)	
				I	II			
20	10	−1	−1	15.8	17.6	16.7	−0.9	0.9
60	10	1	−1	20.3	22.1	21.2	−0.9	0.9
20	30	−1	1	10.8	12.4	11.6	−0.8	0.8
60	30	1	1	5.2	6.4	5.8	−0.6	0.6

the case of electro-conductive yarn experiment. The natural variables are time and temperature. Then, the relationship between coded and natural variable is

$$x_1 = \frac{\text{Time} - (\text{Time}_{low} + \text{Time}_{high})/2}{(\text{Time}_{high} - \text{Time}_{low})/2} = \frac{\text{Time} - (20 + 60)/2}{(60 - 20)/2} = \frac{\text{Time} - 40}{20}$$

$$x_2 = \frac{\text{Temperature} - (\text{Temperature}_{low} + \text{Temperature}_{high})/2}{(\text{Temperature}_{high} - \text{Temperature}_{low})/2}$$

$$= \frac{\text{Temperature} - (10 + 30)/2}{(30 - 10)/2} = \frac{\text{Temperature} - 20}{10}$$

The fitted regression model is

$$\hat{y} = 13.825 + \left(\frac{-0.65}{2}\right)x_1 + \left(\frac{-10.25}{2}\right)x_2 + \left(\frac{-5.15}{2}\right)x_1 x_2$$

where the intercept is the grand average of all eight observations and the regression coefficients are one-half of the corresponding main effect of the factors. The reason for the regression coefficient is one-half of the main effect is stated below. It is known that the regression coefficient measures the effect of a unit change in x on the mean of y. But, the main effect was calculated based on two unit (−1 to +1) change of x. Hence, it is required to take one-half of the main effect as regression coefficient. Using the above equation, the fitted values and residuals are calculated as shown in Table 8.13. The coefficient of determination is found as follows: $R^2 = 0.9805$ and adjusted $R^2 = 0.9659$. It is now necessary to check if the model is adequate or not. Figure 8.3 presents a normal probability plot of the residuals and a plot of the residuals versus the fitted values. The plots appear satisfactory; hence, there is no reason to suspect that the model is not adequate.

Sometimes, it is desirable to express the fitted equation in terms of natural variables. This can be done as follows.

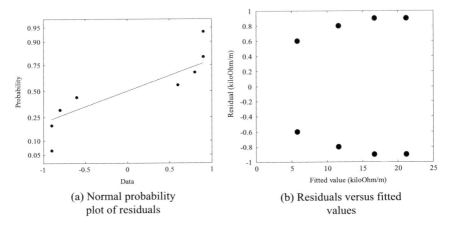

(a) Normal probability
plot of residuals

(b) Residuals versus fitted
values

Fig. 8.3 Plots of residuals for electro-conductive yarn experiment

$$\hat{y} = 13.825 + \left(\frac{-0.65}{2}\right)\left(\frac{\text{Time} - 40}{20}\right) + \left(\frac{-10.25}{2}\right)\left(\frac{\text{Temperature} - 20}{10}\right)$$
$$+ \left(\frac{-5.15}{2}\right)\left(\frac{\text{Time} - 40}{20}\right)\left(\frac{\text{Temperature} - 20}{10}\right)$$
$$= 14.4250 + 0.2413\,\text{Time} + 0.0250\,\text{Temperature} - 0.0129\,\text{TimeTemperature}$$

8.4.5 Response Surface

The regression model

$$\hat{y} = 13.825 + \left(\frac{-0.65}{2}\right)x_1 + \left(\frac{-10.25}{2}\right)x_2 + \left(\frac{-5.15}{2}\right)x_1x_2$$

can be used to generate response surface and contour plots. The response surface plot provides a three-dimensional view of the response surface, while the contour plot is a two-dimensional plot such that all data points that have the same response are connected to produce contour lines of constant responses. These plots are often found to be useful for optimizing the response. Figure 8.4 displays response surface and contour plots of resistivity obtained from the model. As shown, the response surface has a curvature, which is arising from the interaction between the factors. It can be seen that the minimum resistivity can be obtained at higher levels of time and temperature. This can be obtained by employing response surface methodology of analysis.

Fig. 8.4 Response surface and contour plots

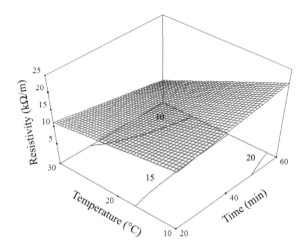

8.5 2^3 Factorial Design

In the last section, we have learnt how to analyze the results of experiments carried out the following 2^2 factorial design. In this section, we will learn 2^3 factorial design.

8.5.1   Display of 2^3 Factorial Design

The 2^3 factorial design involves three factors, say A, B, and C, and each runs at two levels, say high and low, thus yielding eight runs in one replicate. The design matrix of a 2^3 factorial design is displayed in Table 8.14. Like 2^2 factorial design, the low level of a factor is indicated by "$-$" sign and the high level of a factor is denoted by

Table 8.14 Design matrix of 2^3 factorial design

Factor			Replicate				Total
A	B	C	I	II	...	n	
$-$	$-$	$-$			...		$\sum = (1)$
$+$	$-$	$-$			...		$\sum = a$
$-$	$+$	$-$			...		$\sum = b$
$+$	$+$	$-$			...		$\sum = ab$
$-$	$-$	$+$			...		$\sum = c$
$+$	$-$	$+$			...		$\sum = ac$
$-$	$+$	$+$			...		$\sum = bc$
$+$	$+$	$+$			...		$\sum = abc$

Fig. 8.5 Geometric view of 2^3 factorial design

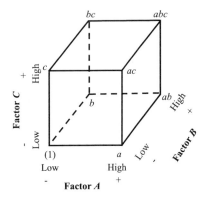

"$+$" sign. In this table, the column under factor A is filled with "$-$" sign and "$+$" sign alternately, the column under factor B is filled with pairs of "$-$" sign and "$+$" sign alternately, and the column under factor C is filled with four "$-$" signs and four "$+$" signs alternately. As shown, there are eight treatment combinations and each treatment combination is replicated n times. Hence, there are $4 \times n$ runs in total. While performing the experiment, these runs are required to be randomized.

Customarily, the summations of the observations obtained at eight different treatment combinations are indicated by (1), a, b, ab, c, ac, bc, and abc, respectively. Here again, it can be seen that the high value of any factor is denoted by the corresponding lowercase letter and that the low value of a factor is indicated by the absence of the corresponding letter. Thus, a represents the treatment combination where A is at high level and B and C are at low level, b represents the treatment combination where B is at high level and A and C are at low level, and c represents the treatment combination where C is at high level and A and B are at low level, ab denotes the treatment combination where A and B are both at high level and C is at low level, ac denotes the treatment combination where A and C are both at high level and B is at low level, bc denotes the treatment combination where B and C are both at high level and A is at low level, and abc denotes the treatment combination where A, B, and C are all at high level. Sometimes, a 2^3 factorial design is represented by its geometric view as shown in Fig. 8.5.

8.5.2 Analysis of Effects in 2^3 Factorial Design

This is carried out in a similar way as it was done earlier in 2^2 factorial design. The main effect of A is defined by the difference between the average of the observations when A is at higher level and the average of the observations when A is at lower level. This is expressed below.

$$A = \bar{y}_{A^+} - \bar{y}_{A^-} = \frac{a + ab + ac + abc}{4n} - \frac{(1) + b + c + bc}{4n}$$

$$= \frac{1}{4n}[a + ab + ac + abc - (1) - b - c - bc]$$

Similarly, the main effects of B and C are expressed as follows

$$B = \bar{y}_{B^+} - \bar{y}_{B^-} = \frac{b + ab + bc + abc}{4n} - \frac{(1) + a + c + ac}{4n}$$

$$= \frac{1}{4n}[b + ab + bc + abc - (1) - a - c - ac]$$

$$C = \bar{y}_{C^+} - \bar{y}_{C^-} = \frac{c + ac + bc + abc}{4n} - \frac{(1) + a + b + ab}{4n}$$

$$= \frac{1}{4n}[c + ac + bc + abc - (1) - a - b - ab]$$

The interaction effect of AB is defined by the difference between the average of the observations when the product of AB is at higher level and the average of the observations when the product of AB is at lower level.

$$AB = \bar{y}_{(AB)^+} - \bar{y}_{(AB)^-} = \frac{(1) + ab + c + abc}{4n} - \frac{a + b + bc + ac}{4n}$$

$$= \frac{1}{4n}[abc - bc + ab - b - ac + c - a + (1)]$$

$$AC = \bar{y}_{(AC)^+} - \bar{y}_{(AC)^-} = \frac{(1) + b + ac + abc}{4n} - \frac{a + ab + c + bc}{4n}$$

$$= \frac{1}{4n}[(1) - a + b - ab - c + ac - bc + abc]$$

$$BC = \bar{y}_{(BC)^+} - \bar{y}_{(BC)^-} = \frac{(1) + a + bc + abc}{4n} - \frac{b + ab + c + ac}{4n}$$

$$= \frac{1}{4n}[(1) + a - b - ab - c - ac + bc + abc]$$

$$ABC = \bar{y}_{(ABC)^+} - \bar{y}_{(ABC)^-} = \frac{a + b + c + abc}{4n} - \frac{ab + bc + ac + (1)}{4n}$$

$$= \frac{1}{4n}[abc - bc - ac + c - ab + b + a - (1)]$$

As known, the contrast of a factor is defined by the total effect of the factor. According to this, the contrasts are expressed as follows.

$$\text{Contrast}_A = [a - (1) + Ab - b + ac - c + abc - bc]$$

$$\text{Contrast}_B = [b + ab + bc + abc - (1) - a - c - ac]$$

$$\text{Contrast}_A = [c + ac + bc + abc - (1) - a - b - ab]$$

$$\text{Contrast}_{AB} = [abc - bc + ab - b - ac + c - a + (1)]$$

$$\text{Contrast}_{AC} = [(1) - a + b - ab - c + ac - bc + abc]$$

$$\text{Contrast}_{BC} = [(1) + a - b - ab - c - ac + bc + abc]$$

$$\text{Contrast}_{ABC} = [abc - bc - ac + c - ab + b + a - (1)]$$

The contrasts are used to calculate the sum of squares. The sum of squares of a factor is equal to the square of the contrast divided by the number of observations in the contrast. According to this, the sum of squares is expressed as follows.

$$SS_A = \frac{[a - (1) + ab - b + ac - c + abc - bc]^2}{8n}$$

$$SS_B = \frac{[b + ab + bc + abc - (1) - a - c - ac]^2}{8n}$$

$$SS_C = \frac{[c + ac + bc + abc - (1) - a - b - ab]^2}{8n}$$

$$SS_{AB} = \frac{[abc - bc + ab - b - ac + c - a + (1)]^2}{8n}$$

$$SS_{AC} = \frac{[(1) - a + b - ab - c + ac - bc + abc]^2}{8n}$$

$$SS_{BC} = \frac{[(1) + a - b - ab - c - ac + bc + abc]^2}{8n}$$

$$SS_{ABC} = \frac{[abc - bc - ac + c - ab + b + a - (1)]^2}{8n}$$

The total sum of squares can be found as usual in the following manner

$$SS_{\text{Total}} = \sum_{i=1}^{2}\sum_{j=1}^{2}\sum_{k=1}^{2}\sum_{l=1}^{n} y_{ijkl}^2 - \frac{y_{....}^2}{4n} \quad \text{where} \quad y_{....} = \sum_{i=1}^{2}\sum_{j=1}^{2}\sum_{k=1}^{2}\sum_{l=1}^{n} y_{ijkl}$$

Table 8.15 ANOVA table

Source of variation	Sum of squares	Degree of freedom	Mean square	F_0
A	SS_A	1	$MS_A = \frac{SS_A}{1}$	$\frac{MS_A}{MS_{Error}}$
B	SS_B	1	$MS_B = \frac{SS_B}{1}$	$\frac{MS_B}{MS_{Error}}$
C	SS_C	1	$MS_C = \frac{SS_C}{1}$	$\frac{MS_C}{MS_{Error}}$
AB	SS_{AB}	1	$MS_{AB} = \frac{SS_{AB}}{1}$	$\frac{MS_{AB}}{MS_{Error}}$
BC	SS_{BC}	1	$MS_{BC} = \frac{SS_{BC}}{1}$	$\frac{MS_{BC}}{MS_{Error}}$
AC	SS_{AC}	1	$MS_{AC} = \frac{SS_{AC}}{1}$	$\frac{MS_{AC}}{MS_{Error}}$
ABC	SS_{ABC}	1	$MS_{ABC} = \frac{SS_{ABC}}{1}$	$\frac{MS_{ABC}}{MS_{Error}}$
Error	SS_{Error}	$8(n-1)$	$MS_{Error} = \frac{SS_{Error}}{8(n-1)}$	
Total	SS_{Total}	$8n - 1$		

Then, the sum of squares due to error can be found as follows

$$SS_{Error} = SS_{Total} - SS_A - SS_B - SS_C - SS_{AB} - SS_{AC} - SS_{BC} - SS_{ABC}$$

The complete analysis of variance can be obtained as shown in Table 8.15.

8.5.3   Yates' Algorithm

There is a quicker method available for analyzing data of a 2^n factorial experiment, where n stands for number of factors. This method is based on Yates' algorithm. The steps of Yates' algorithm are stated below.

Step 1: Write down the results of a 2^n factorial experiment in a standard order. The standard orders of a few 2^n factorial designs are

2^2 factorial design : 1, a, b, ab
2^3 factorial design : 1, a, b, ab, c, ac, bc, abc
2^4 factorial design : 1, a, b, ab, c, ac, bc, abc, d, ad, bd, abd, cd, acd, bcd, $abcd$

Step 2: Write down the response totals for the corresponding treatment combinations in the next column. Label this column as y.

Step 3: Create n number of columns after column y. Here, n stands for number of factors. In the first of these columns (labeled as column C1), the first half of the entries from the top are obtained by adding the pairs from column y and the last half of the entries are obtained by subtracting the top number of each pair from the bottom number of that pair from column y. Just as the entries in column C1 are obtained from column y, the entries in next column (column C2) are obtained from column C1 in the same way, those in next column (column C3) from those in column C2, and so on. This process continues until n number of columns labeled as C1, C2, C3,...,C_n are generated.

Step 4: Create a column just after column C_n, and label it as "factor effect" column. The entries in this column are found by dividing the first entry in column C_n by $(k \times 2^n)$ and the remaining entries by $(k \times 2^{n-1})$, where k stands for number of replicates.

Step 5: Create a column just after "factor effect" column, and label it as "factor sum of squares" column. The entries in this column are obtained by squaring the entries in column C_n and dividing by $(k \times 2^n)$, though no sum of squares is attributable to the first row, which represents overall mean of the data.

Step 6: Create one more column just after "factor sum of squares" column, and label it as "factor." This column identifies the factor combinations corresponding to the sums of squares just calculated. They are found by writing down the factor letters corresponding to the "+" signs in the first n number of column of the table.

8.5.4 A Practical Example

The chemical stability of an enzyme in a buffer solution was studied. The three factors of interest were buffer type, pH, and temperature. Two types of buffer (phosphate and pipes), two levels of pH (6.8 and 7.8), and two levels of temperature (25 and 30 °C) were selected. The percentage decomposition of a solution of the enzyme was measured and considered as response. Let us denote the three factors, namely buffer type, pH, and temperature as A, B, and C, respectively. The phosphate buffer is considered as low level, while the pipes buffer is taken as high level. The pH of 6.8 is considered as low level, while the pH of 7.8 is taken as high level. The temperature of 25 °C is considered as low level, and the temperature of 30 °C is taken as high level. The low level is indicated by "−" sign, and the high level is indicated by "+" sign. A 2^3 factorial design of experiment with three replicates was carried out. Thus, in total, twenty-four runs were carried out. The standard order of the runs is displayed under "standard order" column in Table 8.16. While performing the experiment, the order of these runs was randomized and the sequence of experiment was carried out as mentioned under "run order" column in Table 8.16. The results of experiments are also shown in Table 8.16. The geometric view of this experiment is displayed in Fig. 8.6.

Table 8.16 Standard order, run order, and experimental results for chemical stability experiment

Standard order	Run order	A	B	C	Response
23	1	−	+	+	8.7
7	2	−	+	+	8.1
21	3	−	−	+	12.1
6	4	+	−	+	7.3
14	5	+	−	+	6.1
13	6	−	−	+	13.4
1	7	−	−	−	6.2
3	8	−	+	−	2.4
2	9	+	−	−	0
16	10	+	+	+	1.8
10	11	+	−	−	0.4
20	12	+	+	−	0
11	13	−	+	−	5.3
22	14	+	−	+	8.8
5	15	−	−	+	13
9	16	−	−	−	9.3
4	17	+	+	−	0.8
24	18	+	+	+	0
19	19	−	+	−	3.9
15	20	−	+	+	9.4
17	21	−	−	−	7.2
12	22	+	+	−	0
8	23	+	+	+	3
18	24	+	−	−	0

Fig. 8.6 Geometric view of enzyme stability experiment

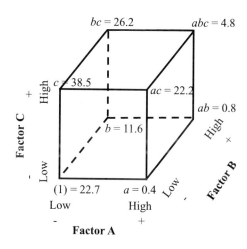

Let us now analyze the main and total effects of the factors A, B, C, AB, AC, BC, and ABC. Using the expressions stated earlier, these main effects are calculated as follows.

$$A = \frac{1}{4 \times 3}[0.4 + 0.8 + 22.2 + 4.8 - 22.7 - 11.6 - 38.5 - 26.2] = -5.9$$

$$B = \frac{1}{4 \times 3}[11.6 + 0.8 + 26.2 + 4.8 - 22.7 - 0.4 - 38.5 - 22.2] = -3.37$$

$$C = \frac{1}{4 \times 3}[38.5 + 22.2 + 26.2 + 4.8 - 22.7 - 0.411.6 - 0.8] = 4.68$$

$$AB = \frac{1}{4 \times 3}[4.8 - 26.2 + 0.8 - 11.622.2 + 38.5 - 0.4 + 22.7] = 0.53$$

$$AC = \frac{1}{4 \times 3}[22.7 - 0.4 + 11.6 - 0.8 - 38.5 + 22.2 - 26.2 + 4.8] = -0.38$$

$$BC = \frac{1}{4 \times 3}[22.7 + 0.4 - 11.6 - 38.5 - 22.2 + 26.2 + 4.8] = -1.52$$

$$ABC = \frac{1}{4 \times 3}[4.8 - 26.2 - 22.2 + 38.5 - 0.8 + 11.6 + 0.4 - 22.7] = -1.38$$

The minus sign before the values for the main effects of A, B, AC, BC, and ABC indicates that the change of levels of these factors from higher to lower results in an increase of the response (percentage decomposition). On the other hand, the plus sign before the values for the main effects of C and AB indicates that the change of levels of these factors from higher to lower results in the decrease of the response (percentage decomposition). It can also be observed that the main effect of factor A is the highest, followed by that of factors C, B, BC, ABC, AB, and AC.

Let us now analyze the total effect (contrast) of the factors A, B, C, AB, AC, BC, and ABC. Using the expressions stated earlier, this is obtained as follows.

$$\text{Contrast}_A = [0.4 + 0.8 + 22.2 + 4.8 - 22.7 - 11.6 - 38.5 - 26.2] = -70.8$$
$$\text{Contrast}_B = [11.6 + 0.8 + 26.2 + 4.8 - 22.7 - 0.4 - 38.5 - 22.2] = -40.44$$
$$\text{Contrast}_C = [38.5 + 22.2 + 26.2 + 4.8 - 22.7 - 0.411.6 - 0.8] = 56.16$$
$$\text{Contrast}_{AB} = [4.8 - 26.2 + 0.8 - 11.622.2 + 38.5 - 0.4 + 22.7] = 6.36$$
$$\text{Contrast}_{AC} = [22.7 - 0.4 + 11.6 - 0.8 - 38.5 + 22.2 - 26.2 + 4.8] = -4.56$$
$$\text{Contrast}_{BC} = [22.7 + 0.4 - 11.6 - 38.5 - 22.2 + 26.2 + 4.8] = -18.24$$
$$\text{Contrast}_{ABC} = [4.8 - 26.2 - 22.2 + 38.5 - 0.8 + 11.6 + 0.4 - 22.7] = -16.56$$

The contrasts are then used to calculate the sum of squares as shown below.

Table 8.17 ANOVA table for chemical stability experiment

Source of variation	Sum of squares	Degree of freedom	Mean square	F_0
A	208.86	1	208.86	159.44
B	68.14	1	68.14	52.02
C	131.41	1	131.41	100.31
AB	1.69	1	1.69	1.29
AC	0.87	1	0.87	0.66
BC	13.86	1	13.86	10.58
ABC	11.43	1	11.43	8.73
Error	21.02	16	1.31	
Total	457.28	23		

$$SS_A = \frac{(-70.8)^2}{(8 \times 3)} = 208.86, \quad SS_B = \frac{(-40.44)^2}{(8 \times 3)} = 68.14$$

$$SS_C = \frac{(56.16)^2}{(8 \times 3)} = 131.41, \quad SS_{AB} = \frac{(6.36)^2}{(8 \times 3)} = 1.69$$

$$SS_{AC} = \frac{(-4.56)^2}{(8 \times 3)} = 0.87, \quad SS_{BC} = \frac{(-18.24)^2}{(8 \times 3)} = 13.86$$

$$SS_{ABC} = \frac{(-16.56)^2}{(8 \times 3)} = 11.43, \quad SS_{Total} = 1131.44 - \frac{(127.2)^2}{(8 \times 3)} = 457.28$$

$$SS_{Error} = 457.28 - 436.26 = 21.02$$

The calculations are summarized in Table 8.17. If the level of significance is taken as 0.05, then the critical region can be adopted as follows

$$F_0 > F_{\alpha, 1, 8(n-1)} = F_{0.05, 1, 16} = 4.49. \text{ (Table A.11)}$$

As the values of the test statistic for factors A, B, C, and interactions between B and C and among A, B, and C fall in the adopted critical region, we conclude that the single-factor effects of buffer type, pH, temperature, two-factor interaction effect between pH and temperature, and three-factor interaction effect among buffer type, pH, and temperature are significant at 0.05 level of significance.

Yates' algorithm can be used to analyze the data of the 2^n factorial experiment. For this, the steps mentioned earlier are followed. At first, a standard order of the design is created in the first three columns of Table 8.18. The low and high levels of each factor are denoted by "−" sign and "−" sign, respectively, in the first three columns of the table headed by the letters A, B, and C denoting the factors. In column A, the "−" and "−" signs appear alternately. In column B, the pairs of "−" and "−" signs alternate. In column C, four "−" signs alternate with four "−" signs. This is how the standard order of a 2^3 factorial design is displayed. The response totals are then mentioned

Table 8.18 Calculations based on Yates' algorithm

A	B	C	y	C1	C2	C3	Factor effect	Factor sum of squares	Factor
−	−	−	22.7	23.1	35.5	127.2	5.3	−	Mean
+	−	−	0.4	12.4	91.7	−70.8	−5.9	208.86	A
−	+	−	11.6	60.7	−33.1	−40.4	−3.37	68.01	B
+	+	−	0.8	31	−37.7	6.4	0.53	1.71	AB
−	−	+	38.5	−22.3	−10.7	56.2	4.68	131.6	C
+	−	+	22.2	−10.8	−29.7	−4.6	−0.38	0.88	AC
−	+	+	26.2	−16.3	11.5	−19	−1.58	15.04	BC
+	+	+	4.8	−21.4	−5.1	−16.6	−1.38	11.48	ABC

in the next column labeled as column y. After this column, three more columns are created and they are labeled as columns C1, C2, and C3. Note that three columns are created as the number of factors involved is three. In column C1, the first four entries are obtained by adding the pairs from column y and the last four entries are obtained by subtracting the top number of each pair from the bottom number of that pair from column y. For example, the first entry from top in this column is obtained as $22.7 + 0.4 = 23.1$, the second entry is obtained as follows: $11.6 + 0.8 = 12.4$, the second last entry is obtained as follows: $22.2 − 38.5 = −16.3$, and the last entry is obtained as follows: $4.8 − 26.2 = −21.4$. Just as the entries in column C1 are obtained from column y, the entries in column C2 are obtained from column C1 in the same way, and those in column C3 are obtained from column C2. The entries in "factor effect" column are found by dividing the first entry in column C3 by (3×2^n) and the remaining entries by $(3 \times 2^{n-1})$. Note that here the divisive factor 3 appears as the number of replicates is 3. The entries in factor sum of squares are obtained by squaring the entries in column C_n and dividing by (3×2^n), though no sum of squares is attributable to the first row, which represents the overall mean of the data. The final column identifies the factor combinations corresponding to the sums of squares just calculated. They are found by writing down the factor letters corresponding to the "+" signs in the first columns of the table. One can check that the factor effects and factor sum of squares calculated in this way are practically the same as calculated earlier for this experiment.

8.6   Blocking and Confounding

There are many situations when it is impossible to carry out all runs in a 2^n factorial design replicated k times under homogeneous conditions. For example, a single batch of raw material is not enough to perform all the runs. In other cases, a single operator cannot perform all the runs or all the runs cannot be performed in a day. In such cases, the design technique used is known as blocking. This is a very useful technique for the fact that it is often desirable to deliberately vary the experimental conditions to ensure that the treatments are equally effective across many situations that are likely to be encountered in practice. In this part of the chapter, we will discuss on blocking in 2^n factorial design.

8.6.1 Replicates as Blocks

Suppose a 2^n factorial design is replicated k times such that each replicate is run in one block which is defined by a set of nonhomogeneous conditions. The order of runs in each block (replicate) would be randomized. The nonhomogeneous conditions include but are not limited to different batches, different suppliers, different operators, etc. Consider the example of electro-conductive yarn experiment discussed in Sect. 8.4.3 Suppose that only four runs can be performed from a single batch of electrically conducting monomer (raw material). Then, two batches of raw materials are required to conduct all the eight runs. The experiment can then be conducted by considering each replicate as one block such that all the runs in one replicate (block) can be performed by using a single batch of raw material. Thus, each batch of raw material corresponds to a block. This is displayed in Table 8.19.

The linear statistical model for this design is stated below

$$y_{ijk} = \mu + \tau_i + \beta_j + (\tau\beta)_{ij} + \delta_k + \varepsilon_{ijkl}; \quad i = 1, 2; \ j = 1, 2; \ k = 1, 2$$

where y_{ijk} is a random variable denoting the ijkth observation, μ is a parameter common to all levels called the overall mean, τ_i is a parameter associated with the ith level effect of factor A, β_j is a parameter associated with jth level effect of factor B, δ_k is a parameter associated with the kth block effect, and ε_{ijk} is a random error component.

Table 8.19 Replicates as blocks in factorial design for electro-conductive yarn experiment

A	B	Response
−	−	$15.8 = (1)$
+	−	$20.3 = a$
−	+	$10.8 = b$
+	+	$5.2 = ab$
−	−	$17.6 = (1)$
+	−	$22.1 = a$
−	+	$12.4 = b$
+	+	$6.4 = ab$

Table 8.20 ANOVA table

Source of variation	Sum of squares	Degree of freedom	Mean square	F_0-value
Blocks	5.12	1	5.12	
A	0.845	1	0.845	0.6450
B	210.125	1	210.125	160.40
AB	53.045	1	53.045	40.49
Error	0.12	3	1.31	
Total	269.255	7		

The calculations for the analysis of variance for this design are shown below. It can be seen that they are very similar to that discussed in Sect. 8.2.3.

$$SS_{\text{Total}} = \sum_{i=1}^{2}\sum_{j=1}^{2}\sum_{k=1}^{2} y_{ijk}^2 - \frac{y_{...}^2}{2 \times 2 \times 2} = 1798.3 - \frac{110.6^2}{8} = 269.255$$

$$SS_A = \frac{1}{2 \times 2}\sum_{i=1}^{2} y_{i..}^2 - \frac{y_{...}^2}{2 \times 2 \times 2} = 1529.89 - \frac{110.6^2}{8} = 0.845$$

$$SS_B = \frac{1}{2 \times 2}\sum_{j=1}^{2} y_{.j.}^2 - \frac{y_{...}^2}{2 \times 2 \times 2} = 1739.17 - \frac{110.6^2}{8} = 210.125$$

$$SS_{AB} = \frac{1}{2}\sum_{i=1}^{2}\sum_{j=1}^{2} y_{ij.}^2 - \frac{y_{...}^2}{2 \times 2 \times 2} - SS_A - SS_B$$

$$= 1793.06 - \frac{110.6^2}{8} - 0.845 - 210.125 = 53.045$$

$$SS_{\text{Block}} = \frac{1}{2 \times 2}\sum_{k=1}^{2} y_{..k}^2 - \frac{y_{...}^2}{2 \times 2 \times 2} = 1534.165 - \frac{110.6^2}{8} = 5.12$$

$$SS_{\text{Error}} = SS_{\text{Total}} - SS_A - SS_B - SS_{AB} - SS_{\text{Block}}$$

$$= 269.255 - 0.845 - 210.125 - 53.045 - 5.12 = 0.12.$$

The results are summarized in Table 8.20.

The F-value obtained from table is $F_{0.05,1,3} = 10.13$ (Table A.11). The conclusion from this analysis is identical to that mentioned earlier and that the block effect is relatively small.

8.6.2   Confounding

It is a design technique for arranging a factorial experiment in blocks, where the block size is smaller than the number of treatment combinations in one replicate. It is very useful when, for example, a single batch of raw material is not large enough to make all the required runs. In the following sections, we will discuss the method of confounding.

Let us take the case of a 2^2 factorial design. There are three possibilities available for confounding: (1) confound A with block, (2) confound B with block, and (3) confound AB with block. Let us analyze these three possibilities one by one. At first, we consider to confound A with blocks. It means all the treatment combinations that have "+" sign on A are assigned to Block 1 and all the treatment combinations that have "+" sign on A are assigned to Block 2. So, the blocks will have the treatment combination as shown below

Block 1	Block 2
a	(1)
ab	b

In this case, the block effect is calculated as follows.

$$\text{Block Effect} = \bar{y}_{\text{Block 1}} - \bar{y}_{\text{Block2}} = \frac{a + ab}{2} - \frac{(1) + b}{2} = \frac{1}{2}[ab + a - b - (1)]$$

Remember that the effect of A was earlier calculated as follows

$$A = \bar{y}_{A+} - \bar{y}_{A-} = \frac{a + ab}{2} - \frac{(1) + b}{2} = \frac{1}{2}[ab + a - b - (1)]$$

It can be thus seen that A is confounded to blocks; that is, the main effect of A is identical (indistinguishable) to the block effect. This is an undesirable situation as the block effect, in principle, should not affect the main effect. Hence, confounding A with blocks is not a good proposition.

Let us then consider the second possibility to confound B with blocks. It means all the treatment combinations that have "+" sign on B are assigned to Block 1 and all the treatment combinations that have "+" sign on B are assigned to Block 2. So, the blocks will have the treatment combination as shown below.

Block 1	Block 2
b	(1)
ab	a

In this case, the block effect is calculated as follows.

$$\text{Block Effect} = \bar{y}_{\text{Block 1}} - \bar{y}_{\text{Block2}} = \frac{b + ab}{2} - \frac{(1) + a}{2} = \frac{1}{2}[ab + b - a - (1)]$$

Remember that the effect of A was earlier calculated as follows

$$B = \bar{y}_{B^+} - \bar{y}_{B^-} = \frac{ab + b}{2n} - \frac{a + (1)}{2n} = \frac{1}{2n}[ab + b - a - (1)]$$

It can be thus seen that B is confounded to blocks; that is, the main effect of B is identical (indistinguishable) to the block effect. This is also not a desirable situation as the block effect, in principle, should not affect the main effect. Hence, confounding B with blocks is not a good proposition.

Let us then consider the third possibility to confound AB with blocks. It means all the treatment combinations that have "+" sign on AB are assigned to Block 1 and all the treatment combinations that have "−" sign on AB are assigned to Block 2. So, the blocks will have the treatment combination as shown below.

Block 1	Block 2
(1)	a
ab	b

In this case, the block effect is calculated as follows.

$$\text{Block Effect} = \bar{y}_{\text{Block 1}} - \bar{y}_{\text{Block2}} = \frac{(1) + ab}{2} - \frac{a + b}{2} = \frac{1}{2}[ab - a - b + (1)]$$

Remember that the effect of AB was earlier calculated as follows

$$AB = \bar{y}_{(AB)^+} - \bar{y}_{(AB)^-} = \frac{(1) + ab}{2n} - \frac{a + b}{2n} = \frac{1}{2n}[ab - a - b + (1)]$$

It can be thus seen that AB is confounded to blocks; that is, the interaction effect of A and B is identical (indistinguishable) to the block effect. This is a desirable situation as the block effect is not affecting the main effects of A and B. Hence, confounding AB with blocks is a good proposition. This is why the usual practice is to confound the highest order interaction with blocks. In case of a 2^2 factorial design, as AB is the highest order interaction, AB needs to be confounded with blocks. Similarly, in case of 2^3 factorial design, as ABC is the highest order interaction, ABC needs to be confounded with blocks.

8.6.3   A Practical Example

Suppose, in Replicate I of electro-conductive yarn experiment, discussed in Sect. 8.4.3 of this chapter, each of the four treatment combinations requires a certain quantity of pyrrole and each batch of pyrrole is only large enough for two treatments; hence, two batches of pyrrole are required for one replicate. If batches of pyrrole are considered

Fig. 8.7 Running two blocks for one replicate

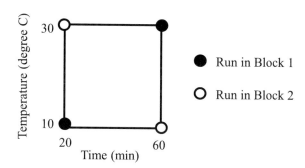

Table 8.21 Experimental results

Replicate I			
Block	Time (min)	Temperature (°C)	Resistivity (kΩ/m)
1	20	10	15.8
	60	30	5.2
2	60	10	20.3
	20	30	10.8
Replicate II			
1	20	10	17.6
	60	30	6.4
2	60	10	22.1
	20	30	12.4

as blocks, then we must assign two of the four treatment combinations to each block. In this way, Replicate I and Replicate II are treated. This is illustrated in Fig. 8.7.

The experimental results are reported in Table 8.21.

The block effect is calculated as follows.

$$\text{Block Effect} = \bar{y}_{\text{Block 1}} - \bar{y}_{\text{Block2}}$$
$$= \frac{(15.8 + 5.2) + (17.6 + 6.4)}{4} - \frac{(20.3 + 10.8) + (22.1 + 12.4)}{4} = -5.15$$

Remember that the interaction effect was calculated earlier as follows.

$$\text{Interaction Effect} = \bar{y}_{(AB)+} - \bar{y}_{(AB)-} = \frac{1}{2 \times 2}[11.6 - 42.4 - 23.2 + 33.4] = -5.15$$

Hence, the block effect is identical to AB interaction effect. That is, AB is confounded with blocks. The sum of squares due to block is calculated as follows

$$SS_{\text{Block}} = \frac{1}{2 \times 2} \sum_{k=1}^{2} y_{..k}^2 - \frac{y_{...}^2}{2 \times 2 \times 2} = \frac{45^2 + 65.6^2}{4} - \frac{110.6^2}{8} = 53.045$$

Table 8.22 ANOVA table for electro-conductive yarn experiment

Source of variation	Sum of squares	Degree of freedom	Mean square	F_0-value
Blocks (AB)	53.045	1		
A	0.845	1	0.845	0.6450
B	210.125	1	210.125	160.40
Error	5.24	4	1.31	
Total	269.255	7		

Thus, obtained analysis of variance is shown in Table 8.22.

The F-value obtained from table is $F_{0.05,1,4} = 7.71$ (Table A.11). The conclusion from this analysis is not identical to that mentioned earlier. This is because the AB interaction effect was very high.

8.7　Two-Level Fractional Factorial Design

As the number of factors increases, the number of runs required for a complete factorial design rapidly outgrows the resources for most of the experimenters. For example, a complete replicate of 2^6 complete factorial experiment requires 64 runs. In this design, only 6 of the 63 degrees of freedom are associated with the main effect and 15 degrees of freedom are corresponding to two-factor interactions. The remaining 42 degrees of freedom are attributed to three-factor and higher-order interactions. If the experimenter can reasonably assume that certain higher-order interactions are negligible, information on the main effects and lower-order interaction effects may be obtained by running a fraction of the complete factorial experiment. Such fractional factorial designs are often used as screening experiments.

8.7.1　Creation of 2^{3-1} Factorial Design

Suppose there are three factors and each factor has two levels, and the experimenter cannot conduct eight runs, but can conduct only four runs. It therefore suggests a one-half fraction of a 2^3 factorial design. As the design contains $2^{3-1} = 4$ treatment combinations, a one-half fraction of the 2^3 design is often called a 2^{3-1} factorial design. Note that a 2^3 factorial design has eight treatment combinations, but a 2^{3-1} factorial design has only four treatment combinations. Then, the obvious question is which four of the eight treatment combinations of a 2^3 factorial design will be

Table 8.23 Plus and minus signs for 2^3 factorial design

Treatment combination	Factorial effect							
	I	A	B	AB	C	AC	BC	ABC
a	$+$	$+$	$-$	$-$	$-$	$-$	$+$	$+$
b	$+$	$-$	$+$	$-$	$-$	$+$	$-$	$+$
c	$+$	$-$	$-$	$+$	$+$	$-$	$-$	$+$
abc	$+$	$+$	$+$	$+$	$+$	$+$	$+$	$+$
ab	$+$	$+$	$+$	$+$	$-$	$-$	$-$	$-$
ac	$+$	$+$	$-$	$-$	$+$	$+$	$-$	$-$
bc	$+$	$-$	$+$	$-$	$+$	$-$	$+$	$-$
(1)	$+$	$-$	$-$	$+$	$-$	$+$	$+$	$-$

Fig. 8.8 Geometric view of a 2^{3-1} factorial design with $I = ABC$

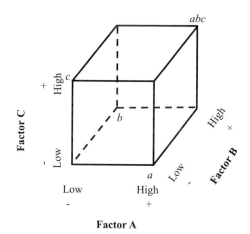

chosen to form a 2^{3-1} factorial design? The treatment combinations of a 2^3 factorial design that yield a plus on the ABC effect constitute a 2^{3-1} factorial design. As shown in Table 8.23, the treatment combinations a, b, c, and abc offer a positive sign on ABC, and they therefore constitute the 2^{3-1} factorial design. Thus, ABC is called the generator of the particular fraction, known as principal fraction. Further, it can be seen that the identity column I always bears plus sign, and hence $I = ABC$ is called the defining relation for this design.

The geometric view of a 2^{3-1} factorial design with $I = ABC$ as the defining relation is displayed in Fig. 8.8.

8.7.2 Analysis of Effects in 2^{3-1} Factorial Design with $I = ABC$

Let us now analyze the main and interaction effects in a 2^{3-1} factorial design. As defined earlier, the main effect of a factor is defined by the difference between the average of the observations when that factor is at higher level and the average of the observations when that factor is at lower level. Thus, we get

$$A = \bar{y}_{A^+} - \bar{y}_{A^-} = \frac{a + abc}{2} - \frac{b + c}{2} = \frac{1}{4n}[a - b - c + abc]$$

$$B = \bar{y}_{B^+} - \bar{y}_{B^-} = \frac{b + abc}{2} - \frac{a + c}{2} = \frac{1}{4n}[-a + b - c + abc]$$

$$A = \bar{y}_{C^+} - \bar{y}_{C^-} = \frac{c + abc}{2} - \frac{a + b}{2} = \frac{1}{2}[-a - b + c + abc]$$

Similarly, the interaction effect of AB is defined by the difference between the average of the observations when the product of AB is at higher level and the average of the observations when the product of AB is at lower level. This is shown below.

$$AB = \bar{y}_{(AB)^+} - \bar{y}_{(AB)^-} = \frac{c + abc}{2} - \frac{a + b}{2} = \frac{1}{2}[-a - b + c + abc]$$

$$BC = \bar{y}_{(BC)^+} - \bar{y}_{(BC)^-} = \frac{a + abc}{2} - \frac{b + c}{2} = \frac{1}{2}[a - b - c + abc]$$

$$AC = \bar{y}_{(AC)^+} - \bar{y}_{(AC)^-} = \frac{b + abc}{2} - \frac{a + c}{2} = \frac{1}{2}[-a + b - c + abc]$$

We observe that the effects of A, B, and AB are, respectively, the same as the effects of BC, AC, and AB. If the effect of a factor is same as the effect of another factor, then the effects are known as alias. Hence, A and BC are aliases, B and AC are aliases, and C and AB are aliases. This can be found out in another way as shown below.

$$A = A.I = A.ABC = (A^2)BC = I.BC = BC$$

$$B = B.I = B.ABC = A(B^2)C = AC$$

$$C = C.I = C.ABC = AB(C^2) = AB$$

Further, the linear combination of observations used to estimate the main effects of A, B, and C can be written as follows.

$$l_A = [a - b - c + abc] = A + BC$$

$$l_B = [-a + b - c + abc] = B + AC$$

$$l_C = [-a - b + c + abc] = C + AB$$

Fig. 8.9 Geometric view of a 2^{3-1} factorial design with $I = -ABC$

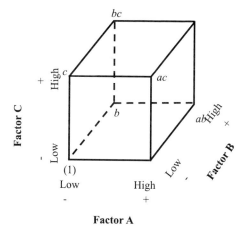

Factor A

8.7.3 Creation of Another 2^{3-1} Factorial Design with $I = -ABC$

It is also possible to select the four treatment combinations of a $\binom{3}{2}$ factorial design that yield a minus on the ABC effect to constitute a $2^3 - 1$ factorial design. As shown in Table 8.23, the treatment combinations ab, ac, bc, and (1) offer a negative sign on ABC and thus constitute another 2^{3-1} factorial design. The identity column I always bears the minus sign, and hence $I = -ABC$ is called the defining relation for this design. The geometric view of this design is shown in Fig. 8.9.

8.7.4 Analysis of Effects in 2^{3-1} Factorial Design with $I = -ABC$

The main and interaction effects in a 2^{3-1} factorial design with $I = -ABC$ are determined as mentioned earlier. They are expressed as follows.

$$A = \bar{y}_{A^+} - \bar{y}_{A^-} = \frac{ab + ac}{2} - \frac{bc + (1)}{2} = \frac{1}{2}[ab + ac - bc - (1)]$$

$$B = \bar{y}_{B^+} - \bar{y}_{B^-} = \frac{ab + bc}{2} - \frac{ac + (1)}{2} = \frac{1}{2}[ab + bc - ac - (1)]$$

$$A = \bar{y}_{C^+} - \bar{y}_{C^-} = \frac{ac + bc}{2} - \frac{ab + (1)}{2} = \frac{1}{2}[-ab + ac + bc - (1)]$$

$$AB = \bar{y}_{(AB)^+} - \bar{y}_{(AB)^-} = \frac{ab + (1)}{2} - \frac{ac + bc}{2} = \frac{1}{2}[ab - ac - bc + (1)]$$

$$BC = \bar{y}_{(BC)^+} - \bar{y}_{(BC)^-} = \frac{bc + (1)}{2} - \frac{ab + ac}{2} = \frac{1}{2}[-ab - ac + bc + (1)]$$

$$AC = \bar{y}_{(AC)^+} - \bar{y}_{(AC)^-} = \frac{ac + (1)}{2} - \frac{ab + bc}{2} = \frac{1}{2}[-ab - bc + ac + (1)]$$

Here, the effects of A, B, and C, respectively, are the same as the effects of $-BC$, $-AC$, and $-AB$. So, A and $-BC$ are aliases, B and $-AC$ are aliases, and C and $-AB$ are aliases. As mentioned earlier, this can be found out in another way.

$$A = A.I = A.(-ABC) = -A.ABC = -(A^2)BC = -I.BC = -BC$$
$$B = B.I = B.(-ABC) = -B.ABC = -A(B^2)C = -A.I.C = -AC$$
$$C = C.I = C.(-ABC) = -C.(ABC) = -AB(C^2) = -AB.I = -AB$$

The linear combination of observations used to estimate the main effects of A, B, and C can be written as follows.

$$l'_A = [ab + ac - bc - (1)] = A - BC$$
$$l'_B = [ab + bc - ac - (1)] = B - AC$$
$$l'_C = [-ab + ac + bc - (1)] = C - AB$$

It is then possible to express the de-aliased estimates of main and interaction effects as follows.

$$\frac{1}{2}(l_A + l'_A) = \frac{1}{2}(A + BC + A - BC) = A$$
$$\frac{1}{2}(l_A - l'_A) = \frac{1}{2}(A + BC - A + BC) = BC$$
$$\frac{1}{2}(l_B + l'_B) = \frac{1}{2}(B + AC + B - AC) = B$$
$$\frac{1}{2}(l_B - l'_B) = \frac{1}{2}(B + AC - B + AC) = AC$$
$$\frac{1}{2}(l_C + l'_C) = \frac{1}{2}(C + AB + C - AB) = C$$
$$\frac{1}{2}(l_C - l'_C) = \frac{1}{2}(C + AB - C + AB) = AB$$

As shown, by combining a sequence of two fractional factorial designs we can isolate both the main effects and the two-factor interactions.

8.7.5   A Practical Example of 2^{3-1} Factorial Design

A 2^{3-1} factorial experiment is carried out to investigate the effects of three factors, namely furnace temperature (A), heating time (B), and transfer time (C) on the surface finish of castings produced in a melting furnace. Each factor is kept at two

Table 8.24 Experimental results of metal furnace experiment

Run	A	B	C	Treatment combinations	Response
1	−	−	−	(1)	40
2	+	+	−	ab	55
3	+	−	+	ac	45
4	−	+	+	bc	65

levels, namely low (−) and high (+). The results of experiment are displayed in Table 8.24.

In this design, the defining relation is $I = -ABC$ and the aliases are

$$A = -BC, \quad B = -AC, \quad C = -AB.$$

Thus, the three main effects (A, B, C) account for the three degrees of freedom for the design. Let us now calculate the main effects of A, B, and C as follows.

$$A = \frac{1}{2}[ab + ac - bc - (1)] = \frac{-5}{2} = -2.5$$
$$B = \frac{1}{2}[ab + bc - ac - (1)] = \frac{35}{2} = 17.5$$
$$C = \frac{1}{2}[-ab + ac + bc - (1)] = \frac{15}{2} = 7.5.$$

The contrasts are calculated as follows

$$\text{Contrast}_A = [ab + ac - bc - (1)] = -5$$
$$\text{Contrast}_B = [ab + bc - ac - (1)] = 35$$
$$\text{Contrast}_C = [-ab + ac + bc - (1)] = 15.$$

The sum of squares is calculated as follows

$$SS_A = \frac{(\text{Contrast}_A)^2}{4} = \frac{(-5)^2}{4} = 6.25$$
$$SS_B = \frac{(\text{Contrast}_B)^2}{4} = \frac{(35)^2}{4} = 306.25$$
$$SS_C = \frac{(\text{Contrast}_C)^2}{4} = \frac{(15)^2}{4} = 56.25.$$

Table 8.25 presents the contribution of factor effects in this experiment. It can be seen that the factor A contributes only 1.7% of total variation in the data. It is therefore considered to be nonimportant. The analysis of variance is shown in Table 8.26.

Table 8.25 Factor contributions for metal furnace experiment

Model term	Effect	Sum of squares	Percent contribution
A	−2.5	6.25	1.70
B	17.5	306.25	83.05
C	7.5	56.25	15.25
Total		368.75	100.00

Table 8.26 ANOVA table for metal furnace experiment

Source of variation	Sum of squares	Degree of freedom	Mean square	F_0-value
B	306.25	1	306.25	49
C	56.25	1	56.25	9
Error	6.25	1	6.25	
Total	368.75	3		

If the level of significance is taken as 0.05, then $F_{0.05,1,1} = 161.4$ (Table A.11), which is higher than the calculated F_0-values (49 and 9) for factor B and factor C, respectively. It is therefore concluded that none of the three factors are statistically significant.

8.7.6   A Practical Example of 2^{4-1} Factorial Design

A 2^{4-1} factorial experiment is carried out to investigate the removal of color from industrial effluent by an electrochemical process. Four factors, each at two levels, are chosen as shown in Table 8.27.

The percentage color removal is taken as response. Here, $I = ABCD$ is the defining relation for this design. The results of experiment are displayed in Table 8.28.

In this design, the aliases are

$$A = BCD, \quad B = ACD, \quad C = ABD, \quad D = ABC, AB = CD, \quad AC = BD, \quad AD = BC$$

Table 8.27 Factor and levels for color removal experiment

Factor	Low level (−)	High level (+)
Current density, mA/cm^2 (A)	14.285	42.857
Dilution, %(B)	10	30
Time of electrolysis, h (C)	2	5
pH (D)	4	9

Table 8.28 Experimental results

Run	A	B	C	D = ABC	Treatment combina-tions	Response
1	−	−	−	−	−1	78.265
2	+	−	−	+	ad	84.325
3	−	+	−	+	bd	72.235
4	+	+	−	−	ab	76.857
5	−	−	+	+	cd	98.25
6	+	−	+	−	ac	93.265
7	−	+	+	−	bc	91.235
8	+	+	+	+	abcd	97.356

Thus, the four main effects (A, B, C, D) and three two-factor interactions (AB, AC, AD) account for the seven degrees of freedom for the design. The main effects are calculated as follows.

$$A = \bar{y}_{A+} - \bar{y}_{A-} = \frac{ad + ab + ac + abcd}{4} - \frac{(1) + bd + cd + bc}{4} = 87.951 - 84.996 = 2.955$$

$$B = \bar{y}_{B+} - \bar{y}_{B-} = \frac{ab + bd + bc + abcd}{4} - \frac{(1) + ad + cd + ac}{4} = 84.421 - 88.526 = -4.105$$

$$C = \bar{y}_{C+} - \bar{y}_{C-} = \frac{cd + ac + bc + abcd}{4} - \frac{(1) + ad + bd + ab}{4} = 95.027 - 77.921 = 17.106$$

$$C = \bar{y}_{C+} - \bar{y}_{C-} = \frac{ad + bd + cd + abcd}{4} - \frac{(1) + ab + ac + bc}{4} = 88.042 - 84.906 = 3.136.$$

It can be seen that the main effect of B is negative, indicating that if the level of B goes from higher to lower, the response increases. The other main effects are positive, indicating that the response decreases if the level of factors A, C, and D changes from higher to lower one. Factor C plays the most important role in deciding the response, followed by B, D, and A.

The interaction effects are calculated as follows.

$$AB = \bar{y}_{(AB)+} - \bar{y}_{(AB)-} = \frac{(1) + ab + cd + abcd}{4} - \frac{ad + bd + ac + bc}{4} = 87.682 - 85.265 = 2.417$$

$$AC = \bar{y}_{(AC)+} - \bar{y}_{(AC)-} = \frac{(1) + bd + ac + abcd}{4} - \frac{ad + ab + cd + bc}{4} = 85.280 - 87.667 = -2.387$$

$$AD = \bar{y}_{(AD)+} - \bar{y}_{(AD)-} = \frac{(1) + ad + bc + abcd}{4} - \frac{bd + ab + cd + ac}{4} = 87.795 - 85.152 = 2.643.$$

It can be seen that the interaction effect of AC is negative, indicating that if the level of AC goes from higher to lower, the response increases. The other interaction effects are positive, indicating that the response decreases if the level of interaction of AB and AD changes from higher to lower one.

The sum of squares is calculated as follows.

Table 8.29 Contribution of factor effects for color removal experiment

Model term	Effect	Sum of squares	Percent contribution
A	2.955	17.464	2.52
B	−4.105	33.703	4.86
C	17.106	585.23	84.43
D	3.136	19.669	2.84
AB	2.417	11.684	1.69
AC	−2.387	11.396	1.64
BC	2.643	13.972	2.02
Total		693.118	100

$$SS_A = \frac{(\text{Contrast}_A)^2}{8} = \frac{(11.82)^2}{8} = 17.464$$

$$SS_B = \frac{(\text{Contrast}_B)^2}{8} = \frac{(-16.42)^2}{8} = 33.703$$

$$SS_C = \frac{(\text{Contrast}_C)^2}{8} = \frac{(68.424)^2}{8} = 585.230$$

$$SS_D = \frac{(\text{Contrast}_D)^2}{8} = \frac{(12.544)^2}{8} = 19.669$$

$$SS_{AB} = \frac{(\text{Contrast}_{AB})^2}{8} = \frac{(9.668)^2}{8} = 11.684$$

$$SS_{AC} = \frac{(\text{Contrast}_{AC})^2}{8} = \frac{(-9.548)^2}{8} = 11.396$$

$$SS_{AD} = \frac{(\text{Contrast}_{AD})^2}{8} = \frac{(10.572)^2}{8} = 13.972$$

The total sum of squares is calculated as follows.

$$SS_{\text{Total}} = 60514.448 - \frac{691.788^2}{8} = 693.118$$

Table 8.29 presents the contribution of factor effects in this experiment. It can be seen that the factor C alone contributes 84.43% of total variation present in the response, whereas all other factors and their interactions contribute only 15.57%. It is therefore concluded that only factor C is important. The resulting analysis of variance is shown in Table 8.30. If the level of significance is taken as 0.05, then $F_{0.05,1,6} = 5.99$, which is less than the calculated F_0-value (32.55) for factor C. It is therefore concluded that the effect of time of electrolysis on the percentage color removal is statistically significant.

Table 8.30 ANOVA table for color removal experiment

Source of variation	Sum of squares	Degree of freedom	Mean square	F_0
C	585.230	1	585.230	32.55
Error	107.888	6	17.981	
Total	693.118	7		

8.7.7 Design Resolution

The concept of design resolution is a useful way to catalogue fractional factorial designs according to the alias patterns they produce.

Resolution I Design: In this design, an experiment of exactly one run involves only one level of a factor. The 2^{1-1} with $I = A$ is a resolution I design. This is not a useful design. We usually employ a Roman numeral subscript to indicate design resolution, and this one-half fraction is a 2_I^{1-1} design.

Resolution II Design: In this design, the main effects are aliased with other main effects. The 2^{2-1} with $I = AB$ is a resolution II design. This is also not a useful design. This is designated as 2_{II}^{2-1} design.

Resolution III Design: In this design, no main effects are aliased with any other main effect, but main effects are aliased with two-factor interactions and some two-factor interactions may be aliased with each other. The 2^{3-1} design with $I = ABC$ is a resolution III design. This is designated as 2_{III}^{3-1} design.

Resolution IV Design: In this design, no main effects are aliased with any other main effect or two-factor interactions, but two-factor interactions are aliased with each other. The 2^{4-1} design with $I = ABCD$ is a resolution IV design. This is designated as 2_{IV}^{4-1} design.

Resolution V Design: In this design, no main effect or two-factor effect is aliased with any other main effect or two-factor interaction, but two-factor interactions are aliased with three-factor interactions. The 2^{5-1} design with $I = ABCDE$ is a resolution V design. This is designated as 2_V^{5-1} design.

Resolution VI Design: In this design, no two-factor interactions are aliased with three-factor interactions. The 2^{6-1} design with $I = ABCDEF$ is a resolution VI design. This is designated as 2_{VI}^{6-1} design.

Note that resolution III and IV designs are particularly suitable in factor screening experiments. Resolution IV design provides good information about main effects and will provide some information about all two-factor interactions.

Problems

8.1 Cauliflowers were grown in a greenhouse under treatments consisting of four types of soils and three types of fertilizers. In order to examine the effects of soil and fertilizer, a factorial experiment was carried out. The results of experiments in terms of the yield (kg) of cauliflowers are shown in Table 8.31.

Analyze the data using $\alpha = 0.05$.

8.2 In order to investigate the effects of type of resin, distance between blade and anvil, and weight fraction of nonvolatile content in a lacquer on the thickness of lacquer film on a substrate, the experiment shown in Table 8.32 was carried out.

Analyze the data using $\alpha = 0.05$.

8.3 A 2^2 factorial experiment was conducted to study the effects of temperature (25 and 65 °C) and catalyst concentration (0.5 and 1.5%) on transesterification of vegetable oil to methanol. The results are shown in Table 8.33.

Table 8.31 Data for Problem 8.1

Type of soil	Type of fertilizer		
	X	Y	Z
A	6.3	7.5	8.6
	6.9	7.8	8.2
B	5.4	3.4	7.5
	5.6	3.2	7.6
C	7.2	5.4	7.2
	7.4	5.8	7.4
D	8.7	6.4	5.2
	8.9	6.8	5.8

Table 8.32 Data for Problem 8.2

Distance (mm) type of soil	Type of resin					
	A			B		
	Weight fraction			Weight fraction		
	0.20	0.30	0.40	0.20	0.30	0.40
1	2.1	1.8	1.5	1.7	1.4	1.2
	2.0	1.7	1.4	1.6	1.3	1.0
3	2.8	2.4	2.1	2.1	1.7	1.4
	2.9	2.3	2.2	2.2	1.6	1.3
5	3.8	3.4	3.0	2.8	2.3	2.0
	3.6	3.2	2.9	2.7	2.1	1.9

Table 8.33 Data for Problem 8.3

Run	Temperature °C	Concentration (%)	Conversion (%)	
1	25	0.5	85.6	86.4
2	65	0.5	98.6	97.6
3	25	1.5	99.5	99.9
4	65	1.5	100	100

Table 8.34 Data for Problem 8.4

Run	pH (−)	Concentration (mM)	Power density (mW/mm^2)	
1	5.8	25	464	460
2	7.4	25	306	282
3	5.8	150	405	370
4	7.4	150	407	380

(a) Display the geometrical view of the aforementioned design.
(b) Analyze the main and interaction effects of temperature and concentration on conversion.
(c) Construct ANOVA for conversion. Which effects are statistically significant at 0.05 level of significance?
(d) Develop an appropriate regression model for conversion.

8.4 In order to study the effects of pH (5.8 and 7.4) and buffer concentration of catholyte (25 and 150 mM) on the power density of a fuel cell, the 2^2 factorial design of experiment was carried out and results are shown in Table 8.34.

(a) Display the geometrical view of the aforementioned design.
(b) Analyze the main and interaction effects of pH and concentration on power density.
(c) Construct ANOVA for power density. Which effects are statistically significant at 0.05 level of significance?
(d) Develop an appropriate regression model for power density.

8.5 An experiment was carried out to investigate the effects of ratio of PP and LLDRE (X_1) and concentration of red mud particles (X_2) on the tensile strength (Y) of red mud filled PP/LLDPE blended composites. The results of experiments are shown in Table 8.35.

(a) Display the geometrical view of the aforementioned design.
(b) Analyze the main and interaction effects of ratio of PP and LLDRE and concentration of red mud particles on the tensile strength.

Table 8.35 Data for Problem 8.5

Run	$X_1(-)$	$X_2(-)$	$Y(MPa)$	
1	0.20	5	10.5	11
2	5	5	24.8	25.2
3	0.20	10	7.6	9.8
4	5	10	18.4	19.8

Table 8.36 Data for Problem 8.6

Run	pH	Buffer	Temperature	Percentage of degraded 1.0×10^{-4} M NADH		
1	6.8	Phosphate	25	6.2	9.3	7.2
2	7.8	Phosphate	25	2.4	5.3	3.9
3	6.8	Pipes	25	0	0.4	0
4	7.8	Pipes	25	0.8	0	0
5	6.8	Phosphate	30	13.0	13.4	12.1
6	7.8	Phosphate	30	8.1	9.4	8.7
7	6.8	Pipes	30	7.3	6.1	8.8
8	7.8	Pipes	30	3.0	1.8	0

(c) Construct ANOVA for tensile strength. Which effects are statistically significant at 0.05 level of significance?

(d) Develop an appropriate regression model.

8.6 An article entitled "Study of NADH stability using ultraviolet-visible spectrophotometric analysis and factorial design" published by L. Rovar et al. in Analytical Biochemistry, 260, 50–55, 1998, reported on the effects of pH, buffer, and temperature on percentage of degraded 1.0×10^{-4} M NADH. In the reported study, two levels of pH (6.8 and 7.8), two buffer solutions (phosphate and pipes), and two levels of temperature (25 and 30 °C) were taken. The results are given in Table 8.36.

(a) Display the geometrical view of the aforementioned design.

(b) Analyze the main and interaction effects of pH, buffer, and temperature on percentage of degraded 1.0 10-4 M NADH.

(c) Construct ANOVA for percentage of degraded 1.0 10-4 M NADH. Which effects are statistically significant at 0.05 level of significance?

8.7 An article entitled "Designed experiments to stabilize blood glucose levels," published by R.E. Chapman and V. Roof in Quality Engineering, 12, 83–87, 1999, reported on the effects of amount of juice intake before exercise (4 oz or 8 oz), amount of exercise (10 min or 20 min), and delay between time of juice intake and beginning of exercise (0 min or 20 min) on the blood glucose level of patients. The data are shown in Table 8.37.

Table 8.37 Data for Problem 8.7

Run	Juice intake (oz)	Exercise	Delay	Blood glucose mg/dL	
1	4	10	0	78	65
2	8	10	0	101	105
3	4	20	0	96	71
4	8	20	0	107	145
5	4	10	20	128	123
6	8	10	20	112	147
7	4	20	20	111	79
8	8	20	20	83	103

Table 8.38 Data for Problem 8.8

Variable	Unit	Low level (−)	High level (+)
Spindle speed (S)	rev/s	17	50
Feed rate (F)	Mm/s	0.09	0.15
Ultrasonic power (P)	%	35	50

(e) Display the geometrical view of the aforementioned design for blood glucose level.

(f) Analyze the main and interaction effects of amount of juice intake, amount of exercise, and delay between juice intake and exercise in determining blood glucose level.

(g) Construct ANOVA for blood glucose level. Which effects are statistically significant at 0.05 level of significance?

(h) Develop an appropriate regression model for blood glucose level.

8.8 An article entitled "Rotary ultrasonic machining of ceramic matrix composites: feasibility study and designed experiments," published by Z.C. Li et al. in International Journal of Machine Tools & Manufacture, 45, 1402–1411, 2005, described the use of a full factorial design to study the effects of rotary ultrasonic machining on the cutting force, material removal rate, and hole quality (in terms of chipping dimensions). The process variables and their levels were taken and are shown in Table 8.38.

The results of experiments are stated in Table 8.39.

(a) Display the geometrical views of the design for cutting force, material removal rate, chipping thickness, and chipping size.

(b) Analyze the main and interaction effects of spindle speed, feed rate, and ultrasonic power in determining cutting force, material removal rate, chipping thickness, and chipping size.

Table 8.39 Data for Problem 8.8

Run	S	F	P	Cutting force (N)		Material removal rate (mm³/s)		Chipping thickness (mm)		Chipping size (mm)	
1	–	–	–	378	398	1.38	1.33	0.49	0.35	4.03	3.24
2	+	–	–	459	472	1.66	1.55	0.61	0.78	4.68	3.27
3	–	+	–	463	352	1.49	1.63	0.43	0.36	3.23	1.89
4	+	+	–	490	394	1.97	1.98	0.40	0.40	1.00	2.13
5	–	–	+	407	332	1.55	1.45	0.52	0.37	3.53	2.01
6	+	–	+	407	332	1.81	2.05	0.82	0.71	4.83	5.53
7	–	+	+	329	397	1.81	1.69	0.36	0.42	1.04	1.93
8	+	+	+	443	465	2.14	2.24	0.54	0.49	2.88	1.45

(c) Construct ANOVA for cutting force, material removal rate, chipping thickness, and chipping size. Which effects are statistically significant at 0.05 level of significance?

(d) Develop appropriate regression models for cutting force, material removal rate, chipping thickness, and chipping size.

8.9 Consider the experiment described in Problem 8.6. Analyze the data using Yates' algorithm. Check if the results obtained are same with those obtained in Problem 8.6.

8.10 Consider the experiment described in Problem 8.7. Analyze the data using Yates' algorithm. Check if the results obtained are same with those obtained in Problem 8.7.

8.11 Consider the experiment described in Problem 8.1. Analyze the experiment assuming that each replicate represents a block and draw conclusions.

8.12 Consider the experiment described in Problem 8.4. Analyze the experiment assuming that each replicate represents a block and draw conclusions.

8.13 Consider the data from the first replicate of Problem 8.6. Suppose that these experiments could not all be run using the same batch of material. Construct a design with two blocks of four observations each with SFP confounded. Analyze the data.

8.14 Consider the data from the first replicate of Problem 8.4. Suppose that these experiments could not all be run using the same batch of material. Suggest a reasonable confounding scheme and analyze the data.

8.15 An article entitled "Screening of factors influencing Cu(II) extraction by soybean oil-based organic solvents using fractional factorial design" published by S.H. Chang et al. in Journal of Environment Management 92, 2580–2585, 2011, described a fractional factorial experiment to study the effect of Cu (II) extraction process on extraction efficiency (η). The process variables and their levels are given in Table 8.40.

The results of experiments are stated in Table 8.41.

Table 8.40 Data for Problem 8.15

Variable	Unit	Low level (−)	High level (+)
Time (A)	min	3	6
Concentration of D2EHPA (B)	mM	50	100
O:A ratio (C)	–	1.0	1.5
Concentration of Na_2SO_4 (D)	mM	200	250
pH (E)	–	4.0	4.5
Concentration of TBP (F)		30	60

Table 8.41 Results of Problem 8.15

Run	A	B	C	D	E	F	$\eta(\%)$
1	6	100	1	250	4	60	87.65
2	3	100	1.5	200	4	30	95.27
3	3	100	1	200	4	60	95.48
4	6	100	1	200	4	30	92.08
5	3	100	1.5	200	4.5	60	99.32
6	6	50	1.5	200	4	30	68
7	6	50	1.5	200	4.5	60	96.58
8	6	50	1	200	4	60	72.07
9	6	100	1	200	4.5	60	98.53
10	6	50	1	250	4	30	79.84
11	3	100	1	250	4.5	30	98.95
12	6	100	1	250	4.5	30	98.49
13	6	50	1	200	4.5	30	96.38
14	3	50	1	200	4	30	68.33
15	3	100	1.5	250	4	60	95.17
16	6	100	1.5	200	4.5	30	98.96
17	3	50	1.5	250	4	30	78.74
18	3	50	1	250	4.5	30	93.97
19	6	100	1.5	250	4.5	60	98.73
20	3	50	1	200	4.5	60	97.11
21	3	50	1.5	200	4.5	30	98.35
22	6	50	1.5	250	4	60	79.64
23	3	50	1.5	250	4.5	60	96.05
24	3	100	1	250	4.5	60	98.54
25	3	100	1	250	4	30	88.1
26	3	100	1.5	250	4.5	30	99.07
27	6	50	1.5	250	4.5	30	97.91
28	3	50	1.5	200	4	60	79.14
29	6	100	1.5	200	4	60	94.71
30	3	50	1	250	4	60	71.5
31	6	100	1.5	250	4	30	92.4
32	6	50	1	250	4.5	60	93.84

Analyze the main and interaction effects of the five process factors, and calculate percent contribution of them. Which effects are statistically significant at 0.05 level of significance? Construct ANOVA for extraction efficiency.

Reference

Leaf GAV (1987) Practical Statistics for the Textile Industry, Part II. The Textile Institute, Manchester, p 60

Chapter 9
Response Surface Methodology

9.1 Introduction

Response surface methodology or in short RSM is a collection of mathematical and statistical tools and techniques that are useful in developing, understanding, and optimizing processes and products. Using this methodology, the responses that are influenced by several variables can be modeled, analyzed, and optimized.

In most of the problems related to RSM, the form of relationship between the response (dependent) and factor (independent) variables is not known. The first step in RSM is therefore to find a suitable relationship between the response and factor variables. For this, an appropriate design of experiments is conducted and the experimental data are used to develop response surface models. Afterward, the response surface is analyzed to locate a direction that goes toward the general vicinity of the optimum. Once the optimum region is found, a more elaborate model—often followed by a new set of designed experiments carried out in the optimum region—is developed and analyzed to locate the optimum. This is why RSM is known as a sequential procedure. This is illustrated in Fig. 9.1 (Montgomery 2007). As stated by Montgomery (2007), the response surface methodology can be thought as "climbing a hill" where the top of the hill represents the point of maximum response, and if the true optimum is a point of minimum response then we may think of "descending into a valley." The ultimate objective of RSM is to optimize a process or product. There are a good number of resources available on response surface methodology. A large number of research articles exist on the development and application of response surface methodology for optimizing processes and products. In addition, there are quite a few textbooks available where a detailed discussion on RSM is presented by authors like Myers et al. (2009), Box et al. (2005), Panneerselvam (2012), to name a few. The interested readers can use these valuable resources to enhance their knowledge and skills on RSM.

© Springer Nature Singapore Pte Ltd. 2018
D. Selvamuthu and D. Das, *Introduction to Statistical Methods,
Design of Experiments and Statistical Quality Control*,
https://doi.org/10.1007/978-981-13-1736-1_9

Fig. 9.1 Illustration of
sequential nature of RSM

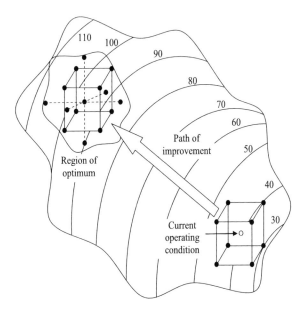

9.2 Response Surface Models

In response surface methodology, the responses are generally modeled by either a
first-order model or a second-order model. Suppose there are n number of independent variables $x_1, x_2, \ldots, x_n$ so that the first-order model can be expressed as
follows

$$y = \beta_0 + \sum_{i=1}^{n} \beta_i x_i + \varepsilon$$

where y denotes the response, βs represent the coefficients, and ε refers to the error.
As this model contains only the main effects, it is sometimes called main effects
model. In this model, the fitted response surface looks like a plane. If there exists an
interaction between the independent variables, then it can be added in the first-order
model as

$$y = \beta_0 + \sum_{i=1}^{n} \beta_i x_i + \sum \sum_{i<j} \beta_{ij} x_i x_j + \varepsilon.$$

This is called a first-order model with interaction. Because of the interaction term
present in the model, the resulting response surface looks curved. Often, the curvature
in the true response surface is so strong that the first-order model even with interaction
term turns out to be inadequate. In such a case, a second-order model of the following
form is found to fit better.

$$y = \beta_0 + \sum_{i=1}^{n} \beta_i x_i + \sum_{i=1}^{n} \beta_{ij} x_i^2 + \sum \sum_{i<j} \beta_{ij} x_i x_j + \varepsilon.$$

It is therefore necessary to check the adequacy of a fitted model. As in the case of regression analysis, this is checked by coefficient of determination, residual analysis, and lack of fit test. The response surface models are developed using multiple linear regression analysis.

9.3 Multiple Linear Regression

Often, we would like to model the effect of more than one independent (regressor) variable on a dependent (response) variable. Then, we talk about multiple regression. In the following, some examples of multiple regression are given.

Let us consider the first-order response surface model involving two factors without any interaction. This is written below.

$$y = \beta_0 + \beta_1 x_1 + \beta_2 x_2 + \varepsilon.$$

Let us now consider the first-order response surface model involving two factors with interaction. This is stated here under

$$y = \beta_0 + \beta_1 x_1 + \beta_2 x_2 + \beta_3 x_1 x_2 + \varepsilon.$$

This expression can also be written as

$$y = \beta_0 + \beta_1 x_1 + \beta_2 x_2 + \beta_3 x_3 + \varepsilon$$

where $x_3 = x_1 x_2$. Further, we consider the second-order response surface model involving two factors, as stated below

$$y = \beta_0 + \beta_1 x_1 + \beta_2 x_2 + \beta_3 x_1^2 + \beta_4 x_2^2 + \beta_5 x_1 x_2 + \varepsilon.$$

This can also be expressed as

$$y = \beta_0 + \beta_1 x_1 + \beta_2 x_2 + \beta_3 x_3 + \beta_4 x_4 + \beta_5 x_5 + \varepsilon$$

where $x_3 = x_1^2$, $x_4 = x_2^2$, $x_5 = x_1 x_2$. In this way, the first as well as the second-order response surface models can be expressed as a multiple linear equation. The statistical technique that explores such a mathematical relationship between dependent and independent variables is known as multiple linear regression analysis.

9.3.1 A Generalized Model

Let us now consider a generalized model as shown below.

$$y_i = \beta_0 + \beta_1 x_{i1} + \beta_2 x_{i2} + \cdots + \beta_k x_{ik} + \varepsilon_i, \ i = 1, 2, \ldots, n.$$

This model can be written in short as

$$\mathbf{y} = \mathbf{x}\boldsymbol{\beta} + \boldsymbol{\varepsilon}$$

where

$$\mathbf{y} = \begin{pmatrix} y_1 \\ y_2 \\ \vdots \\ y_n \end{pmatrix}, \quad \boldsymbol{\beta} = \begin{pmatrix} \beta_0 \\ \beta_1 \\ \vdots \\ \beta_k \end{pmatrix}, \quad \boldsymbol{\varepsilon} = \begin{pmatrix} \varepsilon_1 \\ \varepsilon_2 \\ \vdots \\ \varepsilon_n \end{pmatrix}, \quad \mathbf{x} = \begin{pmatrix} 1 & x_{11} & x_{12} & \ldots & x_{1k} \\ 1 & x_{21} & x_{22} & \ldots & x_{2k} \\ \vdots & \vdots & \vdots & & \vdots \\ 1 & x_{n1} & x_{n2} & \ldots & x_{nk} \end{pmatrix}$$

9.3.2 Estimation of Coefficients: Least Square Method

The least square method for estimating the coefficients of simple linear regression is already discussed in Chap. 7. The same method is used here to estimate the coefficients of multiple linear regression.

The total sum of square errors is given by

$$L = \sum_{i=1}^{n} \varepsilon_i^2 = \boldsymbol{\varepsilon}'\boldsymbol{\varepsilon} = (\mathbf{y} - \mathbf{x}\boldsymbol{\beta})'(\mathbf{y} - \mathbf{x}\boldsymbol{\beta}).$$

Taking partial derivative with respect to β, we have

$$\left. \frac{\partial L}{\partial \boldsymbol{\beta}} \right|_{\hat{\beta}} = 0,$$

$$\frac{\partial}{\partial \boldsymbol{\beta}} \left[(\mathbf{y} - \mathbf{x}\boldsymbol{\beta})'(\mathbf{y} - \mathbf{x}\boldsymbol{\beta}) \right] \big|_{\hat{\beta}} = 0,$$

we get

$$\hat{\boldsymbol{\beta}} = (\mathbf{x}'\mathbf{x})^{-1}\mathbf{x}'\mathbf{y}.$$

Hence, the fitted model is $\mathbf{y} = \mathbf{x}\hat{\boldsymbol{\beta}}$ and residual is $\boldsymbol{\varepsilon} = \mathbf{y} - \mathbf{y}'$.

Table 9.1 Data for strength of parachute joints

y	x_1	x_2
23	5	4
35	5	6
51	5	8
42	10	4
61	10	6
76	10	8

Example 9.1 Let us fit a multiple linear regression model

$$y_i = \beta_0 + \beta_1 x_{i1} + \beta_2 x_{i2} + \varepsilon_i$$

to the data shown in Table 9.1. Here y denotes the strength (kN) of joint in a parachute webbing, x_1 denotes length (mm) of overlap in the joint, and x_2 denotes stitch density (no. of stitches per 25 mm).

Solution:
Following the above-mentioned procedure, we have

$$\mathbf{x'x} = \begin{pmatrix} 1 & 1 & 1 & 1 & 1 & 1 \\ 5 & 5 & 5 & 10 & 10 & 10 \\ 4 & 6 & 8 & 4 & 6 & 8 \end{pmatrix} \begin{pmatrix} 1 & 5 & 4 \\ 1 & 5 & 6 \\ 1 & 5 & 8 \\ 1 & 10 & 4 \\ 1 & 10 & 6 \\ 1 & 10 & 8 \end{pmatrix} = \begin{pmatrix} 6 & 45 & 36 \\ 45 & 375 & 270 \\ 36 & 270 & 232 \end{pmatrix}$$

$$(\mathbf{x'x})^{-1} = \begin{pmatrix} 3.9167 & -0.2000 & -0.3750 \\ -0.2000 & 0.0267 & 0 \\ -0.3750 & 0 & 0.0625 \end{pmatrix},$$

$$\mathbf{x'y} = \begin{pmatrix} 1 & 1 & 1 & 1 & 1 & 1 \\ 5 & 5 & 5 & 10 & 10 & 10 \\ 4 & 6 & 8 & 4 & 6 & 8 \end{pmatrix} \begin{pmatrix} 23 \\ 35 \\ 51 \\ 42 \\ 61 \\ 76 \end{pmatrix} = \begin{pmatrix} 238 \\ 2335 \\ 1852 \end{pmatrix}.$$

This gives

$$\beta = (\mathbf{x'x})^{-1}\mathbf{x'y} = \begin{pmatrix} -33.50 \\ 4.6667 \\ 7.7500 \end{pmatrix}.$$

Hence, the fitted model is

$$\hat{y} = -33.50 + 4.6667x_1 + 7.7500x_2,$$

and the residual is
$$
\begin{pmatrix} \varepsilon_1 \\ \varepsilon_2 \\ \varepsilon_3 \\ \varepsilon_4 \\ \varepsilon_5 \\ \varepsilon_6 \end{pmatrix}
=
\begin{pmatrix} 23 \\ 35 \\ 51 \\ 42 \\ 61 \\ 76 \end{pmatrix}
-
\begin{pmatrix} 20.8335 \\ 36.3335 \\ 51.8335 \\ 44.1670 \\ 59.6670 \\ 75.1670 \end{pmatrix}
=
\begin{pmatrix} 2.1665 \\ -1.3335 \\ -0.8335 \\ -2.1670 \\ 1.3330 \\ 0.8330 \end{pmatrix}.
$$

9.3.3 Estimation of Variance σ^2 of Error Term

We know that the sum of squares of error is

$$SS_E = \sum_{i=1}^{n} \varepsilon_i^2 = \sum_{i=1}^{n} (Y_i - \hat{Y}_i)^2.$$

9.3.4 Point Estimate of Coefficients

The point estimate of coefficients is obtained as follows. It is known that the expected value of SS_E is

$$E(SS_E) = (n - p)\sigma^2.$$

Then, an unbiased estimator of σ^2 is $\hat{\sigma}^2 = \frac{SS_E}{n-p}$. Referring to the previous example, we have

$$\hat{\sigma}^2 = \frac{14.33}{6 - 3} = 4.7778.$$

$$E(\hat{\boldsymbol{\beta}}) = E\left[(\mathbf{x}'\mathbf{x})^{-1}\mathbf{x}'\mathbf{y} \right] = E\left[(\mathbf{x}'\mathbf{x})^{-1}\mathbf{x}'(\mathbf{x}\boldsymbol{\beta} + \boldsymbol{\varepsilon}) \right]$$
$$= E\left[(\mathbf{x}'\mathbf{x})^{-1}\mathbf{x}'\mathbf{x}\boldsymbol{\beta} + (\mathbf{x}'\mathbf{x})^{-1}\mathbf{x}'\boldsymbol{\varepsilon} \right] = \boldsymbol{\beta}$$

Thus $\hat{\boldsymbol{\beta}}$ is an unbiased estimator of $\boldsymbol{\beta}$. It is known that

$$Cov(\hat{\boldsymbol{\beta}}) = \sigma^2 (\mathbf{x}'\mathbf{x})^{-1} = \sigma^2 \mathbf{C}$$

where $\mathbf{C} = (\mathbf{x}'\mathbf{x})^{-1}$ is a symmetric matrix. Then,

$$Cov(\hat{\beta}_i \hat{\beta}_j) = \sigma^2 C_{ij}, \ i \neq j.$$

Hence,

$$V(\hat{\beta}_j) = \sigma^2 C_{jj}, \ j = 0, 1, 2.$$

Thus, the standard error of $\hat{\beta}_j$ equals to $\sqrt{\hat{\sigma}^2 C_{jj}}$. Using the procedure mentioned above, the standard error of the coefficients β_0, β_1, and β_2 in Example 10.1 can be determined as 4.3259, 0.3572, and 0.5465 respectively.

9.3.5 Hypothesis Test for Significance of Regression

Suppose we wish to test the hypothesis that the slope equals to a constant, say $\beta_{1,0}$. Then, our hypotheses are

$$H_0 : \beta_1 = \beta_{1,0} \ \text{against} \ H_1 : \beta_1 \neq \beta_{1,0}.$$

Let us assume that the errors ε_i are normally and independently distributed with mean 0 and variance σ^2. Then, the observations y_i are also normally and independently distributed with mean $\beta_0 + \beta_1 x_i$ and variance σ^2. Since $\hat{\beta}_1$ is a linear combination of independent normal random variables, then $\hat{\beta}_1$ is normally distributed with mean β_1 and variance $\frac{\sigma^2}{S_{xx}}$. In addition, $\frac{(n-2)\hat{\sigma}^2}{\sigma^2}$ is a chi-square distribution with $n - 2$ degree of freedom, and $\hat{\beta}_1$ is independent of $\hat{\sigma}^2$. Then, the statistic $t_0 = \frac{\hat{\beta}_1 - \beta_{1,0}}{\sqrt{\hat{\sigma}^2/S_{xx}}}$ follows t-distribution with n-2 degree of freedom. How can we obtain SS_R and SS_E ? We know that $SS_T = SS_R + SS_E$

$$SS_T = \sum_{i=1}^{n} y_i^2 - \frac{\left(\sum_{i=1}^{n} y_i^2\right)^2}{n}$$

$$SS_E = \sum_{i=1}^{n} \varepsilon_i^2$$

$$= \sum_{i=1}^{n} (y_i - \hat{y}_i)^2$$

Hence,

$$SS_R = SS_T - SS_E.$$

Referring to previous example, we obtain

$$SS_T = 192 \quad SS_E = 14.3333 \quad SS_R = 1777.6667$$

$$F_0 = \frac{SS_R/k}{SS_E/(n-p)} = \frac{1777.6667/2}{14.3333/3} = 186.0353$$

Since the computed value of F_0 is higher than $F_{0.05,2,3} = 19.2$ (Table A.11), we reject H and conclude that the joint strength is linearly related to either the length of overlap in the joint or the stitch density or both. However, this does not imply that the relationship found is an appropriate model for predicting the joint strength as a function of length of overlap in the joint and stitch density. Further tests of model adequacy are required before we can be comfortable in using this model in practice.

9.3.6 Hypothesis Test on Individual Regression Coefficient

Sometimes we may like to determine the potential value of each of the regression coefficients with a view to know if the model would be more effective with the inclusion of additional variables or deletion of one of the regressor variable. Then, the hypotheses are

$$H_0 : \beta_j = 0 \text{ against } H_1 : \beta_j \neq 0.$$

The test statistics is given by

$$|t_0| = \left| \frac{\hat{\beta}_j}{\sqrt{\hat{\sigma}^2 C_{jj}}} \right|$$

We reject H_0 if the computed value of t_0 is greater than the table value of $t_{\frac{\alpha}{2},n-p}$.

Illustration: Hypothesis Test on β_0: The hypotheses are

$$H_0 : \beta_0 = 0 \text{ against } H_1 : \beta_0 \neq 0$$

The test statistics is

$$|t_0| = \left| \frac{\hat{\beta}_0}{\sqrt{\hat{\sigma}^2 C_{00}}} \right| = \left| \frac{-33.5000}{\sqrt{4.7778 \times 3.9167}} \right| = 7.7441$$

Since the computed value of t_0 is greater than the table value of $t_{0.025,3} = 3.1824$ (Table A.10), we reject the hypothesis that $H_0 : \beta_0 = 0$

Illustration: Hypothesis Test on β_1: The hypotheses are

$$H_0 : \beta_1 = 0 \text{ against } H_1 : \beta_1 \neq 0$$

The test statistic is

$$|t_0| = \left| \frac{\hat{\beta}_1}{\sqrt{\sigma^2 C_{11}}} \right| = \left| \frac{4.6667}{\sqrt{4.7778 \times 0.0267}} \right| = 13.0647$$

Since the computed value of t_0 is greater than the table value of $t_{0.025,3} = 3.1824$ (Table A.10), we reject the hypothesis that $H_0 : \beta_1 = 0$. This implies that the length of overlap in the joint contributes significantly to the joint strength.

Illustration: Hypothesis Test on β_2: The hypotheses are

$$H_0 : \beta_2 = 0 \text{ against } H_1 : \beta_2 \neq 0$$

The test statistic is

$$|t_0| = \left| \frac{\hat{\beta}_2}{\sqrt{\sigma^2 C_{22}}} \right| = \left| \frac{7.7500}{\sqrt{4.7778 \times 0.0625}} \right| = 25.9545$$

Since the computed value of t_0 is greater than the table value of $t_{0.025,3} = 3.1824$ (Table A.10), we reject the hypothesis that $H_0 : \beta_2 = 0$. This implies that the stitch density contributes significantly to the joint strength.

9.3.7 Interval Estimates of Regression Coefficients

If the errors ε_i are normally and independently distributed with mean 0 and variance σ^2. Therefore, the observations $\{y_i\}$ are normally and independently distributed with mean

$$\beta_0 + \sum_{j=1}^{k} \beta_j x_{ij} \text{ and variance } \sigma^2.$$

Since the least square estimator $\hat{\beta}$ is a linear combination of the observations, it follows that $\hat{\beta}$ is normally distributed with mean β and covariance $\sigma^2 (\mathbf{x}'\mathbf{x})^{-1}$. The each of the statistics

$$T = \frac{\hat{\beta}_j - \beta_j}{\sqrt{\sigma^2 C_{jj}}}, \quad j = 0, 1, \ldots, k$$

follows t-distribution with $n - p$ degree of freedom. Then, $100(1 - \alpha)\%$ confidence interval for the regression coefficient β_j, $j = 0, 1, 2, \ldots, k$ is

$$\hat{\beta}_j - t_{\frac{\alpha}{2},n-p}\sqrt{\sigma^2 C_{jj}} \leq \beta_j \leq \hat{\beta}_j + t_{\frac{\alpha}{2},n-p}\sqrt{\sigma^2 C_{jj}}.$$

Referring to the previous example, we have confidence intervals for coefficients as

$$-47.2563 \leq \beta_0 \leq -19.7437,$$

$$3.5309 \leq \beta_1 \leq 5.8025,$$

$$6.0123 \leq \beta_2 \leq 9.4877.$$

9.3.8   Point Estimation of Mean

Let us estimate the mean of y at a particular point, say $x_{01}, x_{02}, \ldots, x_{0k}$ The mean response at this point is $\mu_{y|\mathbf{x_0}} = E(y|\mathbf{x_0}) = \mathbf{x_0'}\boldsymbol{\beta}$, where

$$\mathbf{x_0} = \begin{pmatrix} 1 \\ x_{01} \\ x_{02} \\ \vdots \\ x_{0k} \end{pmatrix}$$

This estimator is unbiased, since

$$E(\mathbf{x_0'}\boldsymbol{\beta}) = \mathbf{x_0'}\boldsymbol{\beta} = E(y|\mathbf{x_0}) = \mu_{y|\mathbf{x_0}}$$

The variance of $\hat{\mu}_{y|\mathbf{x_0}}$ is

$$V(\hat{\mu}_{y|\mathbf{x_0}}) = \sigma^2 \mathbf{x_0'}(\mathbf{x'x})^{-1}\mathbf{x_0}.$$

A $100(1-\alpha)\%$ confidence interval on the mean response at the point $x_{01}, x_{02}, \ldots,$ x_{0k} is

$$\hat{\mu}_{y|\mathbf{x_0}} - t_{\frac{\alpha}{2},n-p}\sqrt{\sigma^2 \mathbf{x_0'}(\mathbf{x'x})^{-1}\mathbf{x_0}} \leq \mu_{y|\mathbf{x_0}} \leq \hat{\mu}_{y|\mathbf{x_0}} + t_{\frac{\alpha}{2},n-p}\sqrt{\sigma^2 \mathbf{x_0'}(\mathbf{x'x})^{-1}\mathbf{x_0}}$$

We would like to construct a 95% confidence interval on the mean strength of joint for a parachute webbing with 5 mm length of overlap of joint and 8 stitches per 25 mm.

$$\mathbf{x_0} = \begin{pmatrix} 1 \\ 5 \\ 8 \end{pmatrix}$$

$$\hat{\mu}_{y|x_0} = x_0'\hat{\beta} = (1\ 5\ 8) \begin{pmatrix} -33.50 \\ 4.6667 \\ 7.7500 \end{pmatrix}.$$

The variance of $\hat{\mu}_{y|x_0}$ is estimated by

$$\sigma^2 x_0'(x'x)^{-1}x_0 = 4.7778\,(1\ 5\ 8) \begin{pmatrix} 3.9167 & -0.2000 & -0.3750 \\ -0.2000 & 0.0267 & 0 \\ -0.3750 & 0 & 0.0625 \end{pmatrix} \begin{pmatrix} 1 \\ 5 \\ 8 \end{pmatrix} = 2.7912.$$

Then, $46.5207 \le \mu_{y|x_0} \le 57.1463$

Prediction of New Observation

Let y_0 be the future observation at $x = x_0$. Let the estimator of y_0 be $\hat{y}_0$; $\hat{y}_0 = x_0'\hat{\beta}$, where $x_0' = [1, x_{01}, x_{02}, \ldots, x_{0k}]$.

A $100(1 - \alpha)\%$ prediction interval for this future observation is

$$\hat{y}_0 - t_{\frac{\alpha}{2}, n-p}\sqrt{\sigma^2[1 + x_0'(x'x)^{-1}x_0]} \le y_0 \le \hat{y}_0 + t_{\frac{\alpha}{2}, n-p}\sqrt{\sigma^2[1 + x_0'(x'x)^{-1}x_0]}.$$

We would like to construct a 95% prediction interval on the mean strength of joint for a parachute webbing with 5 mm length of overlap of joint and 5 stitches per 25 mm.

$$x_0 = \begin{pmatrix} 1 \\ 5 \\ 5 \end{pmatrix}, \quad \hat{y}_0 = x_0'\hat{\beta} = (1\ 5\ 5) \begin{pmatrix} -33.5000 \\ 4.6667 \\ 7.7500 \end{pmatrix} = 28.5335.$$

The variance of $\hat{\mu}_{y|x_0}$ is estimated by

$$\sigma^2 x_0'(x'x)^{-1}x_0 = 4.777\,(1\ 5\ 5) \begin{pmatrix} 3.9167 & -0.2000 & -0.3750 \\ -0.2000 & 0.0267 & 0 \\ -0.3750 & 0 & 0.0625 \end{pmatrix} \begin{pmatrix} 1 \\ 5 \\ 5 \end{pmatrix} = 1.8954.$$

Then, $20.3291 \le y_0 \le 36.7379$

9.3.9 Adequacy of Regression Model

The adequacy of a multiple linear regression model is checked by residual analysis and coefficient of multiple determination.

Residual Analysis: Residuals from a regression model are

$$\varepsilon_i = y_i - \hat{y}_i; \quad i = 1, 2, \ldots, n,$$

Table 9.2 Results of transesterification experiment

Temperature (°C)	Concentration (%)	Conversion (%)
25	0.5	86
65	0.5	98.1
25	1.5	99.7
65	1.5	100
45	1	97.7
45	1	97.8
45	1	97.6
45	1	98
45	1.71	100
73.3	1	99.7
16.7	1	96.6
45	0.29	89

where y_i is the actual observation and $\hat{y}_i$ is the corresponding fitted from the regression model. Model is considered to be adequate if the errors are approximately normally distributed with zero mean and constant variance. Plot residuals against $\hat{y}_i$ and x, separately and check for adequacy.

Coefficient of Multiple Determination This is denoted by R^2 and defined by

$$R^2 = \frac{SS_R}{SS_T} = 1 - \frac{SS_E}{SS_T} \quad 0 \le R^2 \le 1$$

For the strength of joint data, we find $R^2 = 0.9920$. Thus, this model accounts for about 99% of the variability in joint strength response. The R^2 statistic is somewhat problematic as a measure of the quality of the fit for a multiple regression model because it always increases with the inclusion of additional variable to a model. So, many regression users prefer to use an adjusted R^2 statistic

$$R^2_{adj} = 1 - \frac{SS_E/(n-p)}{SS_T/(n-1)} \quad R^2_{adj} = 0.9867.$$

Example 9.2 Vicente et al. (1998) made an attempt to study transesterification of sunflower oil with methanol. The effects of operating temperature and catalyst concentration on the conversion of methanol were examined. The experimental results are reported in Table 9.2. Suppose the experimenter fitted the experimental data with a first-order model and obtained the following expression

$$\hat{y}_{[\%]} = 84.17 + 0.1049T_{[°C]} + 7.7891C_{[\%]}.$$

Fig. 9.2 Response surface
and contour plots for
first-order model for
transesterification of oil

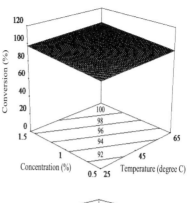

Fig. 9.3 Response surface
and contour plots for
first-order model with
interaction for
transesterification of oil

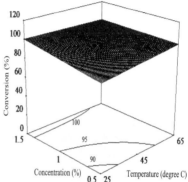

The standard multiple linear regression technique was used. The three-dimensional response surface plot and the two-dimensional contour plot of this model are shown together in Fig. 9.2. As shown, the response surface plot looked like a plane and the contour lines of constant responses were found to be straight. This model, however, gave a poor coefficient of determination (adjusted $R^2 = 0.6466$).

Then the experimenter fitted the experimental data with a first-order model with interaction term and obtained the following expression

$$\hat{y}_{[\%]} = 70.90 + 0.40 T_{[°C]} + 21.06 C_{[\%]} - 0.295 T_{[°C]} C_{[\%]}$$

Here, the last term indicates the interaction between temperature and concentration. Figure 9.3 displays the three-dimensional response surface plot and the two-dimensional contour plot. It can be seen that the contour lines of constant responses were no more straight. This was due to the presence of the interaction term in the model. This model yielded a higher coefficient of determination (adjusted $R^2 = 0.8197$).

Fig. 9.4 Response surface
and contour plots for
second-order model for
transesterification of oil

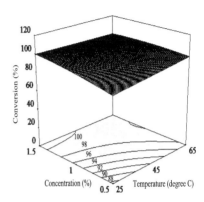

Further, the experimenter fitted the experimental data with a second-order model
and obtained the following expression

$$\hat{y}_{[\%]} = 65.54 + 0.3788T_{[°C]} + 34.91C_{[\%]} - 0.295T_{[°C]}C_{[\%]} + 0.0002T_{[°C]}^2 - 6.925C_{[\%]}^2$$

Here, the last two are the quadratic terms. The three-dimensional response surface
plot and the two-dimensional contour plot of this model are displayed in Fig. 9.4. It
can be seen that the contour lines of constant responses were, as expected, not straight.
This model yielded the highest coefficient of determination (adjusted $R^2 = 0.9301$)
among all the models discussed here.

9.4 Analysis of First-Order Model

One of the most important objectives behind the use of response surface methodology
is to optimize the response. In general, the initial estimate of the optimum process
conditions is very far from the actual optimum ones. In such a situation, the experimenter wishes to move rapidly to the general vicinity of the optimum region. One of
the ways to do so is by following the method of steepest ascent (when maximization
is desired) or the method of steepest decent (when minimization is desired). In the
following, the principles of method of steepest ascent are given. They are similar
in case of steepest decent method. The method of steepest ascent is a procedure for
moving sequentially along the direction of maximum increase in the response. In the
case of first-order model shown below

$$\hat{y} = \hat{\beta}_0 + \sum_{i=1}^{n} \hat{\beta}_i x_i$$

the contours of fitted response ($\hat{y}$) are parallel lines as shown in Fig. 9.5.

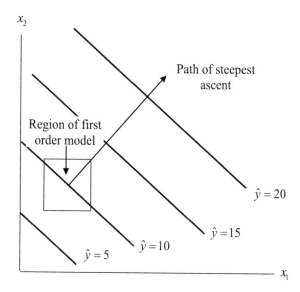

Fig. 9.5 Illustration of method of steepest ascent

The path of steepest ascent denotes the direction in which the fitted response increases most rapidly. This direction is parallel to the normal to the fitted response surface. In general, the path of steepest ascent follows a line which is passing through the center of region of interest and normal to the fitted response surface. The steps along the path of steepest ascent are thus proportional to the regression coefficients $\hat{\beta}$ of the first-order model. This leads to finding the operating conditions for new experiments. Experiments are then conducted along the path of steepest ascent until no further increase in the response is observed. Then, a new first-order model is fitted, and a new path of steepest ascent is determined. This procedure is continued, and the experimenter eventually arrives in the vicinity of the optimum. This is usually indicated by lack of fit of a first-order model. It is indicative that additional experiments be performed to obtain a more precise estimate of the optimum.

The algorithm for the method of steepest ascent is described hereunder. Suppose the first-order model, expressed in coded variables x_i, is shown as

$$\hat{y} = \hat{\beta}_0 + \hat{\beta}_1 x_1 + \hat{\beta}_2 x_2 + \cdots + \hat{\beta}_n x_n.$$

Step 1: Consider $x_1 = x_2 = \cdots = x_n = 0$.

Step 2: Select the independent variable that has the highest absolute value of the regression coefficient. Suppose x_j is the independent variable that has the highest absolute value of regression coefficient $|\hat{\beta}_j|$. Choose a step size Δx_j for x_j.

Step 3: Find out the step size for the other variables as follows

$$\Delta x_i = \frac{\hat{\beta}_i}{\hat{\beta}_j / \Delta x_j}, \quad i = 1, 2, \ldots, n, \; i \neq j.$$

This results in most rapid movement (steepest ascent) toward larger response values.

Step 4: Convert the Δx_i from coded variables to the natural variables.

Let us illustrate this with the help of an example.

Example 9.3 Consider the following first-order model as in Example 9.2

$$\hat{y}_{[\%]} = 84.17 + 0.1049T_{[°C]} + 7.7891C_{[\%]}.$$

This equation is expressed in terms of natural variables of temperature (T) and concentration (C). We need to convert them from natural variables to coded variables. This can be done by considering the following transformations.

$$x_{T_{[-]}} = \frac{T_{[°C]} - \frac{1}{2}(T_{max_{[°C]}} + T_{max_{[°C]}})}{\frac{1}{2}(T_{max_{[°C]}} - T_{max_{[°C]}})} = \frac{T_{[°C]} - 45}{20}$$

$$x_{C_{[-]}} = \frac{C_{[\%]} - \frac{1}{2}(C_{max_{[\%]}} + C_{min_{[\%]}})}{\frac{1}{2}(C_{max_{[\%]}} - C_{min_{[\%]}})} = \frac{C_{[\%]} - 1.0}{0.5}$$

The first-order model, expressed in terms of coded variables, is

$$\hat{y}_{[\%]} = 96.68 + 2.10x_{T_{[°C]}} + 3.89x_{C_{[\%]}}.$$

Then, the following steps are followed.
Step 1: Consider $x_T = x_C = 0$.
Step 2: As $3.89 > 2.10$, it is considered that $\Delta x_C = 1.0$.
Step 3: The step size for temperature is $\Delta x_T = \frac{2.10}{3.89/1.0} = 0.54$.
Step 4: To convert the coded step sizes to the natural units of step sizes, the following relationships are used.

$$\Delta x_{T_{[-]}} = \frac{\Delta T_{[°C]}}{20} \quad \text{and} \quad \Delta x_{C_{[-]}} = \frac{\Delta C_{[\%]}}{0.5}$$

This results in

$$\Delta T_{[°C]} = 20\Delta x_{T_{[-]}} = 20 \times 0.54 = 10.8$$

$$\Delta C_{[\%]} = 0.5\Delta x_{C_{[-]}} = 0.5 \times 1.0 = 0.5$$

The new set of experiments are thus found as follows in Table 9.3. The experimenter needs to perform the experiments and analyzes the responses.

Table 9.3 New set of experiments for Example 10.2

Steps	Coded variable		Natural variables	
Origin	0	0	45	1
Δ	0.54	1	10.8	0.5
Origin+Δ	0.54	1	55.8	1.5
Origin+2Δ	1.08	2	66.6	2
Origin+3Δ	1.62	3	77.4	2.5
$\vdots$	$\vdots$	$\vdots$	$\vdots$	$\vdots$

9.5 Analysis of Second-Order Model

In general, when the experimenter has reached the vicinity of optimum region and a first-order model is exhibiting a lack of fit, then the following second-order model incorporating the curvature of the response is fitted.

$$\hat{y} = \beta_0 + \sum_{i=1}^{n} \beta_i x_i + \sum_{i=1}^{n} \beta_{ij} x_i^2 + \sum \sum_{i<j} \beta_{ij} x_i x_j$$

In this section, we will make an analysis for the fitted second-order model.

9.5.1 Location of Stationary Point

Suppose we wish to find the levels of $x_1, x_2, \ldots, x_n$ that optimize the fitted response. This point, if it exists, can be found out as follows:

$$\frac{\partial \hat{y}}{\partial x_1} = \frac{\partial \hat{y}}{\partial x_2} = \cdots = \frac{\partial \hat{y}}{\partial x_n} = 0$$

This point, say $x_{1,s}, x_{2,s}, \ldots, x_{n,s}$ is called stationary point. It can represent a point of maximum response or a point of minimum response or a saddle point. These three possibilities are displayed in Figs. 9.6a, b, 9.7.

It is possible to obtain a general solution for the location of stationary point. For this, we need to write the second-order model in matrix notation as follows

$$\hat{y} = \hat{\beta}_0 + \mathbf{x}'\mathbf{b} + \mathbf{x}'\mathbf{B}\mathbf{x}$$

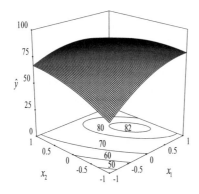 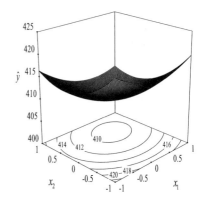

(a) Response surface and contour plot displaying a maximum

(b) Response surface and contour plot displaying a minimum

Fig. 9.6 Response surface and contour plots displaying a maximum and a minimum

Fig. 9.7 Response surface and contour plot displaying a saddle point

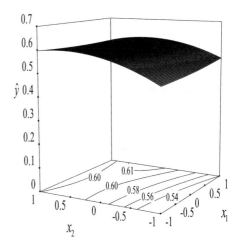

where

$$\mathbf{x} = \begin{bmatrix} x_1 \\ x_2 \\ \vdots \\ x_n \end{bmatrix}, \quad \mathbf{b} = \begin{bmatrix} \hat{\beta}_1 \\ \hat{\beta}_2 \\ \vdots \\ \hat{\beta}_n \end{bmatrix}, \quad \mathbf{B} = \begin{bmatrix} \hat{\beta}_{11} & \frac{\hat{\beta}_{12}}{2} & \cdots & \frac{\hat{\beta}_{1n}}{2} \\ & \hat{\beta}_{22} & \cdots & \frac{\hat{\beta}_{2n}}{2} \\ & & \ddots & \vdots \\ & & & \hat{\beta}_{nn} \end{bmatrix}$$

$\mathbf{x}$ is $(n \times 1)$ vector of independent variables, $\mathbf{b}$ is a $(n \times 1)$ vector of the first-order regression coefficients, and $\mathbf{B}$ is $(n \times n)$ symmetric matrix whose main diagonal elements are the pure quadratic coefficients and the off-diagonal elements are one-half of the mixed quadratic coefficients.

The derivative of $\hat{y}$ with respect to the elements of the vector $\mathbf{x}$ equated to zero is

$$\frac{\partial \hat{y}}{\partial \mathbf{x}} = \mathbf{b} + 2\mathbf{Bx} = 0.$$

The stationary point is the solution to the above equation, that is,

$$\mathbf{x}_S = -\frac{1}{2}\mathbf{B}^{-1}\mathbf{b}.$$

By substituting the expression of stationary point in the second-order model, the predicted response at the stationary point is obtained as follows

$$\hat{y}_S = \hat{\beta}_0 + \frac{1}{2}\mathbf{x}'_S\mathbf{b}.$$

9.5.2 Nature of Stationary Point

After determining the stationary point, it is necessary to know whether the stationary point represents the maximum or minimum response or a saddle point (minimax response). This is done by examining the contour plot of the fitted model, particularly when the model consists of two or three independent variables. As the number of variables increases, the construction and interpretation of the contour plot becomes difficult. Then, a more formal analysis, called canonical analysis, is used. A detailed description of the canonical analysis was given by Myers et al. (2009) and Box et al. (2005). Using this analysis, the nature of response surface (maximum/minimum/saddle point) can be characterized by solving the following equation

$$|\mathbf{B} - \lambda\mathbf{I}| = 0$$

The stationary point corresponds to a maximum, minimum, or minmax response according as the values of $\lambda_1, \lambda_2, \ldots, \lambda_n$ are all negative, positive, or mixed in sign. Let us illustrate this with the help of an example.

Example 9.4 Consider the following second-order model as stated in Example 9.2.

$$\hat{y}_{[\%]} = 65.54 + 0.3788T_{[°C]} + 34.91C_{[\%]} + 0.295T_{[°C]}C_{[\%]} + 0.0002T^2_{[°C]} - 6.925C^2_{[\%]}$$

Let us find the stationary point. In order to do so, the above-stated expression is first converted from natural variables to coded variables as follows.

$$\hat{y}_{[\%]} = 97.77 + 2.10 x_{T_{[-1]}} + 3.89 x_{C_{[-1]}} - 2.95 x_{T_{[-1]}} x_{C_{[-1]}} + 0.094 x_{T_{[-1]}}^2 - 1.73 x_{C_{[-1]}}^2$$

Here,

$$\mathbf{x}_S = \begin{bmatrix} x_{T,s} \\ x_{C,s} \end{bmatrix}, \quad \mathbf{B} = \begin{bmatrix} 0.094 & -1.475 \\ -1.475 & -1.73 \end{bmatrix}, \quad \mathbf{b} = \begin{bmatrix} 2.10 \\ 3.89 \end{bmatrix}.$$

Then,

$$\mathbf{x}_S = -\frac{1}{2} \mathbf{B}^{-1} \mathbf{b} = \begin{bmatrix} -0.3699 & 0.3154 \\ 0.3154 & 0.0201 \end{bmatrix} \begin{bmatrix} 2.10 \\ 3.89 \end{bmatrix} = \begin{bmatrix} 0.4501 \\ 0.7405 \end{bmatrix}$$

Then, the stationary points are

$$x_{T,S} = 0.4501, \quad x_{C,S} = 0.7405.$$

The stationary points in terms of natural variables are

$$x_{T,S} = 0.4501 = \frac{T_{[°C]} - 45}{20} \Rightarrow T_{[°C]} = 54.002 \approx 54$$

$$x_{C,S} = 0.7405 = \frac{C_{[\%]} - 1.0}{0.5} \Rightarrow C_{[\%]} = 1.37 \approx 1.4$$

The predicted response at the stationary point is

$$\hat{y}_{[\%]} = 97.77 + 2.10 x_{T,S_{[-1]}} + 3.89 x_{C,S_{[-1]}} - 2.95 x_{T,S_{[-1]}} x_{C,S_{[-1]}}$$
$$+ 0.094 x_{T,S_{[-1]}}^2 - 1.73 x_{C,S_{[-1]}}^2 = 99.68.$$

Let us now determine the nature of stationary point. For this,

$$|\mathbf{B} - \lambda \mathbf{I}| = 0 \Rightarrow \begin{vmatrix} 0.094 - \lambda & -1.475 \\ -1.475 & -1.73 - \lambda \end{vmatrix} = 0$$

$$\lambda^2 + 1.636\lambda - 2.3382 = 0 \Rightarrow \lambda_1 = 0.9162, \quad \lambda_2 = -2.5522$$

Because λ_1 and λ_2 are mixed in sign, the stationary point refers to the saddle point.

9.6 Response Surface Designs

It is known that appropriate designs of experiments are required for fitting and analyzing of first- and second-order response surface models. It is not true that all designs are corresponding to all models. In what follows below is some of the aspects for selection of appropriate designs for fitting first- and second-order response surface models.

9.6.1 Designs for Fitting First-Order Model

Suppose we wish to fit the following first-order model

$$y = \beta_0 + \sum_{i=1}^{n} \beta_i x_i + \varepsilon$$

The orthogonal first-order designs that minimizes the variance of the regression coefficients are suitable for fitting the first-order model. The first-order orthogonal designs are those designs whose off-diagonal elements of the $(\mathbf{x'x})$ matrix are all zero. The orthogonal first-order designs include two-level full factorial design and two-level fractional factorial designs in which the main effects are not aliased with each other. Special attention needs to be paid while selecting the two-level full factorial design so that they afford an estimate of the experimental error. This can be done by replicating the runs. A common method of including replication in these designs is to augment the designs with several observations at the center. Interestingly, this does not change the orthogonality property of the design. Let us consider following examples.

Example 9.5 Show that 2^2 factorial design is an orthogonal design. Here,

$$\mathbf{x} = \begin{bmatrix} -1 & -1 \\ +1 & -1 \\ -1 & +1 \\ +1 & +1 \end{bmatrix}, \quad \mathbf{x'} = \begin{bmatrix} -1 & +1 & -1 & +1 \\ -1 & -1 & +1 & +1 \end{bmatrix},$$

$$\mathbf{x'x} = \begin{bmatrix} -1 & +1 & -1 & +1 \\ -1 & -1 & +1 & +1 \end{bmatrix} \begin{bmatrix} -1 & -1 \\ +1 & -1 \\ -1 & +1 \\ +1 & +1 \end{bmatrix} = \begin{bmatrix} 4 & 0 \\ 0 & 4 \end{bmatrix}$$

As the off-diagonal elements of the matrix are all zero, the above design is an orthogonal design.

Example 9.6 Show the layout of a 2^2 factorial design augmented with five center points to fit a first-order model. The layout is shown in Table 9.4.

9.6.2 Experimental Designs for Fitting Second-Order Model

Suppose we wish to fit the following second-order model

$$\hat{y} = \beta_0 + \sum_{i=1}^{n} \beta_i x_i + \sum_{i=1}^{n} \beta_{ij} x_i^2 + \sum \sum_{i<j} \beta_{ij} x_i x_j$$

Table 9.4 Layout of 2^2 factorial design augmented with five center points

Run	Factor	
	A	B
1	−1	−1
2	1	−1
3	−1	1
4	1	1
5	0	0
6	0	0
7	0	0
8	0	0
9	0	0

The minimum conditions for response surface designs for fitting the above-stated model are stated below.

- The designs should have at least three levels of each factor.
- The design should have at least $1 + 2n + n(n-1)/2$ distinct design points, where n stands for the number of factors.

Examples of such designs are three-level full factorial designs, central composite design (CCD), and Box Behnken design (BBD).

9.6.2.1 Three-Level Full Factorial Design

The popular three-level full factorial designs are 3^2 full factorial design and 3^3 full factorial design. In a 3^2 full factorial design, there are two factors, each at three levels. The geometrical view of 3^2 and 3^3 full factorial designs is shown in Fig. 9.8a and b, respectively. Here, the low level is indicated by "0", the middle by "1", and high by "2", respectively.

9.6.2.2 Central Composite Design

The central composite design (CCD) is one of the, if not most, popular response surface designs available for fitting the second-order response surface model. This design consists of a 2^n number of factorial runs, $2n$ number of axial runs, and a few center runs. Here, the factorial runs refer to the runs of a 2^n factorial design. The axial runs indicate the points that are lying on the axes of the design. Of course, all the factors are not varied simultaneously on the axial runs, rather they are varied as one-factor-at-a-time. The center runs denote the points that are lying on the center of the design. A two-factor CCD is shown in Table 9.5. The layout of this design is

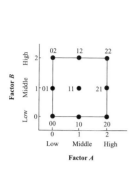

(a) 3^2 full factorial design

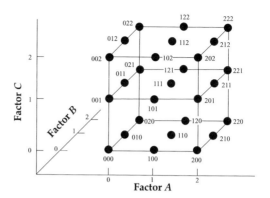

(b) 3^3 full factorial design

Fig. 9.8 Geometrical view of 3^2 and 3^3 full factorial designs

Table 9.5 Layout of two-factor CCD

Run	1	2	3	4	5	6	7	8	9	10	11
Factor A	−1	1	−1	1	α	$-\alpha$	0	0	0	0	0
Factor B	−1	−1	1	1	0	0	α	$-\alpha$	0	0	0

displayed in Fig. 9.9. It can be seen that the first four runs denote the four factorial runs of a 2^2 factorial design. The next four runs refer to the four axial points of this design. Here, α denotes the distance of the axial runs from the design center. There are three choices available for the value of α. If the square region of this design is of interest, then α is chosen as $\alpha = 1$. Such a central composite design is termed as square two-factor CCD. If the circular region of the two-factor CCD is of interest, then α is chosen as $\alpha = \sqrt{n} = \sqrt{2} = 1.414$, where n stands for the number of factors. Such a central composite design is termed as circular two-factor CCD. Further, the two-factor central composite design can be made rotatable when $\alpha = (n_F)^{\frac{1}{4}} = (2^2)^{\frac{1}{4}} = 1.414$, where n_F is the number of factorial runs. As known, rotatability is an important property for the second-order model to provide reasonably consistent and stable variance of the predicted response at points of interest x. This means that the variance of the predicted response is same at all the points x that are lying at the same distance from the design center. Note that a two-factor CCD can be made circular as well as rotatable by choosing α as 1.414. Further, the center runs are important to provide reasonably stable variance of the predicted response. Generally, three to five center runs are recommended for a central composite design. Like a two-factor CCD, a three-factor CCD can be constructed in a similar manner. Table 9.6 displays a three-factor CCD. The layout of a rotatable three-factor CCD is displayed in Fig. 9.10. Here, the values of α for cuboidal, spherical, and rotatable

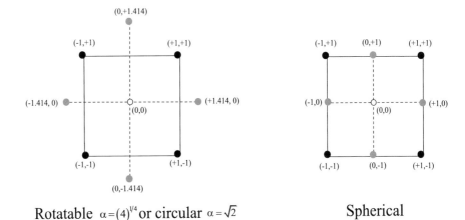

Rotatable $\alpha = (4)^{1/4}$ or circular $\alpha = \sqrt{2}$ Spherical

Fig. 9.9 Geometrical view of two-factor CCD

Table 9.6 Layout of three-factor CCD

Run	1	2	3	4	5	6	7	8	9	10	11	12	13	14	15	16	17
Factor A	−1	1	−1	1	−1	1	−1	1	α	$-\alpha$	0	0	0	0	0	0	0
Factor B	−1	−1	1	1	−1	−1	1	1	0	0	α	$-\alpha$	0	0	0	0	0
Factor C	−1	−1	−1	−1	1	1	1	1	0	0	0	0	α	$-\alpha$	0	0	0

Fig. 9.10 Three-factor CCD

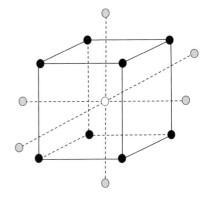

three-factor central composite designs are 1, 1.732, and 1.682, respectively. In this way, central composite designs with more than three factors can be constructed.

9.6.2.3 Box Behnken Design

The Box Behnken design (BBD) is another popular response surface design available for fitting the second-order model. This design is formed by combining 2^n factorials

Table 9.7 Scheme of a balanced incomplete block design

Treatment	Block		
	1	2	3
1	X	X	
2	X		X
3		X	X

Table 9.8 Layout of three-factor BBD

Run	Factor				Run	Factor		
	A	B	C			A	B	C
1	−1	−1	0		13	0	0	0
2	1	−1	0		14	0	0	0
3	−1	1	0		15	0	0	0
4	−1	1	0					
5	−1	0	−1					
6	1	0	−1					
7	−1	0	1					
8	1	0	1					
9	0	−1	−1					
10	0	1	−1					
11	0	−1	1					
12	0	1	1					

with a balanced incomplete block design. Let us illustrate how one can form a three-factor BBD. Table 9.7 shows a scheme of a balanced incomplete block design (BIBD) with three treatments in three blocks. As shown by "X" symbol, treatment 1 is run in blocks 1 and 2, treatment 2 is run in blocks 1 and 3, and treatment 3 is run in blocks 2 and 3 (Table 9.8). Here, each treatment occurs in two blocks (balanced) and all treatment combinations cannot be run on one block (incomplete). Based on the aforesaid BIBD, create a 2^2 factorial design with two blocks keeping the other block at zero and repeat it for all factors you will get a three-factor BBD. This is shown in Table 9.9. The layout of this design is shown in Fig. 9.11. It can be seen that this design does not contain any point at the vertices of the cube and is a spherical design with all points lying on a sphere of radius $\sqrt{2}$. In this way, BBD for more than three factors can also be formed. The reader is instructed to do so. Note that there is no BBD available for two factors. The Box Behnken designs are very efficient in terms of number of runs required. One can compare that a three-factor CCD requires 19 runs including five center runs, whereas a three-factor BBD requires 17 runs including three center runs.

Table 9.9 Factors and levels for corona-charging process

Process factors	Levels		
	−1	0	+1
Applied voltage (kV)	5	10	15
Charging time (min)	15	30	45
Electrode distance (mm)	25	30	35

Fig. 9.11 Geometrical view of three-factor BBD

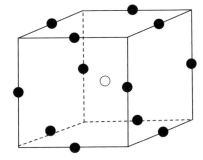

9.7 Multi-factor Optimization

In practice, we often need to optimize many responses simultaneously. Simultaneous optimization of multiple responses first involve development of an appropriate model for each response and then find optimum conditions that optimize all responses together or keep them in desired levels. There are many methods available for the simultaneous optimization of multiple responses, and the most popular ones in response surface methodology are based on overlay of contour plot and desirability function approach. The former works well when there are only a few process variables present, while the latter is able to work with many process variables. The method of overlay of contour plot relies on overlaying the contour plots of different responses to find a region of the optimum operating conditions. Generally, a number of combinations of operating conditions are found to be satisfactory. The other method desirability function approach was popularized by Derringer and Such (1980). In this method, each response y_i is first converted to an individual desirability function d_i such that d_i varies in the range $0 \leq d_i \leq 1$. $d_i = 0$ indicates a completely undesirable response, while $d_i = 1$ denotes the most desirable response. If the goal for the response is to attain a maximum value of target T, then the individual desirability function takes the form

$$d = \begin{cases} 0 & y < L \\ \left(\frac{y-L}{T-L}\right)' & L \leq y \leq T \\ 1 & y > T \end{cases}$$

where L indicates lower limit of the response, T refers to the target, and r denotes the weight of the function. If $r = 1$ then the desirability function is linear. If $r > 1$ then a higher emphasis is paid to the response being close to the target value. If $0 < r < 1$ then the importance is low. If the target for the response is a minimum value then

$$d = \begin{cases} 1 & y < T \\ \left(\frac{U-y}{U-T}\right)^{r} & T \le y \le U \\ 0 & y > U \end{cases}$$

where U indicates the upper limit of the response.

The two-sided desirability function is required when the target T is located in-between the lower (L) and upper (U) limits. This is shown below

$$d = \begin{cases} 0 & y < L \\ \left(\frac{y-L}{T-L}\right)^{r_1} & L \le y \le T \\ \left(\frac{U-y}{U-T}\right)^{r_2} & T \le y \le U \\ 0 & y > U \end{cases}$$

where r_1 and r_2 denote the weights of the function. Then all individual desirabilities are combined into an overall desirability as follows

$$D = (d_1.d_2.\ldots.d_m)^{1/m}$$

where m stands for the number of responses. In this way, the overall desirability becomes a function of the independent variables. The maximization of the overall desirability function generally yields the optimum operating conditions. Let us better understand this with the help of an example.

Example 9.7 In order to study the charge storage capability of a corona-charged electret filter media, a 3^3 factorial experiment is carried out with the following corona-charging process factors and levels.

Initial surface potential (kV) and half-decay time (min) are chosen as responses. The response surface equations are found as follows.

$$Y_1 = 4.09 + 4.57X_1 + 1.29X_2 - 0.21X_3 + 1.00X_1X_2 + 1.82X_1^2$$
$$Y_2 = 4.70 - 0.66X_1 - 0.50X_2 - 0.64X_3 - 1.12X_2^2 - 1.12X_2^2 - 0.60X_3^2$$

where X_1, X_2, and X_3 denote the applied voltage, charging time, and electrode distance, respectively, in terms of coded variables and Y_1 and Y_2 indicate initial surface potential and half-decay time, respectively. The coded variables are found as follows:

$$X_1 = \frac{\text{Applied volatility(kV)} - 10}{5}$$

$$X_2 = \frac{\text{Charging time(min)} - 30}{15}$$

$$X_3 = \frac{\text{Electrode distance(mm)} - 30}{5}.$$

We need to determine the optimum corona-charging process conditions to obtain enhanced initial surface potential and half-decay time simultaneously. We can use the desirability function approach for this purpose. Here, both responses are required to be maximized. Hence, the desirability functions are set as follows

$$d_1 = \frac{Y_1 - Y_{1,min}}{Y_{1,max} - Y_{1,min}} = \frac{Y_1 - 1.11}{12.64 - 1.11}$$

$$= \frac{4.09 + 4.57X_1 + 1.29X_2 - 0.21X_3 + 1.00X_1X_2 + 1.82X_1^2 - 1.11}{12.64 - 1.11},$$

$$d_2 = \frac{Y_2 - Y_{2,min}}{Y_{2,max} - Y_{2,min}} = \frac{Y_1 - 1.57}{6.00 - 1.57}$$

$$= \frac{4.70 - 0.66X_1 - 0.50X_2 - 0.64X_3 - 1.12X_2^2 - 1.12X_2^2 - 0.60X_3^2 - 1.57}{6.00 - 1.57}.$$

where d_1 and d_2 stand for individual desirability functions for initial surface potential and half-decay time, respectively. The minimum and maximum initial surface potential are determined as 1.11 and 12.64 kV, respectively. The minimum and maximum half-decay time are determined as 1.57 and 6.00 min, respectively. The overall desirability function is then found as

$$D = (d_1.d_2)^{1/2} = \left[\left(\frac{4.09 + 4.57X_1 + 1.29X_2 - 0.21X_3 + 1.00X_1X_2 + 1.82X_1^2 - 1.11}{12.64 - 1.11} \right) \right.$$
$$\left. \times \left(\frac{4.70 - 0.66X_1 - 0.50X_2 - 0.64X_3 - 1.12X_2^2 - 1.12X_2^2 - 0.60X_3^2 - 1.57}{6.00 - 1.57} \right) \right].$$

The solution of this expression leads to $X_1 = 1$, $X_2 = 0.06$, $X_3 = -0.536$. They, in terms of natural variables, are as follows: applied voltage = 15 kV, charging time = 30.90 min, and electrode distance = 27.32 mm. At these process conditions, the initial surface potential and half-decay time are predicted as 10.62 kV and 4.18 min, respectively. The overall desirability is found as 0.70. It can be noted that there are many commercial softwares like Design Expert, Minitab can be employed for this purpose.

Problems

9.1 An article entitled "Fermentation of molasses by Zymomonas mobilis: Effects on temperature and sugar concentration on ethanol production" published by M. L. Cazetta et al. in Bioresource Technology, 98, 2824–2828, 2007, described the experimental results shown in Table 9.10.

(a) Name the experimental design used in this study.
(b) Construct ANOVA for ethanol concentration.
(c) Do you think that a quadratic model can be fit to the data? If yes, fit the model and if not, why not?

9.2 Find the path of steepest ascent for the following first-order model

$$\hat{y} = 1000 + 100x_1 + 50x_2$$

where the variables are coded as $-1 \leq x_i \leq 1$.

9.3 In a certain experiment, the two factors are temperature and contact pressure. Two central composite designs were constructed using following ranges on the two factors.

$$\text{Temperature} : 500\,°\text{F} - 1000\,°\text{F}$$
$$\text{Contact pressure} : 15\,\text{psi} - 21\,\text{psi}$$

The designs listed in Table 9.11 are in coded factor levels.

Table 9.10 Data for Problem 9.1

Run	Factors		Response
	Conc. of molasses (g/L)	Temperature (°C)	Concentration of ethanol (g/L)
1	150	25	46.43
2	250	35	42.39
3	150	25	47.73
4	250	35	45.22
5	200	30	55.36
6	200	30	54.31
7	200	30	55.57
8	80	30	28.55
9	200	37	22.83
10	270	30	33.43
11	200	18	7.87

Table 9.11 Data for Problem 9.3

Run	Design 1		Design 2	
	Temperature	Pressure	Temperature	Pressure
1	−1	−1	−1	−1
2	−1	1	−1	1
3	1	−1	1	−1
4	1	1	1	1
5	0	−1	0	−1.5
6	0	1	0	1.5
7	−1	0	−1.5	0
8	1	0	1.5	0
9	0	0	0	0
10	0	0	0	0
11	0	0	0	0

(a) Replace the coded levels with actual factor level.
(b) Is either design rotatable? If not why?
(c) Construct a rotatable central composite design for this.

9.4 A disk-type test rig is designed and fabricated to measure the wear of a textile composite under specified test condition. The ranges of the three factors chosen are as follows:

$$\text{Temperature} : 500 - 1000\,^{\circ}\text{F}$$
$$\text{Contact pressure} : 15 - 21 \text{ psi}$$
$$\text{Sliding speed} : 54 - 60 \text{ ft/sec}$$

Construct a rotatable 3^3 central composite design for this experiment.

9.5 Construct a Box Behnken design for the experiment stated in Problem 9.5.

9.6 In a study to determine the nature of response system that relates yield of electrochemical polymerization (y) with monomer concentration (x_1) and polymerization temperature x_2, the following response surface equation is determined

$$\hat{y} = 79.75 + 10.18x_1 + 4.22x_2 - 8.50x_1^2 - 5.25x_2^2 - 7.75x_1x_2.$$

Find the stationary point. Determine the nature of the stationary point. Estimate the response at the stationary point.

Table 9.12 Data for Problem 9.8

Run	Factors			Response
	Initial pH (–)	Initial conc. of copper (mg/L)	Time (min)	Removal of copper (%)
1	5.5	50	45	93.07
2	3.5	50	90	96
3	4.5	32.5	67.5	93
4	4.5	32.5	67.5	93
5	3.5	15	90	91
6	4.5	61.93	67.5	93.8
7	4.5	32.5	67.5	92.9
8	6.18	32.5	67.5	91.6
9	4.5	32.5	67.5	92.8
10	3.5	50	45	92.7
11	4.5	32.5	67.5	92.9
12	4.5	32.5	29.66	91.24
13	4.5	3.07	67.5	85.5
14	4.5	32.5	105.34	96
15	5.5	50	90	94.16
16	3.5	15	45	85.9
17	5.5	15	45	88.2
18	5.5	15	90	91.2
19	4.5	32.5	67.5	92.9
20	2.82	32.5	67.5	91.4

9.7 Consider the following model

$$\hat{y} = 1.665 - 32 \times 10^{-5}x_1 + 372 \times 10^{-5}x_2 + 1 \times 10^{-5}x_1^2 + 68 \times 10^{-5}x_2^2 - 1 \times 10^{-5}x_1x_2$$

Find the stationary point. Determine the nature of the stationary point. Estimate the response at the stationary point.

9.8 An article entitled "Central composite design optimization and artificial neural network modeling of copper removal by chemically modified orange peel" published by A. Ghosh et al. in Desalination and Water Treatment, 51, 7791–7799, 2013, described the experimental results shown in Table 9.12.

(a) Name the design of experiments used here.
(b) Develop a suitable response surface model and construct ANOVA.
(c) Find out the stationary point and comment on the nature of stationary point.
(d) State the optimum process factors that maximize the percentage removal of copper.

Table 9.13 Data for Problem 9.9

Run	Dopant concentration (M)	Rate of addition (mL/h)	OM ratio (–)	Conductivity (10^2 S/cm)	Yield (%)
1	0.3	10	0.8	11.21	86
2	0.3	30	1.5	15.04	94
3	0.3	50	0.8	1.37	90
4	1.05	50	0.1	0.15	32
5	1.05	10	1.5	1.91	89
6	1.05	30	0.8	153.34	91
7	1.05	30	0.8	253.45	92
8	1.05	50	1.5	6.63	63
9	1.8	10	0.8	0.05	51
10	1.8	50	0.8	0.04	50
11	1.8	30	1.5	0.01	43
12	1.05	30	0.8	191.1	87
13	0.3	30	0.1	0.88	55
14	1.05	10	0.1	0.71	32
15	1.8	30	0.1	0.01	30

9.9 An article entitled "Optimization of the conductivity and yield of chemically synthesized polyaniline using a design of experiments" published by E.J. Jelmy et al. in Journal of Applied Polymer Science, 1047–1057, 2013, described a three-factor Box Behnken design with the results shown in Table 9.13.

(a) Develop an appropriate model and construct ANOVA for conductivity.
(b) Find out the stationary point and comment on the nature of stationary point.
(c) What operating conditions would you recommend if it is important to obtain a conductivity as close as 0.2 S/cm?
(d) Develop an appropriate model and construct ANOVA for yield.
(e) Find out the stationary point and comment on the nature of stationary point.
(f) What operating conditions would you recommend if it is important to obtain a yield as close as 90%?
(g) What operating conditions would you recommend if you wish to maximize both conductivity and yield simultaneously?

References

Box GE, Hunter JS, Hunter WG (2005) Statistics for experimenters. Wiley, New York
Derringer G, Suich R (1980) Simultaneous optimization of several response variables. J Qual Technol 12:214–219

Montgomery DC (2007) Design and analysis of experiments. Wiley, India

Myers RH, Montgomery DC, Anderson-Cook CM (2009) Response surface methodology. Wiley, New York

Panneerselvam R (2012) Design and analysis of experiments. PHI Learning Private Limited, New Delhi

Vicente G, Coteron A, Martinez M, Aracil J (1998) Application of the factorial design of experiments and response surface methodology to optimize biodiesel production. Ind Crops Prod 8:29–35

Chapter 10
Statistical Quality Control

10.1 Introduction

Statistical quality control means application of statistical techniques for checking the quality of products. The products include manufactured goods such as computers, mobile phones, automobiles, clothing and services such as health care, banking, public transportation. The word "quality" is defined in many ways. To some people, quality means fitness for use. To others, quality is inversely proportional to variability. Some people also think that quality means degree of conformance to specifications of products. One can read Montgomery (2001) for a detailed discussion on the meaning of quality. Whatsoever be the definition of quality, there are many statistical methods available for checking the quality of products. In this chapter, we will focus on two important methods: acceptance sampling and quality control charts.

10.2 Acceptance Sampling

In a typical manufacturing set up, the manufacturing industries receive raw materials from vendors and convert them into semifinished or finished products in order to sell them to either suppliers or retailers. The raw materials received by the industries are inspected, and the products received by the suppliers or retailers are also inspected for taking a decision whether to accept or reject the raw materials and the products. This is schematically shown in Fig. 10.1. The statistical technique used for taking such decisions is known as acceptance sampling technique.

The input or output articles are available in lots or batches (population). It is practically impossible to check each and every article of a batch. So we randomly select a few articles (sample) from a batch, inspect them, and then draw conclusion whether the batch is acceptable or not. This procedure is called acceptance sampling.

© Springer Nature Singapore Pte Ltd. 2018
D. Selvamuthu and D. Das, *Introduction to Statistical Methods,
Design of Experiments and Statistical Quality Control*,
https://doi.org/10.1007/978-981-13-1736-1_10

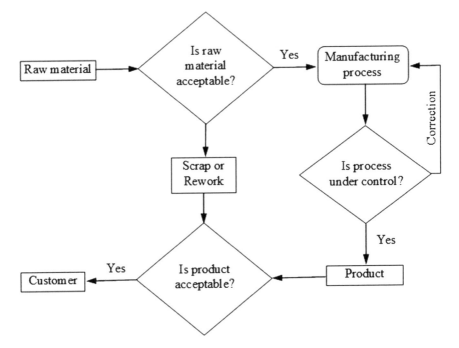

Fig. 10.1 Material and process flow in a typical manufacturing setup

Sometimes the articles inspected are merely classified as defective or nondefective. Then, we deal with acceptance sampling of attributes. In case of acceptance sampling of attributes, whether a batch is accepted or rejected is decided by acceptance sampling plan. Several types of acceptance sampling plans are used in practice. Of them, single sampling plan is the most extensively followed. A more complicated double sampling plan is also used occasionally.

When the property of the articles inspected is actually measured, then we deal with acceptance sampling of variables.

10.3   Single Sampling Plan for Attributes

10.3.1   Definition of a Single Sampling Plan

A single sampling is defined as follows. Suppose a random sample of articles of size n is drawn from a batch of articles of size N such that $n \ll N$. Then, each and every article of the sample is inspected. If the number of defective articles found in the sample is not greater than a certain number c, then the batch is accepted; otherwise, it is rejected.

10.3.2 Operating Characteristic Curve

The performance of an acceptance sampling plan can be seen from a curve, known as operating characteristic curve. This curve plots the probability of acceptance of the batch against the proportion of defective articles present in the batch. It informs us the probability that a batch submitted with a certain fraction of defectives will be accepted or rejected. It thus displays the discriminatory power of the acceptance sampling plan.

The mathematical basis of the OC curve is given below. Let us assume that the proportion of defective articles in the batch is p. Then, when a single article is randomly chosen from a batch, the probability that it will be defective is p. Further, assume that the batch size is sufficiently larger than the sample size n so that this probability is the same for each article in the sample. Thus, the probability of finding exactly r number of defective articles in a sample of size n is

$$P(r) = \frac{n!}{r!(n-r)!} p^r (1-p)^{n-r}$$

Now, the batch will be accepted if $r \leq c$. Then, according to the addition rule of probability (refer to Chap. 2), the probability of accepting the batch is

$$P_a(p) = P(r=0) + P(r=1) + P(r=2) + \cdots + P(r=c)$$
$$= \sum_{r=0}^{c} \frac{n!}{r!(n-r)!} p^r (1-p)^{n-r}$$

This tells that once n and c are known, the probability of accepting a batch depends only on the proportion of defectives in the batch.

In practice, the operating characteristic curve looks like as shown in Fig. 10.2. Nevertheless, an ideal OC curve that discriminates perfectly the good and the bad batches would have looked like as shown in Fig. 10.2. It can be seen that the ideal OC curve runs horizontally at a probability of acceptance $P_a(p) = 1$ until a level of fraction defective which is considered to be "bad" is obtained. At this point, the curve drops vertically to a probability of acceptance $P_a(p) = 0$ and then the curve runs horizontally at $P_a(p) = 0$ for all fraction defectives greater than the undesirable level.

Example 10.1 An automotive company decides to accept or reject a batch of bearings as per the following acceptance sampling plan. A batch is accepted if not more than four bearings are found to be defective in a sample of 100 bearings taken randomly from the batch; otherwise, the batch is rejected. If 100 batches obtained from a process that manufactures 2% defective bearings are submitted to this plan, how many batches the automotive company will expect to accept?

Fig. 10.2 Operating
characteristic (OC) curve

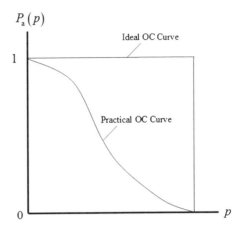

Solution: Here $n = 100, c = 4, p = 0.02$. The probability of acceptance is $P_a(p = 0.02) = \sum_{r=0}^{4} \frac{100!}{r!(100-r)!}(0.02)^r(1-0.02)^{100-r} = 0.9492$. It means that the probability of accepting the batch is approximately 0.95. Hence, the probability of rejecting the batch is $1 - 0.95 = 0.05$. The company will expect to accept 95 out of 100 batches.

10.3.3 Acceptable Quality Level

One of the important characteristics of the OC curve is associated with acceptable quality level (AQL). This represents the poorest level of quality for the producer's process that the consumer would consider to be acceptable as process average say p_1. Ideally, the producer should try to produce lots of quality better than p_1. Assume that there is a high probability, say $1 - \alpha$, of accepting a batch of quality p_1. Then, the probability of rejecting a batch of quality p_1 is α, which is known as producer's risk. This is shown in Fig. 10.3. As shown, when $p = p_1$, $P_a(p_1) = 1 - \alpha$.

10.3.4 Rejectable Quality Level

The other important characteristic of the OC curve is rejectable quality level (RQL). This is otherwise known as lot tolerance proportion defective (LTPD). It represents the poorest level of quality that the consumer is willing to accept in an individual lot. Below this level, it is unacceptable to the consumer. In spite of this, there will be a small chance (probability) β of accepting such a bad batch (with fraction defective p_2) by the consumer; β is known as consumer's risk. This is shown in Fig. 10.4. As shown, when $p > p_2$ then $P_a(p_2) = \beta$.

Fig. 10.3 Acceptable quality level

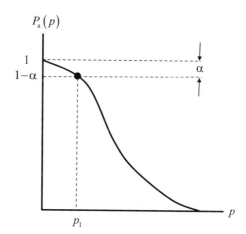

$P_a(p)$

Fig. 10.4 Rejectable quality level

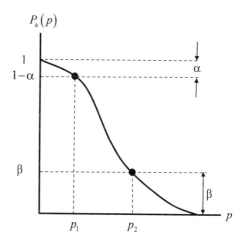

$P_a(p)$

10.3.5 *Designing an Acceptance Sampling Plan*

To design an acceptance sampling plan, it is necessary to know n and c. They can be calculated as follows. It is known that

$$P_a(p_1) = 1 - \alpha = \sum_{r=0}^{c} \frac{n!}{r!(n-r)!} p_1^r (1 - p_1)^{n-r},$$

$$P_a(p_2) = \beta = \sum_{r=0}^{c} \frac{n!}{r!(n-r)!} p_2^r (1 - p_2)^{n-r}.$$

The solution is based on χ^2 distribution with $2(c+1)$ degree of freedom

$$\chi^2_{2(c+1),1-\alpha} = 2np_1, \quad \chi^2_{2(c+1),\beta} = 2np_2 \tag{10.1}$$

$$\frac{\chi^2_{2(c+1),1-\alpha}}{\chi^2_{2(c+1),\beta}} = \frac{2np_1}{2np_2} = \frac{p_1}{p_2}. \tag{10.2}$$

In a sampling plan, p_1, p_2, α, and β are given. Then, the value of c can be found out from Eq. (10.2), and then, the value of n can be found out from Eq. (10.1).

Example 10.2 Design an acceptance sampling plan for which acceptable quality level (AQL) is 0.05, the rejectable quality level (RQL) is 0.15, the producer's risk is 0.05, and the consumer's risk is 0.05.

Solution: Here $p_1 = 0.05$, $p_2 = 0.15$, $\alpha = 0.05$, $\beta = 0.05$.

$$\frac{\chi^2_{2(c+1),1-\alpha}}{\chi^2_{2(c+1),\beta}} = \frac{2np_1}{2np_2} = \frac{p_1}{p_2} = \frac{0.05}{0.15} = 0.3333$$

From χ^2 table (see Table A.8), we find $\chi^2_{18,0.95} = 9.39$ & $\chi^2_{18,0.05} = 28.87$. and hence

$$\frac{\chi^2_{18,0.95}}{\chi^2_{18,0.05}} = \frac{9.39}{28.87} = 0.3253 \approx 0.33$$

$$\chi^2_{18,0.95} = 9.39 = 2np_1 = 0.1n \quad \Rightarrow \quad n = \frac{9.39}{0.1} = 93.9$$

$$\chi^2_{18,0.05} = 28.87 = 2np_2 = 0.3n \quad \Rightarrow \quad n = \frac{28.87}{0.3} = 96.23$$

Then,

$$n = \frac{1}{2}(93.9 + 96.23) = 95.07 \approx 95$$

$$2(c + 1) = 18 \quad \Rightarrow \quad c = 8.$$

Example 10.3 An automotive company receives batches of mirrors and decides to accept or reject a batch according to an acceptance sampling plan for which AQL is 4%, RQL is 16%, producer's risk is 5%, and consumer's risk is 10%. Determine the probability of acceptance of a batch produced by a process that manufactures 10% defective mirrors.

Solution: Here, $p_1 = 0.04$, $p_2 = 0.16$, $\alpha = 0.05$, $\beta = 0.10$,

$$\frac{\chi^2_{2(c+1),1-\alpha}}{\chi^2_{2(c+1),\beta}} = \frac{2np_1}{2np_2} = \frac{p_1}{p_2} = \frac{0.04}{0.16} = 0.25$$

From χ^2 table (see Table A.8), $\chi^2_{10,0.95} = 3.94$ and $\chi^2_{10,0.10} = 15.99$, and we have

$$\frac{\chi^2_{10,0.95}}{\chi^2_{10,0.10}} = \frac{3.94}{15.99} = 0.2464 \approx 0.25.$$

Hence,

$$2(c+1) = 10 \text{ so, } c = 4$$
$$\chi^2_{10,0.95} = 3.94 = 2np_1 = 0.08n \Rightarrow n = 49.25$$
$$\chi^2_{10,0.10} = 15.99 = 2np_2 = 0.32n \Rightarrow n = 49.97$$

Therefore, $n = \frac{1}{2}(49.25 + 49.97) = 49.61 \approx 50$. The required probability of acceptance of a batch is equal to

$$P_a(p = 0.10) = \sum_{r=0}^{4} \binom{50}{r} p^r (1-p)^{50-r} = \binom{50}{0}(0.1)^0(0.9)^{50} + \binom{50}{1}(0.1)^1(0.9)^{49} +$$
$$\binom{50}{2}(0.1)^2(0.9)^{48} + \binom{50}{3}(0.1)^3(0.9)^{47} + \binom{50}{4}(0.1)^4(0.9)^{46}$$
$$= 0.005154 + 0.028632 + 0.077943 + 0.138565 + 0.180905 = 0.431199$$

10.3.6   Effect of Sample Size on OC Curve

It is known that an ideal OC curve discriminates perfectly between good and bad batches. Theoretically, the ideal OC curve can be obtained by 100% inspection. However, in practice, it is almost never realized. Nevertheless, Fig. 10.5 shows that as the sample size n increases, the OC curve becomes more like the idealized OC

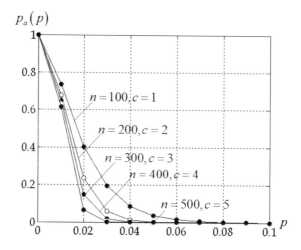

Fig. 10.5 Effect of changing sample size n on OC curve

Fig. 10.6 Effect of changing acceptance number c on OC curve

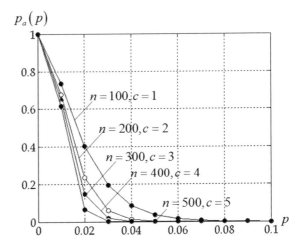

curve. It can be noted that the acceptance number c has been kept proportional to the sample size n. It can be therefore said that the plans with large sample (higher value of n) offer more discriminatory power.

10.3.7 Effect of Acceptance Number on OC Curve

Figure 10.6 displays the effect of acceptance number c on the OC curve. Here, the sample size n is kept constant, but the acceptance number is varied. It can be observed that as the acceptance number c decreases, the OC curve shifts to the left, although the slope of the curve does not change appreciably. It can be thus said that the plans with smaller value of c provide discrimination at lower levels of lot fraction defective than plans with larger values of c do.

10.4 Double Sampling Plan for Attributes

Occasionally, a double sampling plan is used in practice. In this plan, a second sample is inspected if the first sample is not conclusive for either accepting the lot or rejecting the lot. A double sampling plan is described by five parameters N, n_1, n_2, c_1, and c_2. Here, N stands for the lot size from where the samples are taken. n_1 and n_2 denote the sizes of first and second samples, respectively. c_1 refers to the acceptance number for the first sample, whereas c_2 represents the acceptance number for both samples. The above sampling procedure is done as follows. A random sample of size n_1 is selected from the lot of size N. Suppose that the number of defectives found in this sample is d_1. If $d_1 \leq c_1$, then the lot is accepted on the first sample. If $d_1 > c_2$, then the lot is rejected on the first sample. If $c_1 < d_1 \le c_2$, then a second random sample of size n_2 is drawn from the lot. Suppose that the number of defectives found in the second sample is d_2. Then, the combined number of defectives from both the first sample and the second sample, $d_1 + d_2$, is used to determine whether the lot is accepted or rejected. If $d_1 + d_2 \le c_2$, then the lot is accepted. But if $d_1 + d_2 > c_2$ then the lot is rejected.

The performance of a double sampling plan can be seen from its operating characteristic (OC) curve. Let us illustrate the mathematical basis of the OC curve with the help of an example.

Example 10.4 Suppose a double sampling plan is set as follows: $n_1 = 50, c_1 = 1, n_2 = 100, c_2 = 3$. Let $P_a(p)$ be the probability of acceptance on the combined samples and $P_a^I(p)$ and $P_a^{II}(p)$ be the probability of acceptance on the first and the second samples, respectively. Then, the following expression holds true $P_a(p) = P_a^I(p) + P_a^{II}(p)$.
   The expression for $P_a^I(p)$ is

$$P_a^I(p) = \sum_{d_1=0}^{1} \frac{50!}{d_1!(50-d_1)!} p^{d_1}(1-p)^{50-d_1}.$$

If $p = 0.05$ then $P_a^I(p) = 0.279$. Now, a second sample can be drawn only if there are two or three defectives on the first sample. Suppose we find two defectives on the first sample and one or less defectives on the second sample. The probability of this is

$$P_a(d_1 = 2, d_2 \le 1) = P_a(d_1 = 2) \cdot P_a(d_2 \le 1) = \frac{50!}{2!48!}(0.05)^2(0.95)^{48}$$

$$\times \sum_{d_2=0}^{1} \frac{100!}{d_2!(100-d_2)!}(0.05)^{d_2}(0.95)^{100-d_2}$$

$$= 0.261 \times 0.037 = 0.009$$

Suppose we find three defectives on the first sample and no defective on the second sample. The probability of this is

$$P_a(d_1 = 3, d_2 = 0) = P_a(d_1 = 3) \cdot P_a(d_2 = 0) = \frac{50!}{3!47!}(0.05)^3(0.95)^{47}$$

$$\times \frac{100!}{0!100!}(0.05)^0(0.95)^{100}$$

$$= 0.220 \times 0.0059 = 0.001$$

Then, the probability of acceptance on the second sample is

$$P_a^{II}(p) = P_a(d_1 = 2, d_2 \le 1) + P_a(d_1 = 3, d_2 = 0) = 0.009 + 0.001 = 0.010$$

The probability of acceptance of the lot is then

$$P_a(p = 0.05) = P_a^I(p = 0.05) + P_a^{II}(p = 0.05) = 0.279 + 0.010 = 0.289$$

The double sampling plan has certain advantages over the single sampling plan. The principal advantage lies in reduction of total amount of inspection. Suppose that the first sample taken under a double sampling plan is smaller in size than the sample required for a single sampling plan that offers the consumer the same protection. If the decision of accepting or rejecting a lot is taken based on the first sample, then the cost of inspection is less for the double sampling plan than the single sampling plan. Also, it is possible to reject the lot without complete inspection of the second sample. (This is known as curtailment on the second sample.) Hence, the double sampling plan often offers less inspection cost than the single sampling plan. Further, the double sampling plan offers a psychological advantage than the single sampling plan. To a layman, it appears to be unfair to reject a lot on the basis of inspection of one sample and seems to be more convincing to reject the lot based on inspection of two samples. However, there is no real advantage of double sampling plan in this regard. This is because the single and the double sampling plans can be chosen so that they have same OC curves, thus offering same risks of accepting or rejecting lots of given quality. The double sampling plan has a few disadvantages. Sometimes, if curtailment is not done on the second sample, the double sampling plan requires more inspection that the single sampling plan, though both are expected to offer same protection. Further, the double sampling plan is more difficult to administer than the single sampling plan.

10.5 Acceptance Sampling of Variables

10.5.1 Acceptance Sampling Plan

Consider a producer who supplies batches of articles having mean value μ of a variable (length, weight, strength, etc.) and standard deviation σ of the variable. The consumer has agreed that a batch will be acceptable if

$$\mu_0 - T < \mu < \mu_0 + T$$

where μ_0 denotes the critical (nominal) mean value of the variable and T indicates the tolerance for the mean value of the variable. Otherwise, the batch will be rejected. Here, the producer's risk α is the probability of rejecting a perfect batch, for which $\mu = \mu_0$. The consumer's risk β is the probability of accepting an imperfect batch, the one for which $\mu = \mu_0 \pm T$.

Consider that a large number of random samples, each of size n, are prepared from the batch. Let us assume that the probability distribution of mean $\bar{x}$ of samples, each

Fig. 10.7 Producer's risk
condition

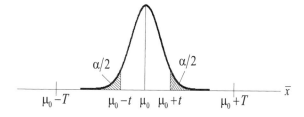

of size n, taken from a batch, is (or tends to) normal with mean μ, where $\mu = \mu_0$
and standard deviation $\frac{\sigma}{\sqrt{n}}$. Then, the batch will be accepted if

$$\mu_0 - t < \bar{x} < \mu_0 + t$$

where t denotes the tolerance for sample mean $\bar{x}$. Otherwise, the batch will be
rejected. Here, the producer's risk α is the probability of rejecting a perfect batch
and the consumer's risk β is the probability of accepting an imperfect batch.

10.5.2   *The Producer's Risk Condition*

The producer's risk α is the probability of rejecting a perfect batch. This is shown in
Fig. 10.7.

Here, $u_{\frac{\alpha}{2}} = \frac{(\mu_0 + t) - \mu_0}{\frac{\sigma}{\sqrt{n}}} = \frac{t\sqrt{n}}{\sigma}$ where $u_{\frac{\alpha}{2}}$ is the standard normal variable corresponding to a tail area $\frac{\alpha}{2}$.

10.5.3   *The Consumer's Risk Condition*

The consumer's risk β is the probability of accepting an imperfect batch. This is
shown in Fig. 10.8.

Here, $u_{\frac{\beta}{2}} = \frac{(\mu_0 - t) - (\mu_0 - T)}{\frac{\sigma}{\sqrt{n}}} = \frac{(T - t)\sqrt{n}}{\sigma}$ where $u_{\frac{\beta}{2}}$ is the standard normal variable
corresponding to a tail area $\frac{\beta}{2}$.

10.5.4   *Designing of Acceptance Sampling Plan*

To design an acceptance sampling plan, it is necessary to know n and t. They can be
calculated as follows. It is known that

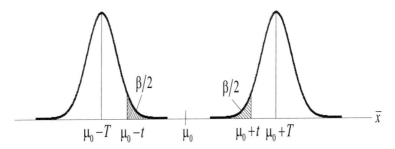

Fig. 10.8 The consumer's risk condition

$$u_{\frac{\beta}{2}} = \frac{(T - t\sqrt{n})}{\sigma} = \frac{T\sqrt{n}}{\sigma} - \frac{t\sqrt{n}}{\sigma} = \frac{T\sqrt{n}}{\sigma} - u_{\frac{\alpha}{2}}$$

Hence,

$$n = \frac{\sigma^2 (u_{\frac{\alpha}{2}} + u_{\frac{\beta}{2}})^2}{T^2}$$

$$u_{\frac{\alpha}{2}} = \frac{t\sqrt{n}}{\sigma} \Rightarrow t = \sigma \frac{u_{\frac{\alpha}{2}}}{\sqrt{n}} = \sigma \frac{u_{\frac{\alpha}{2}}}{\sqrt{\frac{\sigma^2 (u_{\frac{\alpha}{2}} + u_{\frac{\beta}{2}})^2}{T^2}}} = T \frac{u_{\frac{\alpha}{2}}}{(u_{\frac{\alpha}{2}} + u_{\frac{\beta}{2}})}$$

Thus, we get,

$$n = \frac{\sigma^2 (u_{\frac{\alpha}{2}} + u_{\frac{\beta}{2}})^2}{T^2}, \quad t = T \frac{u_{\frac{\alpha}{2}}}{(u_{\frac{\alpha}{2}} + u_{\frac{\beta}{2}})}$$

Example 10.5 A producer (spinner) supplies yarn of nominal linear density equal to be 45 tex. The customer (knitter) accepts yarn if its mean linear density lies within a range of 45 ± 1.5 tex. As the knitter cannot test all the yarns supplied by the spinner, the knitter would like to devise an acceptance sampling scheme with 10% producer's risk and 5% consumer's risk . Assume the standard deviation of count within a delivery is 1.2 tex.

Solution: Here, $\mu_{0[tex]} = 45$, $T_{[tex]} = 1.5$, $\alpha = 0.10$, $\beta = 0.05$, $\sigma_{[tex]} = 1.2$. Assume the mean linear density of yarn samples, each of size n, follows (or tends to follow) normal distribution with mean 45 tex and standard deviation 1.2 tex. Then, the standard normal variable takes the following values

$$u_{\frac{\alpha}{2}} = u_{0.05} = 1.6449, \quad u_{\frac{\beta}{2}} = u_{0.025} = 1.9600.$$

Then, $n_{[-]} = \dfrac{\sigma^2_{[tex]} (u_{\frac{\alpha}{2}[-]} + u_{\frac{\beta}{2}[-]})^2}{T^2_{[tex]}} = \dfrac{1.2^2 (1.6449 + 1.9600)^2}{(1.5)^2} = 8.3 \approx 9$

and $t_{[\text{tex}]} = T_{[\text{tex}]} \frac{u_{\frac{\alpha}{2}[-1]}}{(u_{\frac{\alpha}{2}[-1]} + u_{\frac{\beta}{2}[-1]})} = \frac{(1.5)(1.6449)}{(1.6449 + 1.9600)} = 0.68.$

Thus, the sampling scheme is as follows: Take a yarn sample of size 9, and accept the delivery if the sample mean lies in the range of 45 ± 0.68 tex, that is in between 44.32 tex and 45.68 tex; otherwise, reject the delivery.

Example 10.6 A paper filter manufacturer supplies filters of nominal grammage equal to 455 g/m². A customer accepts filters if the mean grammage lies within a range of 450–460 g/m². As the customer cannot test all the filters supplied by the manufacturer, the customer wishes to devise an acceptance sampling scheme with 5% producer's risk and 5% consumer's risk. Assume the standard deviation of grammage within a delivery is 9 g/m².

Solution: Here, $\mu = 455$ g/m², $T = 5$ g/m², $\alpha = 0.05$, and $\beta = 0.05$. Assume that the grammage of filter samples, each of size n, follows normal distribution with mean 455 g/m² and standard deviation 5 g/m². Then, the standard normal variable takes the following values $u_{\frac{\alpha}{2}} = u_{0.025} = 1.96$ and $u_{\frac{\beta}{2}} = u_{0.025} = 1.96$

Then,

$$n = \frac{\sigma^2 (u_{\frac{\alpha}{2}} + u_{\frac{\beta}{2}})^2}{T^2} = \frac{9^2 (1.96 + 1.96)^2}{5^2} = 49.79 \approx 50$$

and

$$t = T \frac{u_{\frac{\alpha}{2}}}{u_{\frac{\alpha}{2}} + u_{\frac{\beta}{2}}} = \frac{0.5 \times 1.96}{(1.96 + 1.96)} = 0.25.$$

Thus, the sampling scheme is as follows: Take a mirror sample of size 50, and accept the delivery if the sample mean lies in the range of 455 ± 0.25 g/m², that is in between 454.75 g/m² and 455.25 g/m²; otherwise, reject the delivery.

10.6 Control Charts

In a typical manufacturing industry, the input material is processed through a manufacturing process and finally converted to semifinished or finished products. This is shown in Fig. 10.1. In order to achieve the targeted quality of products, the manufacturing process is always kept under control. Whether the manufacturing process is under control or out of control can be found through a technique, called control chart.

10.6.1 Basis of Control Charts

The basis of control charts lies in checking whether the variation in the magnitude of a given characteristic of a manufactured product is arising due to random variation or assignable variation. While the random variation is known as natural variation or allowable variation and it is often small in magnitude, the assignable variation is known as nonrandom variation or preventable variation and it is often relatively high in magnitude. Examples of random variation include slight variation in temperature and relative humidity inside a manufacturing plant, slight vibration of machines, little fluctuation in voltage and current. But the causes of assignable variation include defective raw material, faulty equipment, improper handling of machines, negligence of operators, unskilled technical staff. If the variation is arising due to random variation, the process is said to be under control. But if the variation is arising due to assignable variation, then the process is said to be out of control.

10.6.2   Major Parts of a Control Chart

Figure 10.9 shows a typical control chart. This is a graphical display of a quality characteristic that has been measured or computed from a sample versus the sample number. The three major parts of a control chart are center line (CL), upper control limit (UCL), and lower control limit (LCL). The central line (CL) indicates the average value of the quality characteristic corresponding to the under-control state, desired standard, or the level of the process. The upper control limit (UCL) and lower control limit (LCL) are chosen such that if the process is under control then all the sample points will fall between them.

If m is the underlying statistic so that $E(m) = \mu_m$ and $V(m) = \sigma_m^2$, then

$$CL = \mu_m$$

$$UCL = \mu_m + k\sigma_m$$

Fig. 10.9 Outline of a control chart

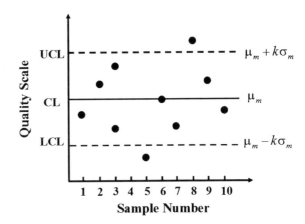

$$LCL = \mu_m - k\sigma_m$$

where E stands for expectation, V stands for variance, μ_m mean of m and σ_m indicates variance of m, and k is the "distance of the control limits from the central line", expressed in standard deviation units. The value of k was first proposed as 3 by Dr. Walter S. Shewhart, and hence, such control charts are known as Shewhart control charts.

10.6.3 Statistical Basis for Choosing k Equal to 3

Let us assume that the probability distribution of the sample statistic m is (or tends to be) normal with mean μ_m and standard deviation σ_m. Then

$$P(\mu_m - 3\sigma_m \leq m \leq \mu_m + 3\sigma_m) = 0.9973.$$

To know more about this, please refer to Chap. 2 of this book. This means the probability that a random value of m falls in between the 3σ limits is 0.9973, which is very high. On the other hand, the probability that a random value of m falls outside of the 3σ limits is 0.0027, which is very low. When the values of m fall in between the 3σ limits, the variations are attributed due to chance variation, then the process is considered to be statistically controlled. But when one or more values of m fall out of the 3σ limits, the variations are attributed to assignable variation, and the process is said to be not under statistical control.

Let us now analyze what happens when $k > 3$ or $k < 3$. In a given situation, two possibilities arise. The chance causes alone are present, or the assignable causes are also present. If the chance causes alone are present, then there are two possible courses of action, namely to accept or reject the process. Needless to say that accepting the process when the chance causes alone are present is the desired correct action, whereas rejecting the process when the chance causes are present is the undesired erroneous action. On the other hand, if the assignable causes are present then also there are two courses of action exist, that is, accept the process or reject the process. Again, needless to say that accepting the process when the assignable causes are present is the undesired erroneous action, whereas rejecting the process when the assignable causes are present is the desired correct action. This is depicted in Table 10.1. Of the two above-mentioned undesired erroneous actions, rejecting a process when chance causes are present (process is in control) is taken as Type I error, while accepting a process when assignable causes are present (process is out of control) is considered as Type II error (Chap. 5 deals with these errors in more detail). When $k > 3$, Type I error decreases but Type II error increases. When $k < 3$, Type I error increases while Type II error decreases.

Table 10.1 Possibilities and courses of action

Possibilities	Courses of action	
Presence of chance causes	Accept a process (desired correct action)	Reject a process (undesired erroneous action) (Type I error)
Presence of assignable causes	Accept a process (undesired erroneous action) (Type II error)	Reject a process (desired correct action)

10.6.4 Analysis of Control Chart

The control charts are analyzed to take a decision whether the manufacturing process is under control or out of control. The following one or more incidents indicate the process to be out of control (presence of assignable variation).

- A point falls outside any of the control limits.
- Eight consecutive points fall within 3σ limits.
- Two out of three consecutive points fall beyond 2σ limits.
- Four out of five consecutive points fall beyond 1σ limits.
- Presence of upward or downward trend.
- Presence of cyclic trend.

Such incidents are displayed in Fig. 10.10. In (a), a point falls outside the upper control limit, thus indicating the process out of control. In (b), eight consecutive points fall within a 3σ limit, though none falls beyond the 3σ limit. Such a pattern is very nonrandom in appearance and, hence, does not indicate statistical control. If the points are truly random, a more even distribution of the points above and below the central line is expected. In (c), two out of three consecutive points fall beyond a 2σ limit, though none falls beyond the 3σ limit. This arrangement of points is known as a run. Since the observations are increasing, this is called a run-up. Similarly, a sequence of decreasing points is called a run down. Runs are an important measure of nonrandom behavior of a control chart and indicates an out-of-control condition. In (d), four out of five consecutive points fall beyond a 1σ limit, though none falls beyond the 3σ limit. This arrangement of run also indicates an out-of-control condition. In (e), an upward trend of points is shown. Sometimes, a downward trend of points can also be seen. In (f), a cyclic trend is shown. Such upward or downward or cyclic trend has a very low probability of occurrence in a random sample of points. Hence, such trends are often taken as a signal of an out-of-control condition.

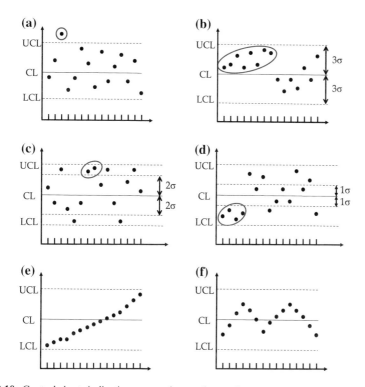

Fig. 10.10 Control charts indicating process is out of control

10.7 Types of Shewhart Control Charts

There are two types of Shewhart control charts often used in practice. They are Shewhart control charts for variables and Shewhart control chart for attributes. The former type includes mean chart (x-bar chart), range chart (R chart), and standard deviation chart (s chart). The latter type includes control chart for fraction defective (p chart), control chart for number of defectives (np chart), and control chart for number of defects per unit (c chart). This is shown in Fig. 10.11.

10.7.1 The Mean Chart

The mean chart (x-bar chart) is constructed to examine whether the process mean is under control or not. This is constructed as follows.

Let x_{ij}, $j = 1, 2, \ldots, n$ be the measurements on ith sample ($i = 1, 2, \ldots, k$). The mean $\bar{x}_i$, range R_i, and standard deviation s_i for ith sample are given by

Fig. 10.11 Types of
Shewhart control charts

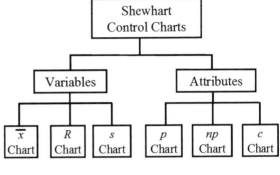

$$\bar{x}_i = \frac{1}{n}\sum_{j=1}^{n} x_{ij}, \quad R_i = \max_{j=1}^{n}(x_{ij}) - \min_{j=1}^{n}(x_{ij}), \quad s_i = \sqrt{\sum_{j=1}^{n}\frac{(x_{ij}-\bar{x}_i)^2}{n}}$$

Then, the mean $\bar{\bar{x}}$ of sample means, the mean $\bar{R}$ of sample ranges, and the mean $\bar{s}$ of
sample standard deviations are given by

$$\bar{\bar{x}} = \frac{1}{k}\sum_{j=1}^{n} \bar{x}_j, \quad \bar{R} = \frac{1}{k}\sum_{j=1}^{n} \bar{R}_j, \quad \bar{s} = \frac{1}{k}\sum_{j=1}^{n} \bar{s}_j$$

Let the mean and standard deviation of the population from which samples are taken
be μ and σ, respectively. Then, the control limits for the ith sample mean $\bar{x}_i$ are
given as

$$\mathrm{CL} = E(\bar{x}_i) = \mu$$

$$\mathrm{UCL} = E(\bar{x}_i) + 3\sqrt{Var(\bar{x}_i)} = \mu + \left(\frac{3}{\sqrt{n}}\right)\sigma = \mu + A\sigma$$

$$\mathrm{LCL} = E(\bar{x}_i) - 3\sqrt{Var(\bar{x}_i)} = \mu - \left(\frac{3}{\sqrt{n}}\right)\sigma = \mu - A\sigma.$$

The values for A for different sample sizes are given in Table 10.2.

When the mean μ and standard deviation σ are not known, the control limits are
given as

$$\mathrm{CL} = \bar{\bar{x}}$$

$$\mathrm{UCL} = \begin{cases} \bar{\bar{x}} + \left(\frac{3}{c_2\sqrt{n}}\right)\bar{s} = \bar{\bar{x}} + A_1\bar{s} \\ \bar{\bar{x}} + \left(\frac{3}{c_2\sqrt{n}}\bar{s}\right) = \bar{\bar{x}} + A_1\bar{s} \end{cases}$$

$$\mathrm{LCL} = \begin{cases} \bar{\bar{x}} - \left(\frac{3}{c_2\sqrt{n}}\right)\bar{s} = \bar{\bar{x}} - A_1\bar{s} \\ \bar{\bar{x}} - \left(\frac{3}{c_2\sqrt{n}}\bar{s}\right) = \bar{\bar{x}} - A_1\bar{s} \end{cases}$$

Table 10.2 Factors for calculation of control charts

Run	Factors			Factors					Factors				
n	A	A_1	A_2	c_2	B_1	B_2	B_3	B_4	d_2	D_1	D_2	D_3	D_4
2	2.121	3.76	1.886	0.5642	0	1.843	0	3.297	1.128	0	3.686	0	3.267
3	1.232	2.394	1.023	0.7236	0	1.858	0	2.568	1.693	0	4.358	0	2.575
4	1.5	1.88	0.729	0.7979	0	1.808	0	2.266	2.059	0	4.698	0	2.282
5	1.342	1.596	0.577	0.8407	0	1.756	0	2.089	2.326	0	4.918	0	2.115
6	1.225	1.41	0.483	0.8686	0.026	1.711	0.03	1.97	2.534	0	5.078	0	2.004
7	1.134	1.277	0.419	0.8882	0.105	1.672	0.118	1.882	2.704	0.205	5.203	0.076	1.924
8	1.061	1.175	0.373	0.9027	0.167	1.638	0.185	1.815	2.847	0.387	5.307	0.136	1.864
9	1	1.094	0.337	0.9139	0.219	1.509	0.239	1.761	2.97	0.546	5.394	0.184	1.816
10	0.949	1.028	0.308	0.9227	0.262	1.584	0.284	1.716	3.078	0.687	5.469	0.223	1.777
11	0.905	0.973	0.285	0.93	0.299	1.561	0.321	1.679	3.173	0.812	5.534	0.256	1.744
12	0.866	0.925	0.266	0.9359	0.331	1.541	0.354	1.646	3.258	0.924	5.592	0.284	1.716
13	0.832	0.884	0.249	0.941	0.359	1.523	0.382	1.618	3.336	1.026	5.646	0.308	1.692
14	0.802	0.848	0.235	0.9453	0.384	1.507	0.406	1.594	3.407	1.121	5.693	0.329	1.671
15	0.775	0.816	0.223	0.9499	0.406	1.492	0.428	1.572	3.472	1.207	5.737	0.348	1.652
16	0.759	0.788	0.212	0.9523	0.427	1.478	0.448	1.552	3.532	1.285	5.779	0.364	1.636
17	0.728	0.762	0.203	0.9951	0.445	1.465	0.466	1.534	3.588	1.359	5.817	0.379	1.621
18	0.707	0.738	0.194	0.9576	0.461	1.454	0.482	1.518	3.64	1.426	5.854	0.392	1.608
19	0.688	0.717	0.187	0.9599	0.477	1.443	0.497	1.503	3.689	1.49	5.888	0.404	1.596
20	0.671	0.697	0.18	0.9619	0.491	1.433	0.51	1.499	3.735	1.548	5.922	0.414	1.586
21	0.655	0.679	0.173	0.9638	0.504	1.424	0.523	1.477	3.778	1.606	5.95	0.425	1.575
22	0.64	0.662	0.167	0.9655	0.516	1.415	0.534	1.466	3.819	1.659	5.979	0.434	1.566
23	0.626	0.647	0.162	0.967	0.527	1.407	0.545	1.455	3.858	1.71	6.006	0.443	1.557
24	0.612	0.632	0.157	0.9684	0.538	1.399	0.555	1.445	3.895	1.759	6.031	0.452	1.548
25	0.6	0.61	0.153	0.9696	0.548	1.392	0.565	1.435	3.931	1.804	6.058	0.459	1.541

The values for A_1 and A_2 for different sample sizes are given in Table 10.2.

10.7.2 The Range Chart

The range chart (R chart) is constructed to examine whether the process variation is under control or out of control. This is constructed as follows.

Let x_{ij}, $j = 1, 2, \ldots, n$ be the measurements on ith sample ($i = 1, 2, \ldots, k$). The range R_i for ith sample is given by

$$R_i = \max_{j=1}^{j=n}(x_{ij}) - \min_{j=1}^{j=n}(x_{ij}).$$

Then, the mean $\overline{R}$ of sample ranges is given by

$$\overline{R} = \frac{1}{k} \sum_{i=1}^{k} R_i.$$

When the standard deviation σ of the population from which samples are taken is known, the control limits for R_i are specified as:

$$\text{CL} = E(R_i) = d_2\sigma$$
$$\text{UCL} = E(R_i) + 3\sqrt{Var(R_i)} = d_2\sigma + 3d_3\sigma = (d_2 + 3d_3)\sigma = D_2\sigma$$
$$\text{LCL} = E(R_i) - 3\sqrt{Var(R_i)} = d_2\sigma - 3d_3\sigma = (d_2 - 3d_3)\sigma = D_1\sigma$$

The values for D_1 and D_2 for different sample sizes are given in Table 10.2. When the standard deviation σ of the population is not known, the corresponding control limits are:

$$\text{CL} = E(R_i) = \overline{R}$$
$$\text{UCL} = E(R_i) + 3\sqrt{Var(R_i)} = \overline{R} + \left(\frac{3d_3}{d_2}\right)\overline{R} = \left(1 + \frac{3d_3}{d_2}\right)\overline{R} = D_4\overline{R}$$
$$\text{LCL} = E(R_i) - 3\sqrt{Var(R_i)} = \overline{R} - \left(\frac{3d_3}{d_2}\right)\overline{R} = \left(1 - \frac{3d_3}{d_2}\right)\overline{R} = D_3\overline{R}.$$

The values of D_3 and D_4 for different sample sizes are given in Table 10.2.

10.7.3   The Standard Deviation Chart (s-Chart)

The standard deviation chart (s chart) is constructed to examine whether the process variation is under control or out of control. Let x_{ij}, $j = 1, 2, \ldots, n$ be the measurements on ith sample ($i = 1, 2, \ldots, k$). The standard deviation s_i for the ith sample is given by

$$s_i = \sqrt{\sum_{j=1}^{n} \frac{(x_{ij} - \overline{x}_i)^2}{n}}$$

Then, the mean $\overline{s}$ of sample standard deviations is given by

$$\overline{s} = \frac{1}{k} \sum_{i=1}^{k} s_i$$

Let us now decide the control limits for s_i . When the standard deviation σ of the population from which samples are taken is known, then

Table 10.3 Yarn strength data

Sample no.	Yarn strength (cN/tex)									
1	14.11	13.09	12.52	13.4	13.94	13.4	12.72	11.09	13.28	12.34
2	14.99	17.97	15.76	13.56	13.31	14.03	16.01	17.71	15.67	16.69
3	15.08	14.41	11.87	13.62	14.84	15.44	13.78	13.84	14.99	13.99
4	13.14	12.35	14.08	13.4	13.45	13.44	12.9	14.08	14.71	13.11
5	13.21	13.69	13.25	14.05	15.58	14.82	14.31	14.92	10.57	15.16
6	15.79	15.58	14.67	13.62	15.9	14.43	14.53	13.81	14.92	12.23
7	13.78	13.9	15.1	15.26	13.17	13.67	14.99	13.39	14.84	14.15
8	15.65	16.38	15.1	14.67	16.53	15.42	15.44	17.09	15.68	15.44
9	15.47	15.36	14.38	14.08	14.08	14.84	14.08	14.62	15.05	13.89
10	14.41	15.21	14.04	13.44	15.85	14.18	15.44	14.94	14.84	16.19

$$CL = E(s_i) = c_2\sigma$$
$$UCL = E(s_i) + 3\sqrt{Var(s_i)} = c_2\sigma + 3c_3\sigma = (c_2 + 3c_3)\sigma = B_2\sigma$$
$$LCL = E(s_i) - 3\sqrt{Var(s_i)}) = c_2\sigma - 3c_3\sigma = (c_2 - 3c_3)\sigma = B_1\sigma.$$

The values for B_1 and B_2 for different sample sizes are given in Table 10.2. When the standard deviation σ of the population is not known, then

$$CL = E(s_i) = \sqrt{s}$$
$$UCL = E(s_i) + 3\sqrt{Var(s_i)} = \sqrt{s} + 3\frac{c_3}{c_2}\sqrt{s} = \left(1 + 3\frac{c_3}{c_2}\right)\sqrt{s} = B_4\sqrt{s}$$
$$LCL = E(s_i) - 3\sqrt{Var(s_i)} = \sqrt{s} - 3\frac{c_3}{c_2}\sqrt{s} = \left(1 - 3\frac{c_3}{c_2}\right)\sqrt{s} = B_3\sqrt{s}.$$

The values for B_3 and B_4 for different sample sizes are given in Table 10.2.

Example 10.7 Table 10.3 displays the experimental data of yarn strength. It is of interest to know whether the yarn manufacturing process was under control or out of control.

Solution: Let us first calculate the mean, range, and standard deviation of the samples of yarn strength. They are shown in Table 10.4.
The mean of sample means is calculated as follows.

$$\overline{\overline{x}}_{cN.tex^{-1}} = 14.41.$$

The mean of sample ranges is calculated as follows.

$$\overline{R}_{cN.tex^{-1}} = 3.11.$$

Table 10.4 Basic statistical characteristics of yarn strength

Sample no. (i)	$\overline{x}_{i[\text{cN.tex}^{-1}]}$	$R_{i[\text{cN.tex}^{-1}]}$	$S_{i[\text{cN.tex}^{-1}]}$
1	12.99	3.02	0.83
2	15.57	4.66	1.54
3	14.19	3.57	0.97
4	13.47	2.36	0.64
5	13.96	5.01	1.36
6	14.55	3.67	1.07
7	14.23	2.09	0.72
8	15.74	2.42	0.69
9	14.59	1.58	0.54
10	14.85	2.75	0.81
Average	14.41	3.11	0.92

The mean of sample standard deviations is calculated as follows.

$$\overline{s}_{\text{cN.tex}^{-1}} = 0.92.$$

Now, the control limits for mean chart are computed as follows. The value of A2 can be obtained from Table 10.2.

$$\text{CL} = \overline{\overline{x}} = 14.41 \text{ cN.tex}^{-1}$$
$$\text{UCL} = \overline{\overline{x}} + A_2\overline{R} = 14.11 + (0.308 \times 3.11) \text{ cN.tex}^{-1} = 15.37 \text{ cN.tex}^{-1}$$
$$\text{LCL} = \overline{\overline{x}} - A_2\overline{R} = 14.11 - (0.308 \times 3.11) \text{ cN.tex}^{-1} = 13.45 \text{ cN.tex}^{-1}.$$

Figure 10.12 displays the mean chart for yarn strength. As shown, there are three points that fall beyond the upper and lower control limits. It is therefore concluded that the process average is out of control.

Now, the control limits for range chart are computed as follows. The values of D_3 and D_4 are obtained from Table 10.2.

$$\text{CL} = \overline{R} = 3.11 \text{ cN.tex}^{-1}$$
$$\text{UCL} = D_4\overline{R} = 1.777 \times 3.11 \text{ cN.tex}^{-1} = 5.5 \text{ cN.tex}^{-1}$$
$$\text{LCL} = D_3\overline{R} = 0.223 \times 3.11 \text{ cN.tex}^{-1} = 0.69 \text{ cN.tex}^{-1}.$$

Figure 10.13 displays the range chart for yarn strength. There is no indication that the range of yarn strength is out of control. It is therefore concluded that the process variation is under control.

Now, the control limits for standard deviation chart are computed as follows. The values of B_3 and B_4 are obtained from Table 10.2.

Fig. 10.12 Mean chart for
yarn strength

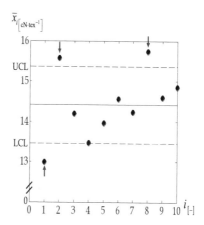

Fig. 10.13 Range chart for
yarn strength

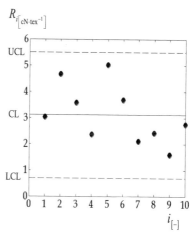

$$CL = \bar{s} = 0.92 \text{ cN.tex}^{-1}$$
$$UCL = B_4\bar{s} = 1.716 \times 0.92 \text{ cN.tex}^{-1} = 1.58 \text{ cN.tex}^{-1}$$
$$LCL = B_3\bar{s} = 0.284 \times 0.92 \text{ cN.tex}^{-1} = 0.26 \text{ cN.tex}^{-1}.$$

Figure 10.14 displays the standard deviation chart for yarn strength. As shown, there
is no indication that the standard deviation of yarn strength is out of control. It is
once again confirmed that the process variation is out of control. It can be overall
concluded that although the process variability is in control, the process cannot be
regarded to be in statistical control since the process average is out of control.

Fig. 10.14 Standard
deviation chart for yarn
strength

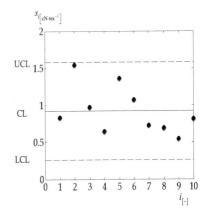

10.8   Process Capability Analysis

When the process is operating under control, we are often required to obtain some information about the performance or capability of the process, that is, whether a controlled process is capable of meeting the specifications. The capability of a statistically controlled process is measured by process capability ratios. One of such ratios is called C_p which is calculated when the process is running at center; that is, the process is centered at nominal dimension. This is shown in Fig. 10.15. C_p is defined as follows

$$C_p = \frac{\text{USL} - \text{LSL}}{6\sigma}$$

where USL and LSL stand for upper specification limit and lower specification limit, respectively, and σ refers to the process standard deviation. $100(1/C_p)$ is interpreted as the percentage of the specifications width used by the process.

Example 10.8 Suppose that the yarn manufacturing process is under control and the specifications of yarn strength are given as 14.50 ± 4 cN.tex^{-1}. As the process standard deviation σ is not given, we need to estimate this as follows

$$\hat{\sigma} = \frac{\overline{R}}{d_2} = \frac{3.11}{3.078} = 1.0104.$$

It is assumed that the sample size is 10 and the yarn strength follows normal distribution with mean at 14.50 cN.tex^{-1} and standard deviation at 1.0104 cN.tex^{-1}. The process capability ratio is calculated as follows.

$$C_p = \frac{18.5 - 10.5}{6 \times 1.0104} = 1.3196$$

That is, 75.78% of the specifications' width is used by the process.

Fig. 10.15 Display of
process centered at nominal
dimension

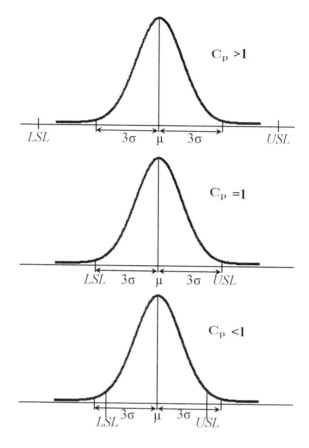

Sometimes, one-sided specification of a process is given; that is, either the upper
specification limit (USL) or the lower specification limit (LSL) is given, but not the
both are given. Then, we speak about two process capability ratios C_{pu} and C_{pl}. The
former is computed when the upper specification limit is known, and the latter is
calculated when the lower specification limit is given. They are defined as follows

$$C_{pu} = \frac{USL - \mu}{3\sigma}$$

$$C_{pl} = \frac{\mu - LSL}{3\sigma}$$

Example 10.9 Suppose the lower specification of yarn strength is given as 10.50
cN.tex^{-1}. The process mean μ and process standard deviation σ were earlier estimated as 14.50 cN.tex^{-1} and 1.0104 cN.tex^{-1}, respectively. Then, the process capability ratio can be computed as follows

Fig. 10.16 Display of
process running off-center

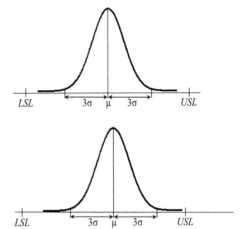

$$C_{pl} = \frac{\mu - \text{LSL}}{3\sigma} = \frac{14.50 - 10.50}{3 \times 1.0104} = 1.32$$

We observed that C_p measures the capability of a centered process. But all processes
are not necessarily always centered at the nominal dimension. That is, processes may
also run off-center (Fig. 10.16), so that the actual capability of noncentered processes
will be less than that indicated by C_p. In the case when the process is running off-
center, the capability of a process is measured by another ratio called C_{pk}. This is
defined below

$$C_{pk} = \min\left[\frac{\text{USL} - \mu}{3\sigma}, \frac{\mu - \text{LSL}}{3\sigma}\right]$$

It is often said that C_{pk} measures actual capability and C_p measures potential capability.

Example 10.10 Suppose the specifications of yarn strength are given as $14 \pm$
4 cN.tex^{-1}.

We assume that the yarn strength follows normal distribution with mean at
$14.50 \text{ cN.tex}^{-1}$ and standard deviation at $1.0104 \text{ cN.tex}^{-1}$. Clearly, the process is
running off-center.

$$C_{pk} = \min\left(\frac{18.5 - 14.5}{3 \times 1.0104}, \frac{14.5 - 10.0}{3 \times 1.0104}\right) = \min(1.1547, 1.4846) = 1.1547$$

But there is a serious problem associated with C_{pk}. Let us illustrate this with the
help of an example. Suppose there are two processes—Process A and Process B.
Process A has a mean at 50 and standard deviation at 5. Process B has a mean at
57.5 and standard deviation at 2.5. The target is given as 50, and the upper and lower
specifications limits are set at 65 and 35, respectively. Then, it can be computed

Table 10.5 Basic statistical characteristics of processes C and D

Process	C	D
Mean	100	110
Standard deviation	3	1

Table 10.6 Process capability ratios of processes C and D

Process capability ratio	C	D
C_p	1.11	3.33
C_{pk}	1.11	0
C_{pm}	1.11	3.33

that Process A has $C_p = C_{pk} = 1$. But Process B has different capability ratios: $C_p = 2$ and $C_{pk} = 1$. It is very surprising to notice that both processes have the same $C_{pk} = 1$; however, Process A is running at center, but Process B is running off-center. This example suggests us that C_{pk} is not an adequate measure of process centering. There is another process capability ratio C_{pm}, which is preferred over C_{pk} for computation of process capability ratio when the process is running off-center. C_{pm} is defined as follows

$$C_{pm} = \frac{\text{USL} - \text{LSL}}{6\tau}$$

where τ is the square root of expected squared deviation from the target $T = \frac{1}{2}(\text{USL} + \text{LSL})$. It can be found that $\tau = 6\sqrt{\sigma^2 + (\mu - T)^2}$. Then, C_{pm} can be expressed as

$$C_{pm} = \frac{\text{USL} - \text{LSL}}{6\sigma\sqrt{1 + (\frac{\mu - T}{\sigma})^2}} = \frac{C_p}{\sqrt{1 + (\frac{\mu - T}{\sigma})^2}}.$$

Based on the above example of two processes A and B, it can be calculated that C_{pm} for Process A is 1 and the same for Process B is 0.63. Clearly, Process B is utilizing the specification width more than Process A. Hence, Process A is preferred to Process B.

Example 10.11 The basic statistical characteristics of two processes (C and D) are given in Table 10.5.

Sample size is 5, and specifications are given at 100 ± 10. The earlier stated process capability ratios are calculated and reported in Table 10.6.

As Process C is utilizing the specification width more than Process D, the former is preferred to the latter.

Worldwide, a few guidelines are available for process capability ratios. They are as follows: (1) $C_p = 1.33$ as a minimum acceptable target for many US companies, (2)

$C_p = 1.66$ as a minimum target for strength, safety, and other critical characteristics for many US companies, and (3) $C_{pk} = 2$ for many internal processes and also for many suppliers.

10.9 Control Chart for Fraction Defectives

The fraction defective is defined as the ratio of the number of defectives in a population to the total number of items in the population. Suppose the production process is operating in a stable manner such that the probability that any item produced will not conform to specifications is p and that successive items produced are independent. Then, each item produced is a realization of a Bernoulli random variable with parameter p. If a random sample of n items of product is selected and if D is the number of items of product that are defectives, then D has a binomial distribution with parameter n and p; that is

$$P(D = x) = {}^nC_x p^x (1 - p)^{n-x}, \quad x = 0, 1, \ldots, n.$$

The mean and variance of the random variable D are np and $np(1 - p)$, respectively. The sample fraction defective is defined as the ratio of the number of defective items in the sample of size n; that is $p' = \frac{D}{n}$. The distribution of the random variable p' can be obtained from the binomial distribution. The mean and variance of p' are p and $\frac{p(1-p)}{n}$, respectively.

When the mean fraction of defectives p of the population from which samples are taken is known, then it follows as mentioned below.

$$\mathrm{CL} = p, \quad \mathrm{UCL} = p + 3\sqrt{\frac{p(1 - p)}{n}}, \quad \mathrm{LCL} = p - 3\sqrt{\frac{p(1 - p)}{n}}$$

When the mean fraction of defectives p of the population is not known then we find out the control limits in the following manner.

Let us select m samples, each of size n. If there are D_i defective items in ith sample, then the fraction defectives in the ith sample are $p_i' = \frac{D_i}{n}$, $i = 1, 2, \ldots, m$. The average of these individual sample fraction defectives is

$$\overline{p}' = \frac{\displaystyle\sum_{i=1}^{m} D_i}{mn} = \frac{\displaystyle\sum_{i=1}^{m} p_i'}{m}$$

$$\mathrm{CL} = \overline{p}', \quad \mathrm{UCL} = \overline{p}' + 3\sqrt{\frac{\overline{p}'(1 - \overline{p}')}{n}}, \quad \mathrm{LCL} = \overline{p}' - 3\sqrt{\frac{\overline{p}'(1 - \overline{p}')}{n}}$$

10.10 Control Chart for the Number of Defectives

It is also possible to base a control chart on the number of defectives rather than the fraction defectives. When the mean number of defectives np of the population from which samples are taken is known, then

$$CL = np, \quad UCL = np + 3\sqrt{np(1 - p)}, \quad LCL = p - 3\sqrt{np(1 - p)}$$

When the mean number of defectives np of the population is not known, then

$$CL = n\overline{p}', \quad UCL = n\overline{p}' + 3\sqrt{n\overline{p}'(1 - \overline{p}')}, \quad LCL = \overline{p}' - 3\sqrt{n\overline{p}'(1 - \overline{p}')}$$

Example 10.12 Table 10.7 refers to the number of defective knitwears in samples of size 180.

Here, $n = 180$ and $\overline{p}' = \frac{423}{30 \times 180} = 0.0783$. The control limits are

$$CL = n\overline{p}' = 14.09, \quad UCL = n\overline{p}' + 3\sqrt{n\overline{p}'(1 - \overline{p}')} = 24.9, \quad LCL = \overline{p}' - 3\sqrt{n\overline{p}'(1 - \overline{p}')} = 3.28$$

Figure 10.17 displays the control chart for number of defectives. It can be seen that the knitwear manufacturing process is out of control.

Table 10.7 Data on defective knitwears

Sample no.	No. of defectives	Sample no.	No. of defectives	Sample no.	No. of defectives
1	5	11	36	21	24
2	8	12	24	22	17
3	10	13	19	23	12
4	12	14	13	24	8
5	12	15	5	25	17
6	29	16	2	26	19
7	25	17	11	27	4
8	13	18	8	28	9
9	9	19	15	29	5
10	20	20	20	30	12

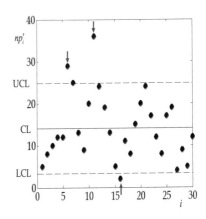

Fig. 10.17 Control chart for
the number of defective
knitwears

10.11   Control Chart for The Number of Defects

Consider the occurrence of defects in an inspection unit of product(s).[1] Suppose that
defects occur in this inspection unit according to Poisson distribution; that is

$$P(x) = \frac{e^{-c}c^x}{x!}, \quad x = 0, 1, 2, \ldots$$

where x is the number of defects and c is known as mean and/or variance of the
Poisson distribution.

When the mean number of defects c in the population from which samples are
taken is known, the control limits for the number of defects per unit of product are

$$CL = c, \quad UCL = c + 3\sqrt{c}, \quad LCL = c - 3\sqrt{c}$$

Note: If this calculation yields a negative value of LCL, then set LCL = 0.

When the mean number of defects c in the population is not known, then it follows
as stated below. Let us select n samples. If there are c_i defects in ith sample, then
the average of these defects in samples of size n is

$$\bar{c}' = \frac{\sum\limits_{i=1}^{n} \bar{c}'_i}{n}, \quad CL = \bar{c}', \quad UCL = \bar{c}' + 3\sqrt{\bar{c}'}, \quad LCL = \bar{c}' - 3\sqrt{\bar{c}'}$$

Note that if this calculation yields a negative value of LCL, then LCL can be set as
follows: LCL = 0.

Example 10.13 The following data set in Table 10.8 refers to the number of holes
(defects) in knitwears.

---

[1] It can be a group of 10 units of products or 50 units of products.

Table 10.8 Data on number of defects in knitwears

Sample no.	No. of holes	Sample no.	No. of holes	Sample no.	No. of holes
1	4	11	3	21	2
2	6	12	7	22	1
3	3	13	9	23	7
4	8	14	6	24	6
5	12	15	10	25	5
6	9	16	11	26	9
7	7	17	7	27	11
8	2	18	8	28	8
9	11	19	9	29	3
10	8	20	3	30	2

Fig. 10.18 Control chart for number of defects in knitwears

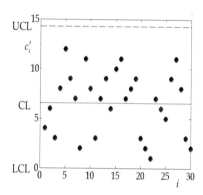

Consider c_i' denote the number of holes in ith sample. The control i limits are calculated as follows.

$$CL = n\bar{c}' = 6.57, \quad UCL = n\bar{c}' + 3\sqrt{n\bar{c}'(1 - \bar{c}')} = 14.26, \quad LCL = \bar{c}' - 3\sqrt{n\bar{c}'(1 - \bar{c}')} = -1.12.$$

Figure 10.18 displays the control chart for number of defects. It can be seen that the knitwear manufacturing process is under control.

10.12 CUSUM Control Chart

The Shewhart control charts discussed earlier have a serious disadvantage. They are known to be relatively insensitive to the small shifts in the process mean, say in the order of about 1.5σ or less. In such cases, a very effective alternative is cumulative sum control chart, or in short, CUSUM control chart.

Let us now obtain the CUSUM control chart for the same data as reported in Table 10.9. The step-by-step procedure to obtain CUSUM control chart is given below.

The cumulative sum (CUSUM) of observations is defined as

$$
C_i = \begin{cases} \sum_{j=1}^{i}(x_j - \mu) = (x_i - \mu) + \sum_{j=1}^{i-1}(x_j - \mu) = (x_i - \mu) + C_{i-1}; & \text{when } i \geq 1 \\ C_0 = 0; & \text{when } i = 0 \end{cases}
$$

When the process remains in control with mean μ, the cumulative sum is a random walk with mean zero. When the mean shifts upward to a value μ_0 such that $\mu > \mu_0$, then an upward or positive drift will be developed in the cumulative sum. When the mean shifts downward with a value μ_0 such that $\mu < \mu_0$, then a downward or negative drift will be developed in the CUSUM.

There are two ways to represent CUSUM, the tabular (algorithmic) CUSUM and V-mask. Of the two, the tabular CUSUM is preferable. We will now present the construction and use of tabular CUSUM.

The tabular CUSUM works by accumulating deviations from μ (the target value) that are above the target with one statistic C^+ and accumulating deviations from μ (the target value) that are below the target with another statistic C^-. These statistics are called as upper CUSUM and lower CUSUM, respectively.

$$
\text{Upper CUSUM: } C_i^+ = \begin{cases} \sum_{j=1}^{i} \max[0, x_i - (\mu + K) + C_{i-1}^+]; & \text{when } i \geq 1 \\ C_0^+ = 0; & \text{when } i = 0 \end{cases}
$$

$$
\text{Lower CUSUM: } C_i^- = \begin{cases} \sum_{j=1}^{i} \max[0, x_i - (\mu + K) + C_{i-1}^-]; & \text{when } i \geq 1 \\ C_0^- = 0; & \text{when } i = 0 \end{cases}
$$

where K is called as reference value or the allowance.

If the shift δ in the process mean value is expressed as

$$
\delta = \frac{|\mu_1 - \mu|}{\sigma}
$$

where μ_1 denotes the new process mean value and μ and σ indicate the old process mean value and the old process standard deviation, respectively, then K is the one-half of the magnitude of shift.

$$
K = \frac{\delta}{2}\sigma = \frac{|\mu_1 - \mu|}{2}.
$$

Fig. 10.19 Shewhart mean
control chart for yarn
strength data of Table 10.9

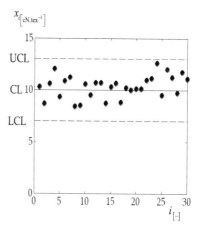

If either C_i^+ or C_i^- exceeds a chosen decision interval H, the process is considered
to be out of control. A reasonable value for H is five times the process standard
deviation, $H = 5\sigma$.

Although we have discussed the development of CUSUM chart for individual
observations ($n = 1$), it is easily extended to the case of averages of samples where
($n > 1$). In such a case, simply replace x_i by $\bar{x}_i$ and σ by $\frac{\sigma}{\sqrt{n}}$.

We have concentrated on CUSUMs for sample averages. However, it is possible
to develop CUSUMs for other sample statistics such as ranges, standard deviations,
number of defectives, proportion of defectives, and number of defects.

Example 10.14 Consider the yarn strength (cN.tex^{-1}) data as shown in Table 10.9.

Let us, for curiosity, obtain the Shewhart mean control chart of the data reported in
Table 10.9. The control limits are calculated as follows.

$$\text{CL} = \mu_{\left[\text{cN.tex}^{-1}\right]} = 10$$

$$\text{UCL} = \mu_{\left[\text{cN.tex}^{-1}\right]} + 3\sigma_{\left[\text{cN.tex}^{-1}\right]} = 10 + (3 \times 1) = 13$$

$$\text{LCL} = \mu_{\left[\text{cN.tex}^{-1}\right]} - 3\sigma_{\left[\text{cN.tex}^{-1}\right]} = 10 - (3 \times 1) = 7.$$

The Shewhart control chart is plotted in Fig. 10.19. It can be observed that the process
mean is under control.

The definition of CUSUM is followed to obtain Table 10.10. Figure 10.20 plots
CUSUM for different samples. It can be seen that the CUSUM increases rapidly after
sample number 20. This information was however not obtained from the Shewhart
control chart.

Let us illustrate how tabular CUSUM is obtained. Table 10.11 reports on the
procedure for obtaining the upper CUSUM. Here, the counter N^+ records the number

Table 10.9 Yarn strength data for CUSUM control chart

Sample no. i	Strength x_i (cN.tex^{-1})	Sample no. i	Strength x_i (cN.tex^{-1})
1	10.29	16	10.65
2	8.66	17	8.8
3	10.61	18	10.2
4	12.03	19	10
5	9.31	20	10.1
6	10.86	21	10.1
7	11.2	22	10.95
8	8.4	23	11.1
9	8.5	24	12.6
10	10.55	25	9.5
11	9.5	26	12
12	10.69	27	11.2
13	10.7	28	9.7
14	8.7	29	11.75
15	10.29	30	11.07

Table 10.10 Calculations of CUSUM

i	x_i (cN.tex^{-1})	C_i (cN.tex^{-1})	i	x_i (cN.tex^{-1})	C_i (cN.tex^{-1})
1	10.29	0.29	16	10.65	0.94
2	8.66	−1.05	17	8.8	−0.26
3	10.61	−0.43	18	10.2	−0.06
4	12.03	1.6	19	10	−0.06
5	9.31	0.91	20	10.1	0.04
6	10.86	1.76	21	10.1	0.14
7	11.2	2.96	22	10.95	1.09
8	8.4	1.36	23	11.1	2.19
9	8.5	−0.14	24	12.6	4.79
10	10.55	0.41	25	9.5	4.29
11	9.5	−0.09	26	12	6.29
12	10.69	0.6	27	11.2	7.49
13	10.7	1.3	28	9.7	7.19
14	8.7	0	29	11.75	8.94
15	10.29	0.29	30	11.07	10.01

Fig. 10.20 Plots of CUSUM
for different samples

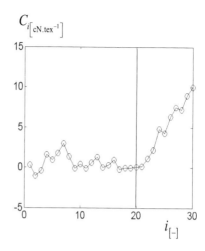

Table 10.11 Calculations of upper CUSUM

$i_{[-]}$	$x_{i(cN.tex^{-1})}$	$C^+_{i(cN.tex^{-1})}$	$N^+_{[-]}$	$i_{[-]}$	$x_{i(cN.tex^{-1})}$	$C^+_{i(cN.tex^{-1})}$	$N^+_{[-]}$
1	10.29	0	0	16	10.65	0.15	1
2	8.66	0	0	17	8.8	0	0
3	10.61	0.11	1	18	10.2	0	0
4	12.03	1.64	2	19	10	0	0
5	9.31	0.45	3	20	10.1	0	0
6	10.86	0.81	4	21	10.1	0	0
7	11.2	1.51	5	22	10.95	0.45	1
8	8.4	0	0	23	11.1	1.05	2
9	8.5	0	0	24	12.6	3.15	3
10	10.55	0.05	1	25	9.5	2.15	4
11	9.5	0	0	26	12	3.65	5
12	10.69	0.19	1	27	11.2	4.35	6
13	10.7	0.39	1	28	9.7	3.55	7
14	8.7	0	0	29	11.75	4.8	8
15	10.29	0	0	30	11.07	5.37	9

of successive points since C^+_i rose above the value of zero. Table 10.12 reports on the procedure for obtaining the lower CUSUM. Here, the counter N_- records the number of successive points since C^-_i rose above the value of zero.

Figure 10.21 plots the CUSUM status chart. It can be observed that the process mean is out of control. Although we have discussed the development of CUSUM chart for individual observations ($n = 1$), it is easily extended to the case of averages of samples where ($n > 1$). Simply replace x_i by $\bar{x}_i$ and σ by $\frac{\sigma}{\sqrt{n}}$. We have concentrated on CUSUMs for sample averages; however, it is possible to develop CUSUMs for

Table 10.12 Calculations of lower CUSUM

$i_{[-]}$	$x_{i(\text{cN.tex}^{-1})}$	$C^+_{i(\text{cN.tex}^{-1})}$	$N^+_{[-]}$	$i_{[-]}$	$x_{i(\text{cN.tex}^{-1})}$	$C^+_{i(\text{cN.tex}^{-1})}$	$N^+_{[-]}$
1	10.29	0	0	16	10.65	0	0
2	8.66	0.84	1	17	8.8	0.7	1
3	10.61	0	0	18	10.2	0	0
4	12.03	0	0	19	10	0	0
5	9.31	0.19	1	20	10.1	0	0
6	10.86	0	0	21	10.1	0	0
7	11.2	0	0	22	10.95	0	0
8	8.4	1.1	1	23	11.1	0	0
9	8.5	2.1	2	24	12.6	0	0
10	10.55	1.05	3	25	9.5	0	0
11	9.5	1.05	4	26	12	0	0
12	10.69	0	0	27	11.2	0	0
13	10.7	0	0	28	9.7	0	0
14	8.7	0.8	1	29	11.75	0	0
15	10.29	0.01	2	30	11.07	0	0

Fig. 10.21 Plot of CUSUM status chart

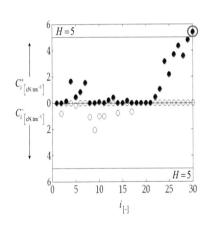

other sample statistics such as ranges, standard deviations, number of defectives, proportion of defectives, and number of defects.

10.13 Exponentially Weighted Moving Average Control Chart

Exponentially weighted moving average (or, in short, EWMA) control chart is another good alternative to Shewhart control chart in detecting small shifts in process mean. As with the CUSUM control chart, the EWMA control chart is typically used for individual observations. And the performance of the EWMA control chart is approximately same as that of the CUSUM control chart.

10.13.1 Basics of EWMA

EWMA is defined by $z_i = \lambda x_i + (1 - \lambda)z_{i-1}$, where i denotes sample number, and $\lambda \in (0, 1]$ is a constant and the starting value which is required with the first sample is the process target so that $z_0 = \mu_0$. Sometimes, the average of preliminary data is used as the starting value of the EWMA, so that $z_0 = \bar{x}$.

Let us now discuss why the EWMA z_i is called the weighted average of all previous sample means. We substitute for z_{i-1} on the right-hand side of the above definition of EWMA and obtain the following

$$z_i = \lambda x_i + (1 - \lambda)z_{i-1} = \lambda x_i + (1 - \lambda)(\lambda x_{i-1} + (1 - \lambda)z_{i-2})$$

$$= \lambda x_i + \lambda(1 - \lambda)x_{i-1} + (1 - \lambda)^2[\lambda x_{i-2} + (1 - \lambda)z_{i-3}] = \lambda \sum_{j=0}^{i-1}(1 - \lambda)^j x_{i-j} + (1 - \lambda)^i z_0.$$

Here, the weights $\gamma(1 - \lambda)^j$ decrease geometrically with the age of sample mean. That is why the EWMA is sometimes also called geometric moving average.

10.13.2 Construction of EWMA Control Chart

EWMA control chart is obtained by plotting z_i against sample number i (or time). The center line and control limits for EWMA control chart are as follows

$$\text{Center line} : \mu_0$$

$$\text{Upper Control Limit}(UCL) : \mu_0 + L\sigma\sqrt{\frac{\lambda}{(2 - \lambda)}\left[1 - (1 - \lambda)^{2i}\right]}$$

$$\text{Lower Control Limit}(LCL) : \mu_0 - L\sigma\sqrt{\frac{\lambda}{(2 - \lambda)}\left[1 - (1 - \lambda)^{2i}\right]}$$

where L is the width of the control limits. We will discuss the effect of choice of the parameters L and λ on the EWMA control chart later on.

Example 10.15 Let us consider the yarn strength data as shown in Table 10.13.

Let us now construct the EWMA control chart with $\lambda = 0.1$ and $L = 2.7$ for the yarn strength (cN.tex^{-1}) data as shown in Table 10.13. We consider that the target value of the mean $\mu_0 = 10$ cN.tex^{-1} and standard deviation $\sigma = 1$ cN.tex^{-1}. The calculations for the EWMA control chart are shown in Table 10.14.

Table 10.13 Yarn strength data

Sample no. i	Strength x_i (cN.tex^{-1})	Sample no. i	Strength x_i (cN.tex^{-1})
1	10.29	16	10.65
2	8.66	17	8.8
3	10.61	18	10.2
4	12.03	19	10
5	9.31	20	10.1
6	10.86	21	10.1
7	11.2	22	10.95
8	8.4	23	11.1
9	8.5	24	12.6
10	10.55	25	9.5
11	9.5	26	12
12	10.69	27	11.2
13	10.7	28	9.7
14	8.7	29	11.75
15	10.29	30	11.07

Table 10.14 Calculations for EWMA control chart

Sample no. i	Strength $x - i$ (cN.tex^{-1})	EWMA z_i (cN.tex^{-1})	Sample no. i	Strength x_i (cN.tex^{-1})	EWMA z_i (cN.tex^{-1})
1	10.29	10.029	16	10.65	10.0442
2	8.66	9.8921	17	8.8	9.9197
3	10.61	9.9639	18	10.2	9.9478
4	12.03	10.1705	19	10	9.953
5	9.31	10.0845	20	10.1	9.9677
6	10.86	10.162	21	10.1	9.9809
7	11.2	10.2658	22	10.95	10.0778
8	8.4	10.0792	23	11.1	10.1801
9	8.5	9.9213	24	12.6	10.422
10	10.55	9.9842	25	9.5	10.3298
11	9.5	9.9358	26	12	10.4969
12	10.69	10.0112	27	11.2	10.5672
13	10.7	10.0801	28	9.7	10.4805
14	8.7	9.9421	29	11.75	10.6074
15	10.29	9.9768	30	11.07	10.6537

To illustrate the calculations, consider the first observation $x_1 = 10.29$. The first value of EWMA is $z_1 = \lambda x_1 + (1 - \lambda)z_0 = (0.1 \times 10.29) + (1 - 0.1) \times 10 = 10.029$. Similarly, the second value of EWMA is $z_2 = \lambda x_2 + (1 - \lambda)z_1 = (0.1 \times 8.66) + (1 - 0.1) \times 10.029 = 9.8921$. In this way, the other values of the EWMA are calculated.

The upper control limits are calculated by using the following formula

$$\mu_0 + L\sigma \sqrt{\frac{\lambda}{(2 - \lambda)} \left[1 - (1 - \lambda)^{2i}\right]}$$

For sample number 1, the upper control limit (UCL) is

$$\mu_0 + L\sigma \sqrt{\frac{\lambda}{(2 - \lambda)} \left[1 - (1 - \lambda)^{2i}\right]} = 10 + 2.7 \times 1 \sqrt{\frac{0.1}{(2 - 0.1)} \left[1 - (1 - 0.1)^{2 \times 1}\right]} = 10.27.$$

and the lower control limit (LCL) is

$$\mu_0 - L\sigma \sqrt{\frac{\lambda}{(2 - \lambda)} \left[1 - (1 - \lambda)^{2i}\right]} = 10 - 2.7 \times 1 \sqrt{\frac{0.1}{(2 - 0.1)} \left[1 - (1 - 0.1)^{2 \times 1}\right]} = 9.73.$$

For sample number 2, the upper control limit (UCL) is

$$\mu_0 + L\sigma \sqrt{\frac{\lambda}{(2 - \lambda)} \left[1 - (1 - \lambda)^{2i}\right]} = 10 + 2.7 \times 1 \sqrt{\frac{0.1}{(2 - 0.1)} \left[1 - (1 - 0.1)^{2 \times 2}\right]} = 10.3632.$$

For sample number 2, the lower control limit (LCL) is

$$\mu_0 - L\sigma \sqrt{\frac{\lambda}{(2 - \lambda)} \left[1 - (1 - \lambda)^{2i}\right]} = 10 - 2.7 \times 1 \sqrt{\frac{0.1}{(2 - 0.1)} \left[1 - (1 - 0.1)^{2 \times 2}\right]} = 9.6368.$$

In this way, the upper and lower control limits for other sample numbers are calculated. The EWMA control chart with $\lambda = 0.1$ and $L = 2.7$ is shown in Fig. 10.22. It can be seen that the control limits increase in width as i increases from $i = 1, 2, \ldots$ until they stabilize later on. The EWMA value at sample number 30 falls beyond the upper control limit; hence, we conclude that the process is out of statistical control.

10.13.3 Choice of L and λ

Let us now see the effects of the choice of L and λ on the EWMA control chart. We fix λ at 0.1 but increase L from 2.7 to 3. We calculated the values of the EWMA and the control limits for sample numbers from 1 to 30. Figure 10.23 displays the EWMA

Fig. 10.22 EWMA control
chart with $\lambda = 0.1$ and
$L = 2.7$

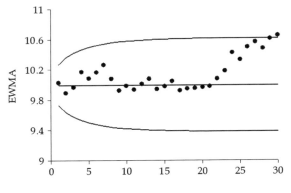

Fig. 10.23 EWMA control
chart with $\lambda = 0.1$ and
$L = 3$

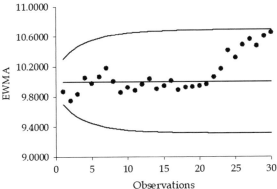

control chart with $\lambda = 0.1$ and $L = 3$. Here, there is no single point of EWMA that falls beyond the control limits. The process is under statistical control. The increase of width of the control charts resulted in statistical control of the process.

We now fix L at 3 and increase λ from $\lambda = 0.1$ to $\lambda = 0.2$. Accordingly, we calculated the values of the EWMA and the control limits for sample numbers from 1 to 30. Figure 10.24 displays the EWMA control chart with $\lambda = 0.2$ and $L = 3$. Here, there is no single point of EWMA that falls beyond the control limits. It can be observed that this resulted in further increase of the width of the control limits.

We then change from $\lambda = 0.2$ to $\lambda = 0.05$. The corresponding EWMA control chart is obtained as shown in Fig. 10.25. As expected, the width of the control limits decreases and there are many points falling beyond the upper and lower control limits. The process is then said to be out of statistical control.

It is thus observed that the choices of λ and L are very critical as far as the EWMA is concerned. In general, it is found that λ works reasonably well in the interval $0.05 \le \lambda \le 0.25$. The popular choices have been $\lambda = 0.05$, $\lambda = 0.10$, and $\lambda = 0.20$. A good rule of thumb is to use smaller values of λ to detect smaller shifts. $L = 3$ (three-sigma limit) works reasonably well, particularly with the larger values of λ. When λ is small, say $\lambda \le 0.1$, there is an advantage in reducing the width of the

Fig. 10.24 EWMA control chart with $\lambda = 0.2$ and $L = 3$

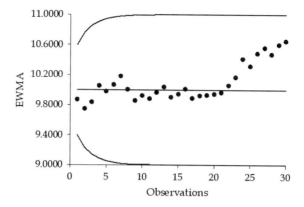

Fig. 10.25 EWMA control chart with $\lambda = 0.05$ and $L = 3$

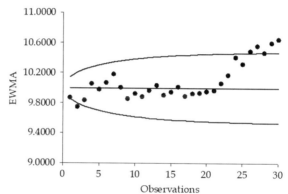

limits by using a value of L between about 2.6 and 2.8. Like the CUSUM, the EWMA performs well against small shifts but it does not react to large shifts as quickly as the Shewhart chart. However, the EWMA is often considered to be superior to the CUSUM for large shifts, particularly if $\lambda > 0.10$.

Problems

10.1 A company uses the following acceptance sampling procedure. A batch is accepted if not more than two items are found to be defective in a sample of 50 items taken randomly from the batch; otherwise, the batch is rejected. If 500 batches obtained from a process that manufactures 2% defective items are submitted to this plan, how many batches are accepted?

10.2 Design an acceptance sampling plan for which AQL is 0.05, RQL is 0.05, α is 0.05, and β is 0.05.

Table 10.15 Data for Problem 10.9

Sample no.	Thickness (mm)				
1	1	1.4	1.5	1.8	1.2
2	1.4	1.3	1.7	1.9	2
3	1.1	1.5	1.6	1.4	1.8
4	1.2	1.4	1.7	1.8	1.4
5	1.2	1.6	1.8	1.4	2
6	1.4	1.2	1.6	1.8	1.5
7	1.8	1.6	1.4	2	1.8
8	1.2	1.8	1.9	2.2	1.8
9	2	1.8	2.4	1.6	1.4
10	1.9	2.4	1.6	1.8	2.4

10.3 Design an acceptance sampling plan for which AQL is 0.05, RQL is 0.05, α is 0.05, and β is 0.10.

10.4 Design an acceptance sampling plan for which AQL is 0.05, RQL is 0.05, α is 0.10, and β is 0.05.

10.5 Design an acceptance sampling plan for which AQL is 0.05, RQL is 0.05, α is 0.05, and β is 0.10.

10.6 Design an acceptance sampling plan for which AQL is 0.05, RQL is 0.10, α is 0.05, and β is 0.05.

10.7 Design an acceptance sampling plan for which AQL is 0.10, RQL is 0.05, α is 0.05, and β is 0.05.

10.8 A ball bearing manufacturing company supplies bearing of inner ring diameter equal to be 50 mm. A customer accepts bearing if its mean diameter lies within a range of 50 ± 0.002 mm. As the customer cannot test all the bearings supplied by the supplier, the customer would like to devise an acceptance sampling scheme with 5% producer's risk and 5% consumer's risk. Consider that the inner ring diameter of the bearing follows normal distribution with standard deviation of 0.001 mm.

10.9 A machine is manufacturing compact disks with thickness shown in Table 10.15. Set up an $\bar{x}$ chart, R chart, and s chart, and comment on whether the manufacturing process is under statistical control.

10.10 The mean, range, and standard deviation of weight of 10 samples of size 9 taken from a manufacturing process are shown in Table 10.16.

(a) Compute 3σ control limits for the mean, range, and standard deviation charts on this process.

Table 10.16 Data for Problem 10.10

Sample number	Mean (g.m^{-2})	Range (g.m^{-2})	Standard deviation (g.m^{-2})
1	253.46	17.12	5.47
2	249.54	19.04	7.05
3	257.76	29.12	9.72
4	250.56	22.56	7.37
5	254.46	20.48	6.44
6	248.77	20.96	8.32
7	250.15	20.8	6.87
8	253.04	22.88	7
9	254.16	14.72	5.18
10	250.34	16.96	6.75

Table 10.17 Data for Problem 10.12

Process	Characteristics	
	Mean (lb.in^{-2})	Standard deviation (lb.in^{-2})
A	50	2
B	55	1

(b) If the specifications on this weight are 250 ± 25 g.m^{-2}, what conclusion would you draw about the capability of this process?

(c) What proportion of materials produced by this process is likely to be within these specifications? Assume the weight of the material follows normal distribution.

10.11 A tire manufacturing company receives an order for a large quantity of tires with mean breaking energy 1000 J and standard deviation of breaking energy 10 J. The company wants to establish a control chart for mean breaking energy of tire based on the aforesaid parameters and samples of size 9 such that the probability of rejecting good tires produced by a statistically controlled process is 0.01. Under this situation can you find out what are the upper and lower control limits for mean breaking energy of tire?

10.12 Consider two manufacturing processes for production of malleable iron castings, and the mean and standard deviation of the tensile strength of the castings resulting from these processes are shown in Table 10.17.

The specifications on the iron casting tensile strength are given as 50 ± 10 lb.in^2.

(a) Estimate the potential capability and the actual capability of these processes.

(b) Which of the two processes would you prefer to use and why?

Table 10.18 Data for Problem 10.13

Sample no.	No. of defectives	Sample no.	No. of defectives	Sample no.	No. of defectives
1	2	11	24	21	2224
2	4	12	14	22	67
3	9	13	9	23	12
4	11	14	14	24	17
5	7	15	2	25	14
6	15	16	1	26	14
7	25	17	1	27	3
8	17	18	18	28	7
9	8	19	5	29	6
10	4	20	18	30	14

Table 10.19 Data for Problem 10.14

Lot no.	1	2	3	4	5	6	7	8	9	10
No. of inspected items	100	200	100	500	800	900	400	700	900	700
No. of defective items	19	16	7	43	77	86	33	53	86	91

10.13 Table 10.18 displays the number of defective bearings found in each sample of size 200.

(a) Compute a 3σ control limits for the number of defectives.

(b) Display the control chart.

(c) Analyze the patterns of data in the control chart, and comment on whether the bearing manufacturing process is under control or out of control.

10.14 Construct an appropriate control chart for the data given in Table 10.19, and comment on whether the process is under statistical control.

10.15 In a production process, a sample of 50 items is inspected on each day. The number of defective found in each sample is as follows: 1, 3, 4, 7, 9, 6, 13, 2, 5, 6. Draw an appropriate control chart and check for control.

10.16 The number of defects in twenty pieces of vertical blind each of 10 m length is as follows: 1, 3, 2, 5, 2, 4, 3, 2, 1, 3, 2, 4, 6, 3, 5, 1, 1, 2, 2, 3. Draw an appropriate control chart, and conclude whether the process can be considered to be under statistical control.

Table 10.20 Data for Problem 10.17

Roll number	1	2	3	4	5	6	7	8	9	10
Number of defects	12	10	8	9	12	10	11	9	11	8

Table 10.21 Data for Problem 10.18

Car no.	1	2	3	4	5	6	7	8	9	10
Number of defects	3	7	5	6	3	1	2	8	8	6

Table 10.22 Data for Problem 10.19

61	66	61	63	64
63	67	68	61	63
62	64	69	68	64
68	68	70	67	62
64	64	65	69	65

10.17 A cloth manufacturer examines ten rolls of clothes each with 100 m length of cloth and counts the number of defects present in each roll. The results are shown in Table 10.20.

Can you design a 3σ control chart for defects and conclude whether the fabrics have been produced by a statistically controlled process?

10.18 An automotive company inspects the defects in ten cars, and the results are shown in Table 10.21. Set up an appropriate control chart, and conclude if the manufacturing process of the cars is under statistical control.

10.19 The data of Table 10.22 represent temperature (°C) of a chemical process recorded every 5 min (read down from left).

The target process mean temperature is 65 °C, and the process standard deviation is 3 °C. Set up a tabular CUSUM for this process, using $H = 5$ and $K = 0.5$.

10.20 The concentration (ppm) of a bath, measured on hourly basis for 20 h, is shown in Table 10.23 (read down from left). The target process mean concentration is 80 ppm, and the process standard deviation is 2.5 ppm. Set up a tabular CUSUM for this process, using $H = 5$ and $K = 0.5$.

Table 10.23 Data for Problem 10.20

84	80	83	78
82	82	84	80
81	78	81	74
79	76	82	79
78	80	80	79

10.21 Consider the data given in Problem No. 11.20 and construct an EWMA control chart with $\lambda = 0.1$ and $L = 2.7$. Compare your results to those obtained with the CUSUM control chart.

Reference

Montgomery DC (2001) Introduction to statistical quality control. Wiley, New York

Appendix A
Statistical Tables

1. Cumulative binomial probabilities (Tables A.1, A.2)
2. Cumulative Poisson probabilities (Tables A.3, A.4, A.5)
3. Cumulative distribution function for the standard normal distribution (Tables A.6, A.7)
4. Critical values for the chi-square distribution (Tables A.8, A.9)
5. Critical values for the student's t distribution (Table A.10)
6. Critical values for the F distribution (Tables A.11, A.12, A.13)
7. Critical values $d_{n,\alpha}$ for the Kolmogorov–Smirnov test (Table A.14)
8. Critical values for the Wilcoxon signed-rank test (Table A.15)
9. Percentage points of the studentized range statistic (Tables A.16, A.17)

© Springer Nature Singapore Pte Ltd. 2018
D. Selvamuthu and D. Das, *Introduction to Statistical Methods,
Design of Experiments and Statistical Quality Control*,
https://doi.org/10.1007/978-981-13-1736-1

Table A.1 Cumulative distribution function for binomial random variable, $B(n, p)$

$n = 5$

x	p														
	0.01	0.05	0.10	0.20	0.25	0.30	0.40	0.50	0.60	0.70	0.75	0.80	0.90	0.95	0.99
0	0.9510	0.7738	0.5905	0.3277	0.2373	0.1681	0.0778	0.0313	0.0102	0.0024	0.0010	0.0003	0.0000		
1	0.9990	0.9774	0.9185	0.7373	0.6328	0.5282	0.3370	0.1875	0.0870	0.0308	0.0156	0.0067	0.0005	0.0000	
2	1.0000	0.9988	0.9914	0.9421	0.8965	0.8369	0.6826	0.5000	0.3174	0.1631	0.1035	0.0579	0.0086	0.0012	0.0000
3		1.0000	0.9995	0.9933	0.9844	0.9692	0.9130	0.8125	0.6630	0.4718	0.3672	0.2627	0.0815	0.0226	0.0010
4			1.0000	0.9997	0.9990	0.9976	0.9898	0.9688	0.9222	0.8319	0.7627	0.6723	0.4095	0.2262	0.0490

$n = 10$

x	p														
	0.01	0.05	0.10	0.20	0.25	0.30	0.40	0.50	0.60	0.70	0.75	0.80	0.90	0.95	0.99
0	0.9044	0.5987	0.3487	0.1074	0.0563	0.0282	0.0060	0.0010	0.0001	0.0000					
1	0.9957	0.9139	0.7361	0.3758	0.2440	0.1493	0.0464	0.0107	0.0017	0.0001	0.0000	0.0000			
2	0.9999	0.9885	0.9298	0.6778	0.5256	0.3828	0.1673	0.0547	0.0123	0.0016	0.0004	0.0001	0.0000		
3	1.0000	0.9990	0.9872	0.8791	0.7759	0.6496	0.3823	0.1719	0.0548	0.0106	0.0035	0.0009	0.0000		
4		0.9999	0.9984	0.9672	0.9219	0.8497	0.6331	0.3770	0.1662	0.0473	0.0197	0.0064	0.0001	0.0000	
5		1.0000	0.9999	0.9936	0.9803	0.9527	0.8338	0.6230	0.3669	0.1503	0.0781	0.0328	0.0016	0.0001	
6			1.0000	0.9991	0.9965	0.9894	0.9452	0.8281	0.6177	0.3504	0.2241	0.1209	0.0128	0.0010	0.0000
7				0.9999	0.9996	0.9984	0.9877	0.9453	0.8327	0.6172	0.4744	0.3222	0.0702	0.0115	0.0001
8				1.0000	1.0000	0.9999	0.9983	0.9893	0.9536	0.8507	0.7560	0.6242	0.2639	0.0861	0.0043
9						1.0000	0.9999	0.9990	0.9940	0.9718	0.9437	0.8926	0.6513	0.4013	0.0956

(continued)

Table A.1 (continued)

$n = 15$

x	0.01	0.05	0.10	0.20	0.25	0.30	0.40	0.50	0.60	0.70	0.75	0.80	0.90	0.95	0.99
0	0.8601	0.4633	0.2059	0.0352	0.0134	0.0047	0.0005	0.0000							
1	0.9904	0.8290	0.5490	0.1671	0.0802	0.0353	0.0052	0.0005	0.0000						
2	0.9996	0.9638	0.8159	0.3980	0.2361	0.1268	0.0271	0.0037	0.0003	0.0000					
3	1.0000	0.9945	0.9444	0.6482	0.4613	0.2969	0.0905	0.0176	0019	0.0001	0.0000				
4		0.9994	0.9873	0.8358	0.6865	0.5155	0.2173	0.0592	0.0093	0.0007	0.0001	0.0000			
5		0.9999	0.9978	0.9389	0.8516	0.7216	0.4032	0.1509	0.0338	0.0037	0.0008	0.0001			
6		1.0000	0.9997	0.9819	0.9434	0.8689	0.6098	0.3036	0.0950	0.0152	0.0042	0.0008			
7			1.0000	0.9958	0.9827	0.9500	0.7869	0.5000	0.2131	0.0500	0.0173	0.0042	0.0000		
8				0.9992	0.9958	0.9848	0.9050	0.6964	0.3902	0.1311	0.0566	0.0181	0.0003	0.0000	
9				0.9999	0.9992	0.9963	0.9662	0.8491	0.5968	0.2784	0.1484	0.0611	0.0022	0.0001	
10				1.0000	0.9999	0.9993	0.9907	0.9408	0.7827	0.4845	0.3135	0.1642	0.0127	0.0006	
11					1.0000	0.9999	0.9981	0.9824	0.9095	0.7031	0.5387	0.3518	0.0556	0.0055	0.0000
12						1.0000	0.9997	0.9963	0.9729	0.8732	0.7639	0.6020	0.1841	0.0362	0.0004
13							1.0000	0.9995	0.9948	0.9647	0.9198	0.8329	0.4510	0.1710	0.0096
14								1.0000	0.9995	0.9953	0.9866	0.9648	0.7941	0.5367	0.1399

Table A.2 Cumulative distribution function for binomial random variable, $B(n, p)$ (Continued)

x	0.01	0.05	0.10	0.20	0.25	0.30	0.40	0.50	0.60	0.70	0.75	0.80	0.90	0.95	0.99
0	0.8179	0.3585	0.1216	0.0115	0.0032	0.0008	0.0000								
1	0.9831	0.7358	0.3917	0.0692	0.0243	0.0076	0.0005	0.0000							
2	0.9990	0.9245	0.6769	0.2061	0.0913	0.0355	0.0036	0.0002							
3	1.0000	0.9841	0.8670	0.4114	0.2252	0.1071	0.0160	0.0013	0.0000						
4		0.9974	0.9568	0.6296	0.4148	0.2375	0.0510	0.0059	0.0003						
5		0.9997	0.9887	0.8042	0.6172	0.4164	0.1256	0.0207	0.0016	0.0000					
6		1.0000	0.9976	0.9133	0.7858	0.6080	0.2500	0.0577	0.0065	0.0003	0.0000				
7			0.9996	0.9679	0.8982	0.7723	0.4159	0.1316	0.0210	0.0013	0.0002	0.0000			
8			0.9999	0.9900	0.9891	0.8887	0.5956	0.2517	0.0565	0.0051	0.0009	0.0001			
9			1.0000	0.9974	0.9861	0.9520	0.7553	0.4119	0.1275	0.0171	0.0039	0.0006			
10				0.9994	0.9961	0.9829	0.8725	0.5881	0.2447	0.0480	0.0139	0.0026	0.0000		
11				0.9999	0.9991	0.9949	0.9435	0.7483	0.4044	0.1133	0.0409	0.0100	0.0001		
12				1.0000	0.9998	0.9987	0.9790	0.8684	0.5841	0.2277	0.1018	0.0321	0.0004		
13					1.0000	0.9997	0.9935	0.9423	0.7500	0.3920	0.2142	0.0867	0.0024	0.0000	
14						1.0000	0.9984	0.9793	0.8744	0.5836	0.3828	0.1958	0.0113	0.0003	
15							0.9997	0.9941	0.9490	0.7625	0.5852	0.3704	0.0432	0.0026	
16							1.0000	0.9987	0.9840	0.8929	0.7748	0.5886	0.1330	0.0159	0.0000
17								0.9998	0.9964	0.9645	0.9087	0.7939	0.3231	0.0755	0.0010
18								1.0000	0.9995	0.9924	0.9757	0.9308	0.6083	0.2642	0.0169
19									1.0000	0.9992	0.9968	0.9885	0.8784	0.6415	0.1821

(continued)

Table A.2 (continued)

$n = 25$

x	0.01	0.05	0.10	0.20	0.25	0.30	0.40	0.50	0.60	0.70	0.75	0.80	0.90	0.95	0.99
0	0.7778	0.2774	0.0718	0.0038	0.0008	0.0001	0.0000								
1	0.9742	0.6424	0.2712	0.0274	0.0070	0.0016	0.0001								
2	0.9980	0.8729	0.5371	0.0982	0.0321	0.0090	0.0004	0.0000							
3	0.9999	0.9659	0.7636	0.2340	0.0962	0.0332	0.0024	0.0001							
4	1.0000	0.9928	0.9020	0.4207	0.2137	0.0905	0.0095	0.0005	0.0000						
5		0.9988	0.9666	0.6167	0.3783	0.1935	0.0294	0.0020	0.0001						
6		0.9998	0.9905	0.7800	0.5611	0.3407	0.0736	0.0073	0.0003						
7		1.0000	0.9977	0.8909	0.7265	0.5118	0.1536	0.0216	0.0012	0.0000					
8			0.9995	0.9532	0.8506	0.6769	0.2735	0.0539	0.0043	0.0001					
9			0.9999	0.9827	0.9287	0.8106	0.4246	0.1148	0.0132	0.0005	0.0000				
10			1.0000	0.9944	0.9703	0.9022	0.5858	0.2122	0.0344	0.0018	0.0002	0.0000			
11				0.9985	0.9893	0.9558	0.7323	0.3450	0.0778	0.0060	0.0009	0.0001			
12				0.9996	0.9966	0.9825	0.8462	0.5000	0.1538	0.0175	0.0034	0.0004			
13				0.9999	0.9991	0.9940	0.9222	0.6550	0.2677	0.0442	0.0107	0.0015			
14				1.0000	0.9998	0.9982	0.9656	0.7878	0.4142	0.0978	0.0297	0.0056	0.0000		
15					1.0000	0.9995	0.9868	0.8852	0.5754	0.1894	0.0713	0.0173	0.0001		
16						0.9999	0.9957	0.9461	0.7265	0.3231	0.1494	0.0468	0.0005		
17						1.0000	0.9988	0.9784	0.8464	0.4882	0.2735	0.1091	0.0023	0.0000	
18							0.9997	0.9927	0.9264	0.6593	0.4389	0.2200	0.0095	0.0002	
19							0.9999	0.9980	0.9706	0.8065	0.6217	0.3833	0.0334	0.0012	
20							1.0000	0.9995	0.9905	0.9095	0.7863	0.5793	0.0980	0.0072	0.0000
21								0.9999	0.9976	0.9668	0.9038	0.7660	0.2364	0.0341	0.0001
22								1.0000	0.9996	0.9910	0.9679	0.9018	0.4629	0.1271	0.0020
23									0.9999	0.9984	0.9930	0.9726	0.7288	0.3576	0.0258
24									1.0000	0.9999	0.9992	0.9962	0.9282	0.7226	0.2222

Table A.3 Cumulative distribution function for Poisson random variable, $P(\mu)$

x	μ									
	0.05	0.10	0.15	0.20	0.26	0.30	0.35	0.40	0.45	0.50
0	0.9512	0.9048	0.8607	0.8187	0.7788	0.7408	0.7047	0.6703	0.6376	0.6065
1	0.9988	0.9953	0.9898	0.9825	0.9735	0.9631	0.9513	0.9384	0.9246	0.9098
2	1.0000	0.9998	0.9995	0.9989	0.9978	0.9964	0.9945	0.9921	0.9891	0.9856
3		1.0000	1.0000	0.9999	0.9999	0.9997	0.9995	0.9992	0.9988	0.9982
4				1.0000	1.0000	1.0000	1.0000	0.9999	0.9999	0.9998
5								1.0000	1.0000	1.0000

x	μ									
	0.55	0.60	0.65	0.70	0.75	0.80	0.85	0.90	0.95	1.00
0	0.5769	0.5488	0.5220	0.4966	0.4724	0.4493	0.4274	0.4066	0.3867	0.3679
1	0.8943	0.8781	0.8614	0.8442	0.8266	0.8088	0.7907	0.7725	0.7541	0.7358
2	0.9815	0.9769	0.9717	0.9659	0.9595	0.9526	0.9451	0.9371	0.9287	0.9197
3	0.9975	0.9966	0.9956	0.9942	0.9927	0.9909	0.9889	0.9865	0.9839	0.9810
4	0.9997	0.9996	0.9994	0.9992	0.9989	0.9986	0.9982	0.9977	0.9971	0.9963
5	1.0000	1.0000	0.9999	0.9999	0.9999	0.9998	0.9997	0.9997	0.9995	0.9994
6		1.0000	1.0000	1.0000	1.0000	1.0000	1.0000	1.0000	0.9999	0.9999
7									1.0000	1.0000

x	μ									
	1.1	1.2	1.3	1.4	1.5	1.6	1.7	1.8	1.9	2.0
0	0.3329	0.3012	0.2725	0.2466	0.2231	0.2019	0.1827	0.1653	0.1496	0.1353
1	0.6990	0.6626	0.6268	0.5918	0.5578	0.5249	0.4932	0.4628	0.4337	0.4060
2	0.9004	0.8795	0.8571	0.8335	0.8088	0.7834	0.7572	0.7306	0.7037	0.6767
3	0.9743	0.9662	0.9569	0.9463	0.9344	0.9212	0.9068	0.8913	0.8747	0.8571
4	0.9946	0.9923	0.9893	0.9857	0.9814	0.9763	0.9704	0.9636	0.9559	0.9473
5	0.9990	0.9985	0.9978	0.9968	0.9955	0.9940	0.9920	0.9896	0.9868	0.9834
6	0.9999	0.9997	0.9996	0.9994	0.9991	0.9987	0.9981	0.9974	0.9966	0.9955
7	1.0000	1.0000	0.9999	0.9999	0.9998	0.9997	0.9996	0.9994	0.9992	0.9989
8			1.0000	1.0000	1.0000	1.0000	0.9999	0.9999	0.9998	0.9998
9						1.0000	1.0000	1.0000	1.0000	1.0000

Table A.4 Cumulative distribution function for Poisson random variable, $P(\mu)$ (Continued)

x	2.1	2.2	2.3	2.4	2.5	2.6	2.7	2.8	2.9	3.0
0	0.1225	0.1108	0.1003	0.0907	0.0821	0.0743	0.0672	0.0608	0.0550	0.0498
1	0.3796	0.3546	0.3309	0.3084	0.2873	0.2674	0.2487	0.2311	0.2146	0.1991
2	0.6496	0.6227	0.5960	0.5697	0.5438	0.5184	0.4936	0.4695	0.4460	0.4232
3	0.8386	0.8194	0.7993	0.7787	0.7576	0.7360	0.7141	0.6919	0.6696	0.6472
4	0.9379	0.9275	0.9162	0.9041	0.8912	0.8774	0.8629	0.8477	0.8318	0.8153
5	0.9796	0.9751	0.9700	0.9643	0.9580	0.9510	0.9433	0.9349	0.9258	0.9161
6	0.9941	0.9925	0.9906	0.9884	0.9858	0.9828	0.9794	0.9756	0.9713	0.9665
7	0.9985	0.9980	0.9974	0.9967	0.9958	0.9947	0.9934	0.9919	0.9901	0.9881
8	0.9997	0.9995	0.9994	0.9991	0.9989	0.9985	0.9981	0.9976	0.9969	0.9962
9	0.9999	0.9999	0.9999	0.9998	0.9997	0.9996	0.9995	0.9993	0.9991	0.9989
10	1.0000	1.0000	1.0000	1.0000	0.9999	0.9999	0.9999	0.9998	0.9998	0.9997
11				1.0000	1.0000	1.0000	1.0000	1.0000	0.9999	0.9999
12									1.0000	1.0000

x	3.1	3.2	3.3	3.4	3.5	3.6	3.7	3.8	3.9	4.0
0	0.0450	0.0408	0.0369	0.0334	0.0302	0.0273	0.0247	0.0224	0.0202	0.0183
1	0.1847	0.1712	0.1586	0.1468	0.1359	0.1257	0.1162	0.1074	0.0992	0.0916
2	0.4012	0.3799	0.3594	0.3397	0.3208	0.3027	0.2854	0.2689	0.2531	0.2381
3	0.6248	0.6025	0.5803	0.5584	0.5366	0.5152	0.4942	0.4735	0.4532	0.4335
4	0.7982	0.7806	0.7626	0.7442	0.7254	0.7064	0.6872	0.6678	0.6484	0.6288
5	0.9057	0.8946	0.8829	0.8705	0.8576	0.8441	0.8301	0.8156	0.8006	0.7851
6	0.9612	0.9554	0.9490	0.9421	0.9347	0.9267	0.9182	0.9091	0.8995	0.8893
7	0.9858	0.9832	0.9802	0.9769	0.9733	0.9692	0.9648	0.9599	0.9546	0.9489
8	0.9953	0.9943	0.9931	0.9917	0.9901	0.9883	0.9863	0.9840	0.9815	0.9786
9	0.9986	0.9982	0.9978	0.9973	0.9967	0.9960	0.9952	0.9942	0.9931	0.9919
10	0.9996	0.9995	0.9994	0.9992	0.9990	0.9987	0.9984	0.9981	0.9977	0.9972
11	0.9999	0.9999	0.9998	0.9998	0.9997	0.9996	0.9995	0.9994	0.9993	0.9991
12	1.0000	1.0000	1.0000	0.9999	0.9999	0.9999	0.9999	0.9998	0.9998	0.9997
13				1.0000	1.0000	1.0000	1.0000	1.0000	0.9999	0.9999
14									1.0000	1.0000

Table A.5 Cumulative distribution function for Poisson random variable, $P(\mu)$ (Continued)

x	5	6	7	8	9	10	15	20	25	30
0	0.0067	0.0025	0.0009	0.0003	0.0001	0.0000				
1	0.0404	0.0174	0.0073	0.0030	0.0012	0.0005				
2	0.1247	0.0620	0.0296	0.0138	0.0062	0.0028	0.0000			
3	0.2650	0.1512	0.0818	0.0424	0.0212	0.0103	0.0002			
4	0.4405	0.2851	0.1730	0.0996	0.0550	0.0293	0.0009	0.0000		
5	0.6160	0.4457	0.3007	0.1912	0.1157	0.0671	0.0028	0.0001		
6	0.7622	0.6063	0.4497	0.3134	0.2068	0.1301	0.0076	0.0003		
7	0.8666	0.7440	0.5987	0.4530	0.3239	0.2202	0.0180	0.0008	0.0000	
8	0.9319	0.8472	0.7291	0.5925	0.4557	0.3328	0.0374	0.0021	0.0001	
9	0.9682	0.9161	0.8305	0.7166	0.5874	0.4579	0.0699	0.0050	0.0002	
10	0.9863	0.9574	0.9015	0.8159	0.7060	0.5830	0.1185	0.0108	0.0006	0.0000
11	0.9945	0.9799	0.9467	0.8881	0.8030	0.6968	0.1848	0.0214	0.0014	0.0001
12	0.9980	0.9912	0.9730	0.9362	0.8758	0.7916	0.2676	0.0390	0.0031	0.0002
13	0.9993	0.9964	0.9872	0.9658	0.9261	0.8645	0.3632	0.0661	0.0065	0.0004
14	0.9998	0.9986	0.9943	0.9827	0.9585	0.9165	0.4657	0.1049	0.0124	0.0009
15	0.9999	0.9995	0.9976	0.9918	0.9780	0.9513	0.5681	0.1565	0.0223	0.0019
16	1.0000	0.9998	0.9990	0.9963	0.9889	0.9730	0.6641	0.2211	0.0377	0.0039
17		0.9999	0.9996	0.9984	0.9947	0.9857	0.7489	0.2970	0.0605	0.0073
18		1.0000	0.9999	0.9993	0.9976	0.9928	0.8195	0.3814	0.0920	0.0129
19			1.0000	0.9997	0.9989	0.9965	0.8752	0.4703	0.1336	0.0219
20				0.9999	0.9996	0.9984	0.9170	0.5591	0.1855	0.0353
21				1.0000	0.9998	0.9993	0.9469	0.6437	0.2473	0.0544
22					0.9999	0.9997	0.9673	0.7206	0.3175	0.0806
23					1.0000	0.9999	0.9805	0.7875	0.3939	0.1146
24						1.0000	0.9888	0.8432	0.4734	0.1572
25							0.9938	0.8878	0.5529	0.2084
26							0.9967	0.9221	0.6294	0.2673
27							0.9983	0.9475	0.7002	0.3329
28							0.9991	0.9657	0.7634	0.4031
29							0.9996	0.9782	0.8179	0.4757
30							0.9998	0.9865	0.8633	0.5484
31							0.9999	0.9919	0.8999	0.6186
32							1.0000	0.9953	0.9285	0.6845
33								0.9973	0.9502	0.7444
34								0.9985	0.9662	0.7973
35								0.9992	0.9775	0.8426
36								0.9996	0.9854	0.8804
37								0.9998	0.9998	0.9110
38								0.9999	0.9943	0.9352

(continued)

Table A.5 (continued)

x	μ									
	5	6	7	8	9	10	15	20	25	30
39								0.9999	0.9966	0.9537
40								1.0000	0.9980	0.9677
41									0.9988	0.9779
42									0.9993	0.9852
43									0.9996	0.9903
44									0.9998	0.9937

Table A.6   Cumulative distribution function for the standard normal random variable

This table contains values of the cumulative distribution function for the standard normal random variable $\Phi(z) =$

$$P(Z \le z) = \int_{-\infty}^{z} \frac{1}{\sqrt{2\pi}} e^{-s^2/2} \, dz.$$

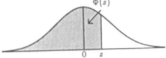

z	.00	.01	.02	.03	.04	.05	.06	.07	.08	.09
-3.4	.0003	.0003	.0003	.0003	.0003	.0003	.0003	.0003	.0003	.0002
-3.3	.0005	.0005	.0005	.0004	.0004	.0004	.0004	.0004	.0004	.0003
-3.2	.0007	.0007	.0006	.0006	.0006	.0006	.0006	.0005	.0005	.0005
-3.1	.0010	.0009	.0009	.0009	.0008	.0008	.0008	.0008	.0007	.0007
-3.0	.0013	.0013	.0013	.0012	.0012	.0011	.0011	.0011	.0010	.0010
-2.9	.0019	.0018	.0018	.0017	.0016	.0016	.0015	.0015	.0014	.0014
-2.8	.0026	.0025	.0024	.0023	.0023	.0022	.0021	.0021	.0020	.0019
-2.7	.0035	.0034	.0033	.0032	.0031	.0030	.0029	.0028	.0027	.0026
-2.6	.0047	.0045	.0044	.0043	.0041	.0040	.0039	.0038	.0037	.0036
-2.5	.0062	.0060	.0059	.0057	.0055	.0054	.0052	.0051	.0049	.0048
-2.4	.0082	.0080	.0078	.0075	.0073	.0071	.0069	.0068	.0066	.0064
-2.3	.0107	.0104	.0102	.0099	.0096	.0094	.0091	.0089	.0087	.0084
-2.2	.0139	.0136	.0132	.0129	.0125	.0122	.0119	.0116	.0113	.0110
-2.1	.0179	.0174	.0170	.0166	.0162	.0158	.0154	.0150	.0146	.0143
-2.0	.0228	.0222	.0217	.0212	.0207	.0202	.0197	.0192	.0188	.0183
-1.9	.0287	.0281	.0274	.0268	.0262	.0256	.0250	.0244	.0239	.0233
-1.8	.0359	.0351	.0344	.0336	.0329	.0322	.0314	.0307	.0301	.0294
-1.7	.0446	.0436	.0427	.0418	.0409	.0401	.0392	.0384	.0375	.0367
-1.6	.0548	.0537	.0526	.0516	.0505	.0495	.0485	.0475	.0465	.0455
-1.5	.0668	.0655	.0643	.0630	.0618	.0606	.0594	.0582	.0571	.0559
-1.4	.0808	.0793	.0778	.0764	.0749	.0735	.0721	.0708	.0694	.0681
-1.3	.0968	.0951	.0934	.0918	.0901	.0885	.0869	.0853	.0838	.0823
-1.2	.1151	.1131	.1112	.1093	.1075	.1056	.1038	.1020	.1003	.0985
-1.1	.1357	.1335	.1314	.1292	.1271	.1251	.1230	.1210	.1190	.1170
-1.0	.1587	.1562	.1539	.1515	.1492	.1469	.1446	.1423	.1401	.1379
-0.9	.1841	.1814	.1788	.1762	.1736	.1711	.1685	.1660	.1635	.1611
-0.8	.2119	.2090	.2061	.2033	.2005	.1977	.1949	.1922	.1894	.1867
-0.7	.2420	.2389	.2358	.2327	.2296	.2266	.2236	.2206	.2177	.2148
-0.6	.2743	.2709	.2676	.2643	.2611	.2578	.2546	.2514	.2483	.2451
-0.5	.3085	.3050	.3015	.2981	.2946	.2912	.2877	.2843	.2810	.2776
-0.4	.3446	.3409	.3372	.3336	.3300	.3264	.3228	.3192	.3156	.3121
-0.3	.3821	.3783	.3745	.3707	.3669	.3632	.3594	.3557	.3520	.3483
-0.2	.4207	.4168	.4129	.4090	.4052	.4013	.3974	.3936	.3897	.3859
-0.1	.4602	.4562	.4522	.4483	.4443	.4404	.4364	.4325	.4286	.4247
-0.0	.5000	.4960	.4920	.4880	.4840	.4801	.4761	.4721	.4681	.4641

Table A.7 Cumulative distribution function for the standard normal random variable (Continued)

Z	0.00	0.01	0.02	0.03	0.04	0.05	0.00	0.07	0.08	0.09
0.0	0.5000	0.5040	0.5080	0.5120	0.5160	0.5199	0.5239	0.5279	0.5319	0.5359
0.1	0.5398	0.5438	0.5478	0.5517	0.5557	0.5596	0.5636	0.5675	0.5714	0.5753
0.2	0.5793	0.5832	0.5871	0.5910	0.5948	0.5987	0.6026	0.6064	0.6103	0.6141
0.3	0.6179	0.6217	0.6255	0.6293	0.6331	0.6368	0.6406	0.6443	0.6480	0.6517
0.4	0.6554	0.6591	0.6628	0.6664	0.6700	0.6736	0.6772	0.6808	0.6844	0.6879
0.5	0.6915	0.6950	0.6985	0.7019	0.7054	0.7088	0.7123	0.7157	0.7190	0.7224
0.6	0.7257	0.7291	0.7324	0.7357	0.7389	0.7422	0.7454	0.7486	0.7517	0.7549
0.7	0.7580	0.7611	0. 7642	0.7673	0.7704	0.7734	0.7764	0.7794	0.7823	0.7852
0.8	0.7881	0.7910	0.7939	0.7967	0.7995	0.8023	0.8051	0.8078	0.8106	0.8133
0.9	0.8159	0.8186	0.8212	0.8238	0.8264	0.8289	0.8315	0.8340	0.8365	0.8389
1.0	0.8413	0.8438	0.8461	0.8485	0.8508	0.8531	0.8554	0.8577	0.8599	0.8621
1.1	0.8643	0.8665	0.8686	0.8708	0.8729	0.8749	0.8770	0.8790	0.8810	0.8830
1.2	0.8849	0.8889	0.8888	0.8907	0.8925	0.8944	0.8962	0.8980	0.8997	0.9015
1.3	0.9032	0.9049	0.9066	0.9082	0.9099	0.9115	0.9131	0.9147	0.9162	0.9177
1.4	0.9192	0.9207	0.9222	0.9236	0.9251	0.9265	0.9279	0.9292	0.9306	0.9319
1.5	0.9332	0.9345	0.9357	0.9370	0.9382	0.9394	0.9406	0.9418	0.9429	0.9441
1.6	0.9452	0.9463	0.9474	0.9484	0.9495	0.9505	0.9515	0.9525	0.9535	0.9545
1.7	0.9554	0.9564	0.9573	0.9582	0.9591	0.9599	0.9608	0.9616	0.9625	0.9633
1.8	0.9641	0.9649	0.9656	0.9664	0.9671	0.9678	0.9686	0.9693	0.9699	0.9706
1.9	0.9713	0.9719	0.9726	0.9732	0.9738	0.9744	0.9750	0.9756	0.9761	0.9767
2.0	0.9772	0.9778	0.9783	0.9788	0.9793	0.9798	0.9803	0.9808	0.9812	0.9817
2.1	0.9821	0.9826	0.9830	0.9834	0.9838	0.9842	0.9846	0.9850	0.9854	0.9857
2.2	0.9861	0.9864	0.9868	0.9871	0.9875	0.9878	0.9881	0.9884	0.9887	0.9890
2.3	0.9893	0.9896	0.9898	0.9901	0.9904	0.9906	0.9909	0.9911	0.9913	0.9916
2.4	0.9918	0.9920	0.9922	0.9925	0.9927	0.9929	0.9931	0.9932	0.9934	0.9936
2.5	0.9938	0.9940	0.9941	0.9943	0.9945	0.9946	0.9948	0.9949	0.9951	0.9952
2.6	0.9953	0.9955	0.9956	0.9957	0.9959	0.9960	0.9961	0.9962	0.9963	0.9964
2.7	0.9965	0.9966	0.9967	0.9968	0.9969	0.9970	0.9971	0.9972	0.9973	0.9974
2.8	0.9974	0.9975	0.9976	0.9977	0.9977	0.9978	0.9979	0.9979	0.9980	0.9981
2.9	0.9981	0.9982	0.9982	0.9983	0.9984	0.9984	0.9985	0.9985	0.9986	0.9986
3.0	0.9987	0.9987	0.9987	0.9988	0.9988	0.9989	0.9989	0.9989	0.9990	0.9990
3.1	0.9990	0.9991	0.9991	0.9991	0.9992	0.9992	0.9992	0.9992	0.9993	0.9993
3.2	0.9993	0.9993	0.9994	0.9994	0.9994	0.9994	0.9994	0.9995	0.9995	0.9995
3.3	0.9995	0.9995	0.9995	0.9996	0.9996	0.9996	0.9996	0.9996	0.9996	0.9997
3.4	0.9997	0.9997	0.9997	0.9997	0.9997	0.9997	0.9997	0.9997	0.9997	0.9998

Critical values, $P(Z \geq z_\alpha) = \alpha$

α	0.10	0.05	0.025	0.01	0.005	0.001	0.0005	0.0001		
z_α	1.2816	1.6449	1.9600	2.3263	2.5758	3.0902	3.2905	3.7190		
α	0.00009	0.00008	0.00007	0.00006	0.00005	0.00004	0.00003	0.00002	0.00001	
z_α	3.7455	3.7750	3.8082	3.8461	3.8906	3.9444	4.0128	4.1075	4.2649	

Table A.8 Critical values for the chi-square distribution (This table contains critical values $\chi^2_{\alpha,\nu}$ for the chi-square distribution defined by $P(\chi^2 \geq \chi^2_{\alpha,\nu}) = \alpha$)

ν	α							
	0.9999	0.9995	0.999	0.995	0.99	0.975	0.95	0.90
1	0.0^7157	0.0^6393	0.0^5157	0.0^4393	0.0002	0.0010	0.0039	0.0158
2	0.0002	0.0010	0.0020	0.0100	0.0201	0.0506	0.1026	0.2107
3	0.0052	0.0153	0.0243	0.0717	0.1148	0.2158	0.3518	0.5844
4	0.0284	0.0639	0.0908	0.2070	0.2971	0.4844	0.7107	10.0636
5	0.0822	0.1581	0.2102	0.4117	0.5543	0.8312	1.1455	1.6103
6	0.1724	0.2994	0.3811	0.6757	0.8721	1.2373	1.6354	2.2041
7	0.3000	0.4849	0.05985	0.9893	1.2390	1.6899	2.1673	2.8331
8	0.4636	0.7104	0.8571	1.3444	1.6465	2.1797	2.7326	3.4895
9	0.6608	0.9717	1.1519	1.7349	2.0879	2.7004	3.3251	4.1682
10	0.8889	1.2650	1.4787	2.1559	2.5582	3.2470	3.9403	4.8652
11	1.1453	1.5868	1.8339	2.6032	3.0535	3.8157	4.5748	5.5778
12	1.4275	1.9344	2.2142	3.0738	3.5706	4.4038	5.2260	6.3038
13	1.7333	2.3051	2.6172	3.5650	4.069	5.0088	5.8919	7.0415
14	2.0608	2.6967	3.0407	4.0747	4.6604	5.6287	6.5706	7.7895
15	2.4082	3.1075	3.4827	4.6009	5.2293	6.2621	7.2609	8.5468
16	2.7739	3.5358	3.9416	5.1422	5.8122	6.9077	7.9616	9.3122
17	3.1567	3.9802	4.4161	5.6972	6.4078	7.5642	8.6718	10.0852
18	3.5552	4.4394	4.9048	6.2648	7.0149	8.2307	9.3905	10.8649
19	3.9683	4.9123	5.4068	6.8440	7.6327	8.9065	10.1170	11.6509
20	4.3952	5.3981	5.9210	7.4338	8.2604	9.5908	10.8508	12.4426
21	4.8348	5.8957	6.4467	8.0337	8.8972	10.2829	11.5913	13.2395
22	5.2865	6.4045	6.9830	8.6427	9.5425	10.9823	12.3380	14.0415
23	5.7494	6.9237	7.5292	9.2604	10.1957	11.6886	13.0905	14.8480
24	6.2230	7.4527	8.0849	9.8862	10.8564	12.4012	13.8484	15.6587
25	6.7066	7.9910	8.6493	10.5197	11.5240	13.1197	14.6114	16.4734
26	7.1998	8.5379	9.2221	11.1602	12.1981	13.8439	15.3792	17.2919
27	7.7019	9.0932	9.8028	11.8076	12.8785	14.5734	16.1514	18.1139
28	8.2126	9.6563	10.3909	12.4613	13.5647	15.3079	16.9279	18.9392
29	8.7315	10.2268	10.9861	13.1211	14.2565	16.0471	17.7084	19.7677
30	9.2581	10.8044	11.5880	13.7867	14.9535	16.7908	18.4927	20.5992
31	9.7921	11.3887	12.1963	14.4578	15.6555	17.5387	19.2806	21.4336
32	10.3331	11.9794	12.8107	15.1340	16.3622	18.2908	20.0719	22.2706
33	10.8810	12.5763	13.4309	15.8153	17.0735	19.0467	20.8665	23.1102
34	11.4352	13.1791	14.0567	16.5013	17.7891	19.8063	21.6643	23.9523
35	11.9957	13.7875	14.6878	17.1918	18.5089	20.5694	22.4650	24.7967
36	12.5622	14.4012	15.3241	17.8867	19.2327	21.3359	23.2686	25.6433

(continued)

Table A.8 (continued)

ν	α							
	0.9999	0.9995	0.999	0.995	0.99	0.975	0.95	0.90
37	13.1343	15.0202	15.9653	18.5658	19.9602	22.1056	24.0749	26.4921
38	13.7120	15.6441	16.6112	19.2889	20.6914	22.8785	24.8839	27.3430
39	14.2950	16.2729	17.2616	19.9959	21.4262	23.6543	25.6954	28.1958
40	14.8831	16.9062	17.9164	20.7065	22.1643	24.4330	26.5093	29.0505
50	21.0093	23.4610	24.6739	27.9907	29.7067	32.3574	34.7643	37.6886
60	27.4969	30.3405	31.7383	35.5345	37.4849	40.4817	43.1880	46.4589
70	34.2607	37.4674	39.0364	43.2752	45.4417	48.7576	51.7393	55.3289
80	41.2445	44.7910	46.5199	51.1719	53.5401	57.1532	60.3915	64.2778
90	48.4087	52.2758	54.1552	59.1963	61.7541	65.6466	69.1260	73.2911
100	55.7246	59.8957	61.9179	67.3276	70.0649	74.2219	77.9295	82.3581

Table A.9 Critical values for the chi-square distribution

ν	α							
	0.10	0.05	0.025	0.01	0.005	0.001	0.0005	0.0001
1	2.7055	3.8415	5.0239	6.6349	7.8794	10.8276	12.1157	15.1367
2	4.6052	5.9915	7.3778	9.2103	10.5966	13.8155	15.2013	18.4207
3	6.2514	7.8147	9.3484	11.3449	12.8382	16.2662	17.7300	21.1075
4	7.7794	9.4877	11.1433	13.2767	14.8603	18.4668	19.9974	23.5127
5	9.2364	11.0705	12.8325	15.0863	16.7496	20.5150	22.1053	25.7418
6	10.6446	12.5916	14.4494	16.8119	18.5476	22.4577	24.1028	27.8563
7	12.0170	14.0671	16.0128	18.4753	20.2777	24.3219	26.0178	29.8775
8	13.3616	15.5073	17.5345	20.0002	21.9550	26.1245	27.8680	31.8276
9	14.6837	16.9190	19.0228	21.6660	23.5894	27.8772	29.6658	33.7199
10	15.9872	18.3070	20.4832	23.2093	25.1882	29.5883	31.4198	35.5640
11	17.2750	19.6751	21.9200	24.7250	26.7568	31.2641	33.1366	37.3670
12	18.5493	21.0261	23.3367	26.2170	28.2995	32.9095	34.8213	39.1344
13	19.8119	22.3620	24.7356	27.6882	29.8195	34.5282	36.4778	40.8707
14	21.0641	23.6848	26.1189	29.1412	31.3193	36.1233	38.1094	42.5793
15	22.3071	24.9958	27.4884	30.5779	32.8013	37.6973	39.7188	44.2632
16	23.5418	26.2962	28.8454	31.9999	34.2672	39.2524	41.3081	45.9249
17	24.7690	27.5871	30.1910	33.4087	35.7185	40.7902	42.8792	47.5664
18	25.9894	28.8693	31.5264	34.8053	37.1565	42.3124	44.4338	49.1894
19	27.2036	30.1435	32.8523	36.1909	38.5823	43.8202	45.9731	50.7955
20	28.4120	31.4104	34.1696	37.5662	39.9968	45.3147	47.4985	52.3860
21	29.6151	32.6706	35.4789	38.9322	41.4011	46.7970	49.0108	53.9620
22	30.8133	33.9244	36.7807	40.2894	42.7957	48.2679	50.5111	55.5246

(continued)

Table A.9 (continued)

ν	α 0.10	0.05	0.025	0.01	0.005	0.001	0.0005	0.0001
23	32.0069	35.1725	38.0756	41.6384	44.1813	49.7282	52.0002	57.0746
24	33.1962	36.4150	39.3641	42.9798	45.5585	51.1786	53.4788	58.6130
25	34.3816	37.6525	40.6465	44.3141	46.9279	52.6197	54.9475	60.1403
26	35.5632	38.8851	41.9232	45.6417	48.2899	54.0520	56.4069	61.6573
27	36.7412	40.1133	43.1945	46.9629	49.6449	55.4760	57.8576	63.1645
28	37.9159	41.3371	44.4608	48.2782	50.9934	56.8923	59.3000	64.6624
29	39.0875	42.5570	45.7223	49.5879	52.3356	58.3012	60.7346	66.1517
30	40.2560	43.7730	46.9792	50.8922	53.6720	59.7031	62.1619	67.6326
31	41.4217	44.9853	48.2319	52.1914	55.0027	61.0983	63.5820	69.1057
32	42.5847	46.1943	49.4804	53.4858	56.3281	62.4872	64.9955	70.5712
33	43.7452	47.3999	50.7251	54.7755	57.6484	63.8701	66.4025	72.0296
34	44.9032	48.6024	51.9660	56.0609	58.9639	63.2472	67.8035	73.4812
35	46.0588	49.8018	53.2033	57.3421	60.2748	66.6188	69.1986	74.9262
36	47.2122	50.9985	54.4373	58.6192	61.5812	67.9852	70.5881	76.3650
37	48.3634	52.1923	55.6680	59.8925	62.8833	69.3465	71.9722	77.7977
38	49.5126	53.3835	56.8955	61.1621	64.1814	70.7029	73.3512	79.2247
39	50.6598	54.5722	58.1201	62.4281	65.4756	72.0547	74.7253	80.6462
40	51.8051	55.7585	59.3417	63.6907	66.7660	73.4020	76.0946	82.0623
50	63.1671	67.5048	71.4202	76.1539	79.4900	86.6608	89.5605	95.9687
60	74.3970	79.0819	83.2977	88.3794	91.9517	99.6072	102.6948	109.5029
70	85.5270	90.5312	95.0232	100.4252	104.2149	112.3169	115.5776	122.7547
80	96.5782	101.8795	106.6286	112.3288	116.3211	124.8392	128.2613	135.7825
90	107.5650	113.1453	118.1359	124.1163	128.2989	137.2084	140.7823	143.6273
100	118.4980	124.3421	129.5612	135.8067	140.1695	149.4493	153.1670	161.3187

Table A.10 Critical values for the student's t distribution (This table contains critical values $t_{\nu,\alpha}$ for the student's t distribution defined by $P(T \geq t_{\nu,\alpha}) = \alpha$)

ν	α 0.20	0.10	0.05	0.025	0.01	0.005	0.001	0.0005	0.0001
1	1.3764	3.0777	6.3138	12.7062	31.8205	63.6567	313.3088	636.6192	3183.0988
2	1.0607	1.8856	2.9200	4.3027	6.9646	9.9248	22.3271	31.5991	70.7001
3	0.9785	1.6377	2.3534	3.1824	4.5407	5.8409	10.2145	12.9240	22.2037
4	0.9410	1.5332	2.1318	2.7764	3.7469	4.6041	7.1732	8.6103	13.0337
5	0.9195	1.4759	2.0150	2.5706	3.3649	4.0321	5.8934	6.8688	9.6776
6	0.9057	1.4398	1.9432	2.4469	3.1427	3.7074	5.2076	5.9588	8.0248
7	0.8960	1.4149	1.8946	2.3646	2.9980	3.4995	4.7853	5.4079	7.0634
8	0.8889	1.3968	1.8595	2.3060	2.8965	3.3554	4.5008	5.0413	6.4420
9	0.8834	1.3830	1.8331	2.2622	2.8214	3.2493	4.2968	4.7809	6.0101
10	0.8791	1.3722	1.8125	2.2281	2.7638	3.1693	4.1437	4.5869	5.6938
11	0.8755	1.3634	1.7959	2.2010	2.7181	3.1058	4.0247	4.4370	5.4528
12	0.8726	1.3562	1.7823	2.1788	2.6810	3.0545	3.9296	4.3178	5.2633
13	0.8702	1.3502	1.7709	2.1604	2.6503	3.0123	3.8520	4.2208	5.1106
14	0.8681	1.3450	1.7613	2.1448	2.6245	2.9768	3.7874	4.1405	4.9850
15	0.8662	1.3406	1.7531	2.1314	2.6025	2.9467	3.7328	4.0728	4.8800
16	0.8647	1.3368	1.7459	2.1199	2.5835	2.9208	3.6862	4.0150	4.7909
17	0.8633	1.3334	1.7396	2.1098	2.5669	2.8982	3.6458	3.9651	4.7144
18	0.8620	1.3304	1.7341	2.1009	2.5524	2.8784	3.6105	3.9216	4.6480
19	0.8610	1.3277	1.7291	2.0930	2.5395	2.8609	3.5794	3.8834	4.5899
20	0.8600	1.3253	1.7247	2.0860	2.5280	2.8453	3.5518	3.5495	4.5385
21	0.8591	1.3232	1.7207	2.0796	2.5176	2.8314	3.5271	3.8192	4.4929
22	0.8583	1.3212	1.7171	2.0739	2.5083	2.8187	3.5050	3.7921	4.4520
23	0.8575	1.3195	1.7139	2.0687	2.4999	2.8073	2.4850	3.7676	4.4152
24	0.8569	1.3178	1.7109	2.0639	2.4922	2.7969	3.4008	3.7454	4.3819
25	0.8562	1.3163	1.7081	2.0595	2.4351	2.7874	3.4502	3.7251	4.3517
26	0.8557	1.3150	1.7056	2.0555	2.4786	2.7787	3.4350	3.7066	4.3240
27	0.8551	1.3137	1.7033	2.0518	2.4727	2.7707	3.4210	3.6896	4.2987
28	0.8546	1.3125	1.7011	2.0484	2.4671	2.7633	3.4081	3.6739	4.2754
29	0.8542	1.3114	1.6991	2.0452	2.4620	2.7564	3.3962	3.6594	4.2539
30	0.8538	1.3104	1.6973	2.0423	2.4573	2.7500	3.3852	3.6460	4.2340
40	0.8507	1.3031	1.6839	2.0211	2.4233	2.7045	3.3069	3.5510	4.0942
50	0.8489	1.2987	1.6759	2.0086	2.4033	2.6778	3.2614	3.4960	4.0140
60	0.8477	1.2958	1.6706	2.0003	2.3901	2.6603	3.2317	3.4602	3.9621
120	0.8446	1.2886	1.6577	1.9799	2.3578	2.6174	3.1595	3.3735	3.8372
∞	0.8416	1.2816	1.6449	1.9600	2.3263	2.5758	3.0902	3.2905	3.7190

Table A.11 Critical values for the F distribution (This table contains critical values F_{α,ν_1,ν_2} for the F distribution defined by $P(F \geq F_{\alpha,\nu_1,\nu_2}) = \alpha$)

$\alpha = 0.05$

ν_2 \ ν_1	1	2	3	4	5	6	7	8	9	10	15	20	30	40	60	120	∞
1	161.45	199.50	215.71	224.58	230.16	233.99	236.77	238.88	240.54	241.88	245.95	248.01	250.10	251.14	252.20	253.25	254.25
2	18.51	19.00	19.16	19.25	19.30	19.33	19.35	19.37	19.38	19.40	19.43	19.45	19.46	19.47	19.48	19.49	19.50
3	10.13	9.55	9.28	9.12	9.01	8.94	8.89	8.85	8.81	8.79	8.70	8.66	8.62	8.59	8.57	8.55	8.53
4	7.71	6.94	6.59	6.39	6.26	6.16	6.09	6.04	6.00	6.96	5.86	5.80	5.75	5.72	5.69	5.66	5.63
5	6.61	5.79	5.41	5.19	5.05	4.95	4.88	4.82	4.77	4.74	4.62	4.56	4.50	4.46	4.43	4.40	4.37
6	5.99	5.14	4.76	4.53	4.39	4.28	4.21	4.15	4.10	4.06	3.94	3.87	3.81	3.77	3.74	3.70	3.67
7	5.59	4.74	4.35	4.12	3.97	3.87	3.79	3.73	3.68	3.64	3.51	3.44	3.38	3.34	3.30	3.27	3.23
8	5.32	4.46	4.07	3.84	3.69	3.58	3.50	3.44	3.39	3.35	3.22	3.15	3.08	3.04	3.01	2.97	2.93
9	5.12	4.26	3.86	3.63	3.48	3.37	3.29	3.23	3.18	3.14	3.01	2.94	2.86	2.83	2.79	2.75	2.71
10	4.96	4.10	3.71	3.48	3.33	3.22	3.14	3.07	3.02	2.98	2.85	2.77	2.10	2.66	2.62	2.58	5.54
11	4.84	3.98	3.59	3.36	3.20	3.09	3.01	2.95	2.90	2.85	2.72	2.65	2.51	2.53	2.49	2.45	2.41
12	4.75	3.89	3.49	3.26	3.11	3.00	2.91	2.85	2.80	2.75	2.62	2.54	2.47	2.43	2.38	2.34	2.30
13	4.67	3.81	3.41	3.18	3.03	2.92	2.83	2.77	2.71	2.67	2.53	2.46	2.38	2.34	2.30	2.25	2.21
14	4.60	3.74	3.34	3.11	2.96	2.85	2.76	2.70	2.65	2.60	2.46	2.39	2.31	2.27	2.22	2.18	2.13
15	4.54	3.68	3.29	3.06	2.90	2.79	2.71	2.64	2.59	2.54	2.40	2.33	2.25	2.20	2.16	2.11	2.07
16	4.49	3.63	3.24	3.01	2.85	2.74	2.66	2.59	2.54	2.49	2.35	2.28	2.19	2.15	2.11	2.06	2.01
17	4.45	3.59	3.20	2.96	2.81	2.70	2.61	2.55	2.49	2.45	2.31	2.23	2.15	2.10	2.06	2.01	1.96
18	4.41	3.55	3.16	2.93	2.77	2.66	2.58	2.51	2.46	2.41	2.27	2.19	2.11	2.06	2.02	1.97	1.92
19	4.38	3.52	3.13	2.90	2.74	2.63	2.54	2.48	2.42	2.38	2.23	2.16	2.07	2.03	1.98	1.93	1.88
20	4.35	3.49	3.10	2.87	2.71	2.60	2.51	2.45	2.39	2.35	2.20	2.12	2.04	1.99	1.95	1.90	1.85
21	4.32	3.47	3.07	2.84	2.68	2.57	2.49	2.42	2.37	2.32	2.18	2.10	2.01	1.96	1.92	1.87	1.82
22	4.30	3.44	3.05	2.82	2.66	2.55	2.46	2.40	2.34	2.30	2.15	2.07	1.98	1.94	1.89	1.84	1.79
23	4.28	3.42	3.03	2.80	2.64	2.53	2.44	2.37	2.32	2.27	2.13	2.05	1.96	1.91	1.86	1.81	1.76

(continued)

Table A.11 (continued)

$\alpha = 0.05$

$\nu_2 \backslash \nu_1$	1	2	3	4	5	6	7	8	9	10	15	20	30	40	60	120	∞
24	4.26	3.40	3.01	2.78	2.62	2.51	2.42	2.36	2.30	2.25	2.11	2.03	1.94	1.89	1.84	1.79	1.74
25	4.24	3.39	2.99	2.76	2.60	2.49	2.40	2.34	2.28	2.24	2.09	2.01	1.92	1.87	1.82	1.77	1.71
30	4.17	3.32	2.92	2.69	2.53	2.42	2.33	2.27	2.21	2.16	2.01	1.93	1.84	1.79	1.74	1.68	1.63
40	4.08	3.23	2.84	2.61	2.45	2.34	2.25	2.18	2.12	2.08	1.92	1.84	1.74	1.69	1.64	1.58	1.51
50	4.03	3.18	2.79	2.56	2.40	2.29	2.20	2.13	2.07	2.03	1.87	1.78	1.69	1.63	1.58	1.51	1.44
60	4.00	3.15	2.76	2.53	2.37	2.25	2.17	2.10	2.04	1.99	1.84	1.75	1.65	1.59	1.53	1.47	1.39
120	3.92	3.07	2.68	2.45	2.29	2.18	2.09	2.02	1.96	1.91	1.75	1.66	1.55	1.50	1.43	1.35	1.26
∞	3.85	3.00	2.61	2.38	2.22	2.10	2.01	1.94	1.88	1.84	1.67	1.58	1.46	1.40	1.32	1.23	1.00

Table A.12 Critical values for the F distribution (Continued)

$\alpha = 0.01$

v_1																	
v_1	1	2	3	4	5	6	7	8	9	10	15	20	30	40	60	120	∞
2	98.50	99.00	99.17	99.25	99.30	99.33	99.36	99.37	99.39	99.40	99.43	99.45	99.47	99.47	99.48	99.49	99.50
3	34.12	30.82	29.46	28.71	28.24	27.91	27.67	27.49	27.35	27.23	26.87	26.69	26.50	26.41	26.32	26.22	26.13
4	21.20	18.00	16.69	15.98	15.52	15.21	14.98	11.80	14.66	14.55	14.20	14.02	13.84	13.75	13.65	13.56	13.47
5	16.26	13.27	12.06	11.39	10.97	10.67	10.46	10.29	10.16	10.05	9.72	9.55	9.38	9.29	9.20	9.11	9.03
6	13.75	10.92	9.78	9.15	8.75	8.47	8.26	8.10	7.98	7.87	7.56	7.40	7.23	7.14	7.06	6.97	6.89
7	12.25	9.55	8.45	7.85	7.46	7.19	6.99	6.84	6.72	6.62	6.31	6.16	5.99	5.91	5.82	5.74	5.65
8	11.26	8.65	7.59	7.01	6.63	6.37	6.18	6.03	5.91	5.81	5.52	5.36	5.20	5.12	5.03	4.95	4.86
9	10.56	8.02	6.99	6.42	6.06	5.80	5.61	5.47	5.35	5.26	4.96	4.81	4.65	4.57	4.48	4.40	4.32
10	10.04	7.56	6.55	5.99	5.64	5.39	5.20	5.06	4.94	4.85	4.56	4.41	4.25	4.17	4.08	4.00	3.91
11	9.65	7.21	6.22	5.67	5.32	5.07	4.89	4.74	4.63	4.54	4.25	4.10	3.94	3.86	3.78	3.69	3.61
12	9.33	6.93	5.95	5.41	5.06	4.82	4.64	4.50	4.39	4.30	4.01	3.86	3.70	3.62	3.54	3.45	3.37
13	9.07	6.70	5.74	5.21	4.86	4.62	4.44	4.30	4.19	4.10	3.82	3.66	3.51	3.43	3.34	3.25	3.17
14	8.86	6.51	5.56	5.04	4.69	4.46	4.28	4.14	4.03	3.94	3.66	3.51	3.35	3.27	3.18	3.09	3.01
15	8.68	6.36	5.42	4.89	4.56	4.32	4.14	4.00	3.89	3.80	3.52	3.37	3.21	3.13	3.05	2.96	2.87
16	8.53	6.23	5.29	4.77	4.44	4.20	4.03	3.89	3.78	3.69	3.41	3.26	3.10	3.02	2.93	2.84	2.76
17	8.40	6.11	5.19	4.67	4.34	4.10	3.93	3.79	3.63	3.59	3.31	3.16	3.00	2.92	2.83	2.75	2.66
18	8.29	6.01	5.09	4.58	4.25	4.01	3.84	3.71	3.60	3.51	3.23	3.08	2.92	2.84	2.75	2.66	2.57
19	8.18	5.93	5.01	4.50	4.17	3.94	3.77	3.63	3.52	3.43	3.15	3.00	2.84	2.76	2.67	2.58	2.50
20	8.10	5.85	4.94	4.43	4.10	3.87	3.70	3.56	3.46	3.37	3.09	2.94	2.78	2.69	2.61	2.52	2.43
21	8.02	5.78	4.87	4.37	4.04	3.81	3.64	3.51	3.40	3.31	3.03	2.88	2.72	2.64.	2.55	2.46	2.37
22	7.95	5.72	4.82	4.31	3.99	3.76	3.59	3.45	3.35	3.26	2.98	2.83	2.67	2.58	2.50	2.40	2.31

(continued)

Table A.12 (continued)

$\alpha = 0.01$	ν_1																
ν_1	1	2	3	4	5	6	7	8	9	10	15	20	30	40	60	120	∞
23	7.88	5.66	4.76	4.26	3.94	3.71	3.54	3.41	3.30	3.21	2.93	2.78	2.62	2.54	2.45	2.35	2.26
24	7.82	5.61	4.72	4.22	3.90	3.67	3.50	3.36	3.26	3.17	2.89	2.74	2.58	2.49	2.40	2.31	2.22
25	7.77	5.57	4.68	4.18	3.85	3.63	3.46	3.32	3.22	3.13	2.85	2.70	2.54	2.45	2.36	2.27	2.18
30	7.56	5.39	4.51	4.02	3.70	3.47	3.30	3.17	3.07	2.98	2.70	2.55	2.39	2.30	2.21	2.11	2.01
40	7.31	5.18	4.31	3.83	3.51	3.29	3.12	2.99	2.89	2.80	2.52	2.37	2.20	2.11	2.02	1.92	1.81
50	7.17	5.06	4.20	3.72	3.41	3.19	3.02	2.89	2.78	2.70	2.42	2.27	2.10	2.01	1.91	1.80	1.69
60	7.08	4.98	4.13	3.65	3.34	3.12	2.95	2.82	2.72	2.63	2.35	2.20	2.03	1.94	1.84	1.73	1.61
120	6.85	4.79	3.95	3.48	3.17	2.96	2.79	2.66	2.56	2.47	2.19	2.03	1.86	1.76	1.66	1.53	1.39
∞	6.65	4.62	3.79	3.33	3.03	2.81	2.65	2.52	2.42	2.33	2.05	1.89	1.71	1.60	1.48	1.34	1.00

Table A.13 Critical values for the F distribution (Continued)

$\alpha = 0.01$

ν_2 / ν_1	1	2	3	4	5	6	7	8	9	10	15	20	30	40	60	120	∞
2	998.50	999.00	999.17	999.25	999.30	999.33	999.36	999.37	999.39	999.40	999.43	999.45	999.47	999.47	999.48	999.49	999.50
3	167.03	148.50	141.11	137.10	134.58	132.85	131.58	130.62	129.86	129.25	127.37	126.42	125.45	124.96	124.17	123.97	123.50
4	74.14	61.25	56.18	53.44	51.71	50.53	49.66	49.00	48.47	48.05	46.76	46.10	45.43	45.09	44.75	44.40	44.07
5	47.18	37.12	33.20	31.09	29.75	28.83	28.16	27.65	27.24	26.92	25.91	25.39	24.87	24.60	24.33	24.06	23.80
6	35.51	27.00	23.70	21.92	20.80	20.03	19.46	19.03	18.69	18.41	17.56	17.12	16.67	16.44	16.21	15.98	15.76
7	29.25	21.69	18.77	17.20	16.21	15.52	15.02	14.63	14.33	14.08	13.32	12.93	12.53	12.33	12.12	11.91	11.71
8	25.41	18.49	15.83	14.39	13.48	12.86	12.40	12.05	11.77	11.54	10.84	10.48	10.11	9.92	9.73	9.53	9.35
9	22.86	16.39	13.90	12.56	11.71	11.13	10.70	10.37	10.11	9.89	9.24	8.90	8.55	8.37	8.19	8.00	7.82
10	21.04	14.91	12.55	11.28	10.48	9.93	9.52	9.20	8.96	8.75	8.13	7.80	7.47	7.30	7.12	6.94	6.77
11	19.69	13.81	11.56	10.35	9.58	9.05	8.66	8.35	8.12	7.92	7.32	7.01	6.68	6.52	6.35	6.18	6.01
12	18.64	12.97	10.80	9.63	8.89	8.38	8.00	7.71	7.48	7.29	6.11	6.40	6.09	5.93	5.76	5.59	5.43
13	17.82	12.31	10.21	9.07	8.35	7.86	7.49	7.21	6.98	6.80	6.23	5.93	5.63	5.47	5.30	5.14	4.98
14	17.14	11.78	9.73	8.62	7.92	7.44	7.08	6.80	6.58	6.40	5.35	5.56	5.25	5.10	4.94	4.77	4.61
15	16.59	11.34	9.34	8.25	7.57	7.09	6.74	6.47	6.26	6.08	5.54	5.25	4.95	4.80	4.64	4.47	4.32
16	16.12	10.97	9.01	7.94	7.27	6.80	6.46	6.19	5.98	5.81	5.27	4.99	4.70	4.54	4.39	4.23	4.07
17	15.72	10.66	8.73	7.68	7.02	6.56	6.22	5.96	5.75	5.58	5.05	4.78	4.48	4.33	4.18	4.02	3.86
18	15.38	10.39	8.49	7.46	6.81	6.35	6.02	5.76	5.56	5.39	4.87	4.59	4.30	4.15	4.00	3.84	3.68
19	15.08	10.16	8.28	7.27	6.62	6.18	5.85	5.59	5.39	5.22	4.70	4.43	4.14	3.99	3.84	3.68	3.52
20	14.82	9.95	8.10	7.10	6.46	6.02	5.69	5.44	5.24	5.08	4.56	4.29	4.00	3.86	3.70	3.54	3.39
21	14.59	9.77	7.94	6.95	6.32	5.88	5.56	5.31	5.11	4.95	4.44	4.17	3.88	3.74	3.58	3.42	3.27
22	14.38	9.61	7.80	6.81	6.19	5.76	5.44	5.19	4.99	4.83	4.33	4.06	3.78	3.63	3.48	3.32	3.16
23	14.20	9.47	7.67	6.70	6.08	5.65	5.33	5.09	4.89	4.73	4.23	3.96	3.68	3.53	3.38	3.22	3.07

(continued)

Table A.13 (continued)

$\alpha = 0.01$

v_2 \ v_1	1	2	3	4	5	6	7	8	9	10	15	20	30	40	60	120	∞
24	14.03	9.34	7.55	6.59	5.98	5.55	5.23	4.99	4.80	4.64	4.14	3.87	3.59	3.45	3.29	3.14	2.98
25	13.88	9.22	7.45	6.49	5.89	5.46	5.15	4.91	4.71	4.56	4.06	3.79	3.52	3.37	3.22	3.06	2.90
30	13.29	8.77	7.05	6.12	5.53	5.12	4.82	4.58	4.39	4.24	3.75	3.49	3.22	3.07	2.92	2.76	2.60
40	12.61	8.25	6.59	5.70	5.13	4.73	4.44	4.21	4.02	3.87	3.40	3.14	2.87	2.73	2.57	2.41	2.24
50	12.22	7.96	6.34	5.46	4.90	4.51	4.22	4.00	3.82	3.67	3.20	2.95	2.68	2.53	2.38	2.21	2.04
60	11.97	7.77	6.17	5.31	4.76	4.37	4.09	3.86	3.69	3.54	3.08	2.83	2.55	2.41	2.25	2.08	1.90
120	11.38	7.32	5.78	4.95	4.42	4.04	3.77	3.55	3.38	3.24	2.78	2.53	2.26	2.11	1.95	1.77	1.56
∞	10.86	6.93	5.44	4.64	4.12	3.76	3.49	3.28	3.11	2.97	2.53	2.28	2.01	1.85	1.68	1.47	1.00

Table A.14 Critical values $d_{n,\alpha}$ for the Kolmogorov–Smirnov test

n/α	0.2	0.1	0.05	0.02	0.01	n/α	0.2	0.1	0.05	0.02	0.01
1	0.900	0.950	0.975	0.990	0.995	16	0.258	0.295	0.327	0.366	0.392
2	0.684	0.776	0.842	0.900	0.929	17	0.250	0.286	0.318	0.355	0.381
3	0.565	0.636	0.708	0.785	0.829	18	0.244	0.0279	0.309	0.346	0.371
4	0.493	0.565	0.624	0.689	0.734	19	0.237	0.271	0.301	0.337	0.361
5	0.447	0.509	0.563	0.627	0.669	20	0.232	0.265	0.294	0.329	0.352
6	0.410	0.468	0.519	0.577	0.617	21	0.226	0.259	0.287	0.321	0.344
7	0.381	0.436	0.483	0.538	0.576	22	0.221	0.253	0.281	0.314	0.337
8	0.358	0.410	0.454	0.507	0.542	23	0.216	0.247	0.275	0.307	0.330
9	0.339	0.387	0.430	0.480	0.513	24	0.212	0.242	0.264	0.301	0.323
10	0.323	0.369	0.409	0.457	0.489	25	0.208	0.238	0.264	0.295	0.317
11	0.308	0.352	0.391	0.437	0.468	26	0.204	0.233	0.259	0.290	0.311
12	0.296	0.338	0.375	0.419	0.449	27	0.200	0.229	0.254	0.284	0.305
13	0.285	0.325	0.361	0.404	0.432	28	0.197	0.225	0.250	0.279	0.300
14	0.275	0.314	0.349	0.390	0.418	29	0.193	0.221	0.246	0.275	0.295
15	0.266	0.304	0.338	0.377	0.404	30	0.190	0.218	0.242	0.270	0.281

Table A.15 Critical values for the Wilcoxon signed-rank test $P(S_+ \geq c)$ when H_0 is true

n	c_1	$p_0(S_+ \geq c_1)$	n	c_1	$p_0(S_+ \geq c_1)$
3	6	0.125		78	0.011
4	9	0.125		79	0.009
	10	0.062		81	0.005
5	13	0.094	14	73	0.108
	14	0.062		74	0.097
	15	0.031		79	0.052
6	17	0.109		84	0.025
	19	0.047		89	0.010
	20	0.031		92	0.005
	21	0.016	15	83	0.104
7	22	0.109		84	0.094
	24	0.055		89	0.053
	26	0.023		90	0.047
	28	0.008		95	0.024
8	28	0.098		100	0.011
	30	0.055		101	0.009
	32	0.027		104	0.005
	34	0.012	16	93	0.106
	35	0.008		94	0.096
	36	0.004		100	0.052
9	34	0.102		106	0.025
	37	0.049		112	0.011
	39	0.027		113	0.009
	42	0.010		116	0.005
	44	0.004	17	104	0.103
10	41	0.097		105	0.095
	44	0.053		112	0.049
	47	0.024		118	0.025
	50	0.010		125	0.010
	52	0.005		129	0.005
11	48	0.103	18	116	0.098
	52	0.051		124	0.049
	55	0.027		131	0.024
	59	0.009		138	0.010
	61	0.005		143	0.005
12	56	0.102	19	128	0.098
	60	0.055		136	0.052
	61	0.046		137	0.048
	64	0.026		144	0.025
	68	0.010		152	0.010
	71	0.005		157	0.005
13	64	0.108	20	140	0.101
	65	0.095		150	0.049
	69	0.055		158	0.024
	70	0.047		167	0.010
	74	0.024		172	0.005

Table A.16 Percentage points of the studentized range statistic $q_{0.01}(p, f)$

f \ p	2	3	4	5	6	7	8	9	10	11	12	13	14	15	16	17	18	19	20
1	90.0	135	164	186	202	216	227	237	246	253	260	266	272	272	282	286	290	294	298
2	14.0	19.0	22.3	24.7	26.6	28.2	29.5	30.7	31.7	32.6	33.4	31.4	34.8	35.4	36.0	36.5	37.0	37.5	37.9
3	8.26	10.6	12.2	13.3	14.2	15.0	15.6	16.2	16.7	17.1	17.5	17.9	18.2	18.5	18.8	19.1	19.3	19.5	19.8
4	6.51	8.12	9.17	9.96	10.6	11.1	11.5	11.9	12.3	12.6	12.8	13.1	13.3	13.5	13.7	13.9	14.1	14.2	14.4
5	5.70	6.97	7.80	8.42	8.91	9.32	9.67	9.97	10.24	10.48	10.70	10.89	11.08	11.24	11.40	11.55	11.68	11.81	11.93
6	5.24	6.33	7.03	7.56	7.97	8.32	8.61	8.87	9.10	9.30	9.49	9.65	9.81	9.95	10.08	10.21	10.32	10.43	10.54
7	4.95	5.92	6.54	7.01	7.37	7.68	7.94	8.17	8.37	8.55	8.71	8.86	9.00	9.12	9.24	9.35	9.46	9.55	9.65
8	4.74	5.63	6.20	6.63	6.96	7.24	7.47	7.68	7.87	8.03	8.18	8.31	8.44	8.55	8.66	8.76	8.85	8.94	9.03
9	4.60	5.43	5.96	6.35	6.66	6.91	7.13	7.32	7.49	7.65	7.78	7.91	8.03	8.13	8.23	8.32	8.41	8.49	8.57
10	4.48	5.27	5.77	6.14	6.43	6.67	6.87	7.05	7.21	7.36	7.48	7.60	7.71	7.81	7.91	7.99	8.07	8.15	8.22
11	4.39	5.14	5.62	5.97	6.25	6.48	6.67	6.84	6.99	7.13	7.25	7.36	7.46	7.56	7.65	7.73	7.81	7.88	7.95
12	4.32	5.04	5.50	5.84	6.10	6.32	6.51	6.67	6.81	6.94	7.06	7.17	7.26	7.36	7.44	7.52	7.59	7.66	7.73
13	4.26	4.96	5.40	5.73	5.98	6.19	6.37	6.53	6.67	6.79	6.90	7.01	7.10	7.19	7.27	7.34	7.42	7.48	7.55
14	4.21	4.89	5.32	5.63	5.88	6.08	6.26	6.41	6.54	6.66	6.77	6.87	6.96	7.05	7.12	7.20	7.27	7.33	7.39
15	4.17	4.83	5.25	5.56	5.80	5.99	6.16	6.31	6.44	6.55	6.66	6.76	6.84	6.93	7.00	7.07	7.14	7.20	7.26
16	4.13	4.78	5.19	5.49	5.72	5.92	6.08	6.22	6.35	6.46	6.56	6.66	6.74	6.82	6.90	6.97	7.03	7.09	7.15
17	4.10	4.74	5.14	5.43	5.66	5.85	6.01	6.15	6.27	6.38	6.48	6.57	6.66	6.73	6.80	6.87	6.94	7.00	7.05
18	4.07	4.70	5.09	5.38	5.60	5.79	5.94	6.08	6.20	6.31	6.41	6.50	6.58	6.65	6.72	6.79	6.85	6.91	6.96

(continued)

Table A.16 (continued)

f	p																		
	2	3	4	5	6	7	8	9	10	11	12	13	14	15	16	17	18	19	20
19	4.05	4.67	5.05	5.33	5.55	5.73	5.89	6.02	6.14	6.25	6.34	6.43	6.51	6.58	6.65	6.72	6.78	6.84	6.89
20	4.02	4.64	5.02	5.29	5.51	5.69	5.84	5.97	6.09	6.19	6.29	6.37	6.45	6.52	6.59	6.65	6.71	6.76	6.82
24	3.96	4.54	4.91	5.17	5.37	5.54	5.69	5.81	5.92	6.02	6.11	6.19	6.26	6.33	6.39	6.45	6.51	6.56	6.61
30	3.89	4.45	4.80	5.05	5.24	5.40	5.54	5.65	5.76	5.85	5.93	6.01	6.08	6.14	6.20	6.26	6.31	6.36	6.41
40	3.82	4.37	4.70	4.93	5.11	5.27	5.39	5.50	5.60	5.69	5.77	5.84	5.90	5.96	6.02	6.07	6.12	6.17	6.21
60	3.76	4.28	2.60	4.82	4.99	5.13	5.25	5.36	5.45	5.53	5.60	5.67	5.73	5.79	5.84	5.89	5.93	5.98	6.02
120	3.70	4.20	4.50	4.71	4 87	5.01	5.12	5.21	5.30	5.38	5.44	5.51	5.56	5.61	5.66	5.71	5.75	5.79	5.83
∞	3.64	4.12	4.40	4.60	4.76	4.88	4.99	5.08	5.16	5.23	5.29	5.35	5.40	5.45	5.49	5.54	5.57	5.61	5.65

f = degrees of freedom

Table A.17 Percentage points of the studentized range statistic (Continued) $q_{0.05}(p, f)$

f \\ p	2	3	4	5	6	7	8	9	10	11	12	13	14	15	16	17	18	19	20
1	18.1	26.7	32.8	37.2	40.5	43.1	45.4	47.3	49.1	50.6	51.9	53.2	54.3	55.4	56.3	57.2	58.0	58.8	59.6
2	6.09	8.28	9.80	10.89	11.73	12.43	13.03	13.54	13.99	14.39	14.75	15.08	15.38	15.65	15.91	16.14	16.36	16.57	16.77
3	4.50	5.88	6.83	7.51	8.04	8.47	8.85	9.18	9.46	9.72	9.95	10.16	10.35	10.52	10.69	10.84	10.98	11.12	11.24
4	3.93	5.00	5.76	6.31	6.73	7.06	7.35	7.60	7.83	8.03	8.21	8.37	8.52	8.67	8.80	8.92	9.03	9.14	9.24
5	3.64	4.60	5.22	5.67	6.03	6.33	6.58	6.80	6.99	7.17	7.32	7.47	7.60	7.72	7.83	7.93	8.03	8.12	8.21
6	3.46	4.34	4.90	5.31	5.63	5.89	6.12	6.32	6.49	6.65	6.79	6.92	7.04	7.14	7.24	7.34	7.43	7.51	7.59
7	3.34	4.16	4.68	5.06	5.35	5.59	5.80	5.99	6.15	6.29	6.42	6.54	6.65	6.75	6.84	6.93	7.01	7.08	7.16
8	3.26	4.04	4.53	4.89	5.17	5.40	5.60	5.77	5.92	6.05	6.18	6.29	6.39	6.48	6.57	6.65	6.73	6.80	6.87
9	3.20	3.95	4.42	4.76	5.02	5.24	5.43	5.60	5.74	5.87	5.98	6.09	6.19	6.28	6.36	6.44	6.51	6.58	6.65
10	3.15	3.88	4.33	4.66	4.91	5.12	5.30	5.46	5.60	5.72	5.83	5.93	6.03	6.12	6.20	6.27	6.34	6.41	6.47
11	3.11	3.82	4.26	4.58	4.82	5.03	5.20	5.35	5.49	5.61	5.71	5.81	5.90	5.98	6.06	6.14	6.20	6.27	6.33
12	3.08	3.77	4.20	4.51	4.75	4.95	5.12	5.27	5.40	5.51	5.61	5.71	5.80	5.88	5.95	6.02	6.09	6.15	6.21
13	3.06	3.73	4.15	4.46	4.69	4.88	5.05	5.19	5.32	5.43	5.53	5.63	5.71	5.79	5.86	5.93	6.00	6.06	6.11
14	3.03	3.70	4.11	4.41	4.64	4.83	4.99	5.13	5.25	5.36	5.46	5.56	5.64	5.72	5.79	5.86	5.92	5.98	6.03
15	3.01	3.67	4.08	4.37	4.59	4.78	4.94	5.08	5.20	5.31	5.40	5.49	5.57	5.65	5.72	5.79	5.85	5.91	5.96
16	3.00	3.65	4.05	4.34	4.56	4.74	4.90	5.03	5.15	5.26	5.35	5.44	5.52	5.59	5.66	5.73	5.79	5.84	5.90
17	2.98	3.62	4.02	4.31	4.52	4.70	4.86	4.99	5.11	5.21	5.31	5.39	5.47	5.55	5.61	5.68	5.74	5.79	5.84
18	2.97	3.61	4.00	4.28	4.49	4.67	4.83	4.96	5.07	5.17	5.27	5.35	5.43	5.50	5.57	5.63	5.69	5.74	5.79
19	2.96	3.59	3.98	4.26	4.47	4.64	4.79	4.92	5.04	5.14	5.23	5.32	5.39	5.46	5.53	5.59	5.65	5.70	5.75

(continued)

Table A.17 (continued)

f	p 2	3	4	5	6	7	8	9	10	11	12	13	14	15	16	17	18	19	20
20	2.95	3.58	3.96	4.24	4.45	4.62	4.77	4.90	5.01	5.11	5.20	5.28	5.36	5.43	5.50	5.56	5.61	5.66	5.71
24	2.92	3.53	3.90	4.17	4.37	4.54	4.68	4.81	4.92	5.01	5.10	5.18	5.25	5.32	5.38	5.44	5.50	5.55	5.59
30	2.89	3.48	3.84	4.11	4.30	4.46	4.60	4.72	4.83	4.92	5.00	5.08	5.15	5.21	5.27	5.33	5.38	5.43	5.48
40	2.86	3.44	3.79	4.04	4.23	4.39	4.52	4.63	4.74	4.82	4.90	4.98	5.05	5.11	5.17	5.22	5.27	5.32	5.36
60	2.83	3.40	3.74	3.98	4.16	4.31	4.44	4.55	4.65	4.73	4.81	4.88	4.94	5.00	5.06	5.11	5.15	5.20	5.24
120	2.80	3.36	3.69	3.92	4.10	4.24	4.36	4.47	4.56	4.64	4.71	4.78	4.84	4.90	4.95	5.00	5.04	5.09	5.13
∞	2.77	3.32	3.63	3.86	4.03	4.17	4.29	4.39	4.47	4.55	4.62	4.68	4.74	4.80	4.84	4.98	4.93	4.97	5.01

Index

A

Acceptance number, 360
Acceptance sampling, 353
  of attributes, 354
  of variables, 354
  plan, 355
  technique, 353
Actual capability, 378
Aliases, 302
Allowance, 384

B

Basu's theorem, 122
Bayes' theorem, 24, 122
Block, 240, 254
Blocking, 239, 295
Borel measurable function, 35

C

Canonical analysis, 337
Center line, 366
Central limit theorem, 55
Central tendency, 79
Characteristic function, 42
Chart
  bar, 70
  control, 365
  mean, 369
  pie, 71
  quality control, 353
  range, 371
  Shewhart control, 367
  standard deviation, 372

Coefficient
  mixed quadratic, 337
  of determination, 283
  of kurtosis, 88
  of multiple determination, 330
  of skewness, 88
  of variation, 87
  Pearson's correlation, 194
  pure quadratic, 337
  regression, 238
  Spearman's correlation, 195
Confidence
  interval, 131
Confounding, 297, 315
Consumer's risk, 356
Contrasts, 305
Control chart
  CUSUM, 383
  EWMA, 389
  for number of defectives, 381
  for number of defects, 382
  fraction defective, 380
Control limit
  lower, 366
  upper, 366
Correlation
  multiple, 201
  partial, 203
Covariance, 193
  analysis of, 239
Cramér and Rao inequality, 114
Critical region, 148, 231
Cuboidal, 341

© Springer Nature Singapore Pte Ltd. 2018
D. Selvamuthu and D. Das, *Introduction to Statistical Methods,
Design of Experiments and Statistical Quality Control*,
https://doi.org/10.1007/978-981-13-1736-1

Printed in the United States
By Bookmasters